D0207636

Chemistry and Analysis of Volatile Organic Compounds in the Environment

Chemistry and Analysis of Volatile Organic Compounds in the Environment

Edited by

H.J. Th. BLOEMEN
Research Scientist, Air Quality

and

J. BURN
Science Associate, International Air Quality

Dutch Institute of Public Health and Environmental Protection (RIVM)
The Netherlands

BLACKIE ACADEMIC & PROFESSIONAL
An Imprint of Chapman & Hall
London · Glasgow · New York · Tokyo · Melbourne · Madras

TD
196
.073
C48
1993

Published by Blackie Academic & Professional, an imprint of Chapman & Hall
Wester Cleddens Road, Bishopbriggs, Glasgow G64 2NZ

Chapman & Hall, 2–6 Boundary Row, London SE1 8HN, UK

Blackie Academic & Professional, Wester Cleddens Road, Bishopbriggs, Glasgow G64 2NZ, UK

Chapman & Hall Inc., One Penn Plaza 41st Floor, New York, NY 10019, USA

Chapman & Hall Japan, Thomson Publishing Japan, Hirakawacho Nemoto Building, 6F, 1-7-11 Hirakawa-cho, Chiyoda-ku, Tokyo 102, Japan

DA Book (Aust.) Pty Ltd, 648 Whitehorse Road, Mitcham 3132, Victoria, Australia

Chapman & Hall India, R. Seshadri, 32 Second Main Road, CIT East, Madras 600 035, India

First edition 1993

© 1993 Chapman & Hall

Typeset in 10/12pt Times by Thomson Press (India) Ltd, New Delhi
Printed in Great Britain by St Edmundsbury Press, Bury St Edmunds, Suffolk

ISBN 0 7514 0000 9

Apart from any fair dealing for the purposes of research or private study, or criticism or review, as permitted under the UK Copyright Designs and Patents Act, 1988, this publication may not be reproduced, stored, or transmitted, in any form or by any means, without the prior permission in writing of the publishers, or in the case of reprographic reproduction only in accordance with the terms of the licences issued by the Copyright Licensing Agency in the UK, or in accordance with the terms of licences issued by the appropriate Reproduction Rights Organization outside the UK. Enquiries concerning reproduction outside the terms stated here should be sent to the publishers at the Glasgow address printed on this page.

The publisher makes no representation, express or implied, with regard to the accuracy of the information contained in this book and cannot accept any legal responsibility or liability for any errors or omissions that may be made.

A catalogue record for this book is available from the British Library

Library of Congress Cataloging-in-Publication data

Chemistry and analysis of voltaile organic compounds in the
environment / edited by H.J. Th. Bloemen and J. Burn.—1st ed.
p. cm.
Includes bibliographical references and index.
ISBN 0-7514-0000-9 (alk. paper)
1. Volatile organic compounds—Environmental aspects. 2. Volatile organic compounds—Analysis. 3. Volatile organic compounds—Health aspects. I. Bloemen, H.J. Th. (Henk J. Th.) II. Burn. J. (James)
TD196.073C48 1993 93-9751
CIP

Printed on acid-free text paper, manufactured in accordance with ANSI/NISO Z39.48-1992 (Permanence of Paper).

LONGWOOD COLLEGE LIBRARY
FARMVILLE, VIRGINIA 23901

Contributors

Dr H.J.Th. Bloemen Air Quality Laboratory (LLO), Dutch National Institute of Public Health and Environmental Protection (RIVM), PO Box I-NL 3720 BA, Bilthoven, The Netherlands

Mr J. Burn Air Quality Laboratory (LLO), Dutch National Institute of Public Health and Environmental Protection (RIVM), PO Box 1-NL 3720 BA, Bilthoven, The Netherlands

Dr P. Ciccioli Instituto sull' Inquinamento Atmospherico del CNR, Area della Ricerca di Roma, Via Salaria km. 29,300, CP10, 00016 Monterotondo Scalo, Italy

Dr J.J.G. Kliest Research Coordination Environmental Incidents, Dutch National Institute of Public Health and Environmental Protection (RIVM), PO Box 1-NL 3720 BA, Bilthoven, The Netherlands

Dr D. Kotzias Commision of the European Communities, Joint Research Centre, Ispra Environment Institute, 21020 Ispra (VA), Italy

Dr W.A. McClenny US Environmental Protection Agency, Research Triangle Park, North Carolina 27711, USA

Dr.Ir G. de Mik Substances and Risks Division, Dutch National Institute of Public Health and Environmental Protection (RIVM), PO Box 1-NL 3720 BA, Bilthoven, The Netherlands

Dr C. Spartà Commission of the European Communities, Joint Research Centre, Ispra Environment Institute, 21020 Ispra (VA), Italy

Dr G.J.A. Speijers Laboratory of Toxicology, Dutch National Institute of Public Health and Environmental Protection (RIVM), PO Box 1-NL 3720 BA, Bilthoven, The Netherlands

Dr L.A. Wallace US Environmental Protection Agency, Building 166 Bicher Raod, Vint Hill Farms Station, Warrenton, VA 22091, USA.

For Monique and Ingrid

Contents

Editorial introduction

Volatile organic compounds (VOCs) and their reaction products are increasingly regarded as posing unacceptable risks to public and occupational health, as well as to biological and physical environments. VOCs may be identified and measured in all compartments of the environment and on all scales ranging from the individual to the global. Their ubiquity reflects partly their volatility—a vapour pressure greater than 0.13 kPa is conventionally used to distinguish them from less volatile organics[1]—and partly their involvement in comparably ubiquitous living processes—some VOCs may be referred to as biogenic. However, the key role of VOCs in the industries and economies of today's fossil fuel-based societies, together with the quantities of VOCs involved, are the prime contributing factors. Anthropogenic VOCs have now also appeared in abundance.

While some biogenic VOCs perform various biophysical and biochemical functions within living cells and mircoorganisms, others are released by living organisms—sometimes in great quantities—into the surroundings. Dimethyl sulphide is emitted into seawater and thence to the atmosphere by marine microorganisms throughout the world; isoprene and terpenes are released into the air by vegetation as diverse as the woodlands of Central Europe and the prairies of North America, and, during the rice-growing season, vast amounts of methane are produced and released into the atmosphere from Asian paddy fields. This methane may be biogenic in origin, but its occurrence and quantity are anthropogenically determined, being dependent both on rice-growing and on specific paddy cultivation techniques. Similarly, most of the VOCs referred to as anthropogenic are, or are believed to be, of biological origin, but their presence, especially at current quantities, is the result of social, economic and industrial activities.

In pre-industrial societies sources of VOCs such as wood and 'natural' oils, tars and waxes were exploited largely as fuels, primarily on a domestic scale. When the Industrial Revolution gained momentum in the 19th century, emissions of organic compounds increased exponentially. In pre-war Germany the chemical industry manufactured numerous compounds, among them novel VOCs. In this emerging industry, still based on coal, VOCs were usually formed after processing the heavier starting material. Exploration of the vast

[1]Subdivisions of the group, excluding certain species or distinguishing classes according to criteria other than volatility, may be encountered if indicated by the application.

oil fields in the USA and the Middle East transformed pre-war communities into oil-based societies. Nowadays, light oil products, obtained from crude oil directly or after cracking heavy fractions, make up the essential ingredients of modern transportation, energy generation and the petrochemical industries. The seemingly unlimited capacity of the chemical industry has created a wide range of new and sometimes exotic compounds. Ever-expanding product applications are accompanied by seemingly inevitable spills. Only in a few cases are VOCs not released into the environment; in most cases they are flushed into aquatic systems, evaporated into the air or dumped in the soil. In addition to 'regular' spills (the end result of malcontrolled chemical or physical processes), accidental releases (e.g. uncontrolled emissions from industrial plants or wrecked tankers at sea) contribute to the increasing levels of a wide range of compounds. This happens not only near industrial and urban areas, but also in remote areas such as the Arctic stratosphere, where stable organic compounds are found in elevated concentrations.

The mere presence of VOCs in the environment, even in large quantities, would simply be a scientifically interesting phenomenon, were it not for the range of adverse effects known, or suspected, to be associated with them. The exposure of the environment and living organisms to VOCs has altered the delicate balance of ecosystems and has caused adverse health effects. The relationship between these health effects and exposure to VOCs was first recognized in the chemical industry, where high concentrations of a limited number of compounds existed. Toxicological research revealed that many VOCs have various reversible and irreversible effects on the human body, ranging from acute anaesthesia to long-term effects such as induction of carcinomas. Ecotoxicological research has indicated effects on the environment; these range from changes in the population of terrestrial and aquatic ecosystems to the extinction of vulnerable species. All of these effects can be considered to be directly due to VOCs, although the active agent may be one or more metabolites produced by the target species.

A current, extremely relevant, indirect effect of VOCs was recognized in the 1950s. Heavy air pollution in the Los Angeles basin was found to be created by the photo-oxidation of VOCs in the presence of nitrogen oxides. The main constituent of this type of smog was found to be ozone, now known to have reversible and irreversible effects on the lungs. Furthermore, the effects are not limited to humans and animals; crops have also been proven to be damaged after exposure to this type of air pollution. Recent research has revealed that this 'urban soup' also contains a multitude of other reactive compounds, a number of which are suspected to be biologically and ecologically active.

In the 1980s two further major indirect effects of VOCs on a global scale were discovered. First, after sophisticated satellite measurement of the Antarctic ozone column and through examination and re-evaluation of the data obtained, it was shown that inert VOCs, until then thought to be harmless, were related to dramatic ozone depletion in the stratosphere above the Antarctic. A second indirect effect is global climatic change. The presence of, again, relatively unreactive compounds in the upper atmosphere is thought to be responsible for trapping surplus solar radiation. Although the process is not completely understood, current evidence indicates changes for some global regions that are dramatic enough to warrant strong recommendations of dramatic measures.

The presence of VOCs in the indoor environment also poses a threat to public health. Modern building materials and household products are sources of many different VOCs. The increasing effort to reduce heat loss from buildings has initiated nationwide programmes for insulating walls and windows and recirculating air in air-conditioned buildings. This has led to a reduction of forced and natural ventilation and so to the build-up of concentrations of compounds emitted by indoor sources. It is clear that VOCs have increasingly affected the world we live in and that a better understanding of their occurrence and behaviour is of eminent importance to counteract their effects.

Scientific and public awareness of the threat that VOCs pose to humans and the environment has alerted regulatory bodies. Many initiatives have been launched to reduce the emission of VOCs or to reduce their adverse impact on humans and the environment, especially in the industrial nations where local and regional levels and effects are most pronounced. The intrinsic conflict arising from the basic role of VOCs in modern society and the need to prevent the environment from becoming hostile has complicated and delayed, but not obstructed, the ratification of various agreements on reduction of VOCs. Wherever possible, the air toxic, benzene, is banned from industrial processes and solvents. The Montreal Protocol to limit the use of and, ultimately, to ban chlorofluorocarbons is ratified by many countries. The ECE VOC Protocol to reduce ozone levels by the reduction of VOC emission is already having some effect. The amendments to the US Clean Air Act deal with the reduction of VOCs from industrial and mobile sources. These are some major illustrations of the intention to preserve the delicate balances so important to life.

All of these initiatives are based on the work of researchers from several disciplines in many institutions worldwide. Their research is quite diverse, including public health, studies on the physics and chemistry in the upper stratosphere, and research on ecosystems at the bottom of the sea. As in many other fields, communication between researchers of various disciplines on the one hand, and between researchers, policy makers, research programme officers and the public on the other, although essential, is hampered by the complexity and limited knowledge of the issues.

This book can be seen as a contribution to improving communication between the various participants in the search for solutions to the many problems created by the presence of VOCs. Scientists from the main fields of VOC investigation highlight aspects relevant to the impact assessment of VOCs. In addition to the biochemistry of VOCs in the human body, and the chemistry and physics of VOCs in environmental compartments, special attention is given to both analytical and modelling techniques, for defining identities and assessing levels of VOCs. Distinction is made between direct effects on human health, and mainly ecological, but also some indirect human health effects encountered in environmental compartments. The first aspect is dealt with in chapters 1 and 2. In chapters 3, 4 and 5 the main compartments of the environment—air, water and soil, are explored. Assessing the occurrence of VOCs by means of analytical techniques is considered to be one of the driving forces in this area of research, therefore chapter 6 is devoted to future developments in monitoring techniques for VOCs. Finally, chapter 7 deals solely with occupational health, probably the most developed research area.

The compilation of facts presented in this book cannot, and is not intended to be complete but will certainly paint a broad picture of what is happening in the many institutes concerned with increasing our knowledge of the role VOCs. The aim of the book is to enable researchers, policy makers and programme officers concerned with public health and environmental quality to familiarize themselves with the insights and approaches presented, and so to improve their understanding of the uncertainties and the scope of current knowledge. This may prove to be essential for progress between disciplines.

We would like to express our gratitude to the individual contributors presenting so many facts, insights and approaches from the day-to-day work done in their particular fields. Compiling this book, in which such a wide area of research has been covered, has only been possible through their valuable contributions. Finally, we thank the Board of Directors of the National Institute of Public Health and Environmental Protection in Bilthoven for the facilities necessary to complete this book.

Henk J.Th. Bloemen
James Burn

1 VOCs and the environment and public health—exposure

L.A. WALLACE

1.1 Introduction

Volatile organic compounds (VOCs) are present in essentially all natural and synthetic materials, from gasoline to flowers, from water to wine. The uses of these versatile compounds are innumerable, but include fuels, solvents, fragrances, biocides and flavor constituents. VOCs often exist as vapors or liquids at room temperature but may also be in the form of solids (e.g. naphthalene and para-dichlorobenzene, used as mothballs and bathroom deodorants).

For most VOCs, air is the main route of exposure. In some special cases, drinking water is an important route (e.g. chloroform from chlorinated water). For some natural VOCs, food and beverages are important routes (e.g. limonene, a natural component of citrus fruits but also a synthetic addition to soft drinks, teas, etc.). Travel through soil has been hypothesized to be important in the case of homes or schools built on contaminated landfills.

Two main health effects are of interest: cancer and acute irritative effects (e.g. *sick building syndrome*). The latter may have as great or a greater economic effect than the former.

1.2 Measurement methods

1.2.1 Air

1.2.1.1 Sampling methods

Activated charcoal. Historically, the measurement method of choice for *occupational* exposures (generally 1–100 parts per million, or ppm, for a given VOC) has been to pump a sample of air across a sorbent (usually activated charcoal) in order to concentrate the VOCs. They are then recovered by a solvent such as carbon disulphide.

In the early 1980s, passive badges employing activated charcoal were developed for use in occupational sampling. The badges operate on the principle of diffusion, and are often operated over an 8 h workday to provide an integrated average exposure for comparison to the occupational standards (e.g. the threshold limit value, or TLV). However, the manufacturing process for these badges leaves residues of VOCs on the activated carbon, which makes the badges unsuitable for short-term sampling at *environmental* concentrations, which are usually at part per billion (ppb) levels. However, the high background contamination on the badges can be overcome by extending the time of sampling to a week or more, and several studies of indoor air pollution have adopted this technique (Seifert and Abraham, 1983; Mailahn *et al.*, 1987).

Tenax. The background problems associated with activated charcoal, as well as problems in obtaining reliable recoveries of sorbed chemicals, led to a search for a more suitable sorbent. A polymer known as Tenax was widely adopted during the 1970s as a more reliable sorbent than charcoal for ppb levels (Barkley *et al.*, 1980; Krost *et al.*, 1982). Tenax, properly cleaned, has very low background contamination. It is also stable at temperatures up to 250 °C, allowing thermal desorption instead of solvent desorption. (Solvent desorption involves a redilution of the VOCs, thus partially negating the concentration made possible by the sorbent.) Drawbacks include artefact formation of several chemicals (e.g. benzaldehyde, phenol) and an inability to retain very volatile organic chemicals (e.g. vinyl chloride, methylene chloride). Although most uses of Tenax sorbent have been with active (pumped) samplers, a passive badge containing Tenax has also been developed (Countant *et al.*, 1986).

Multisorbent systems. In the late 1980s, attempts were made to combine the best attributes of charcoal and Tenax into a multisorbent system. Newer types of activated charcoal (Spherocarb, Carbosieve) were developed to provide more reliable recoveries. Tandem systems employing Tenax as the first sorbent and activated charcoal as the second, or backup, sorbent were employed. The Tenax collected the bulk of the VOCs and the activated charcoal collected those more volatile VOCs that 'broke through' the Tenax. Systems were also developed using three sorbents, such as Tenax, Ambersorb, and Spherocarb or Carbosieve (Hodgson *et al.*, 1986). All such systems allow collection of a broader range of chemical types and volatilities.

Direct (whole air) sampling. This method, first developed in the 1970s for upper atmosphere sampling, avoids the sorption–desorption step, which should theoretically allow less chance for contamination. (However, it requires great sensitivity on the part of the detection instruments.) The method may involve real-time sampling in mobile laboratories, with direct injection of the air sample into a cold trap attached to a gas chromatogram;

or sampling in evacuated electropolished aluminum canisters for later laboratory analysis (Oliver *et al.*, 1986).

Comparison of sampling methods. No single method of sampling VOCs in the atmosphere or indoors has become a standard or reference method. In the US, the two preferred methods are Tenax and evacuated canisters. These two methods were compared under controlled conditions in an unoccupied house (Spicer *et al.*, 1986). Ten chemicals were injected at nominal levels of about 3, 9 and 27 $\mu g/m^3$. The results showed that the two methods were in excellent agreement, with precisions of better than 10% for all chemicals at all spiked levels.

In Europe, the two most common methods are Tenax and activated charcoal. One study employing both methods side by side (Skov *et al.*, 1990) found consistently higher levels of total VOCs on the charcoal sorbent. The difference may be due to very volatile organics such as pentane and isopentane, which are collected by charcoal but which break through Tenax readily.

The sorbent methods lend themselves to personal monitoring—a small battery-powered pump is worn for an 8 h or 12 h period to provide a time-integrated sample. However, at present, the whole-air methods employ bags or canisters that are too bulky or heavy to be used as personal monitors.

1.2.1.2 Analysis. Samples are usually analyzed by first separating the components using gas chromatography (GC). Three detection methods in common use are flame ionization detection (FID), electron capture detection (ECD) and mass spectrometry (MS). Only GC–MS has the ability to unambiguously identify many chemicals. Neither GC–FID nor GC–ECD is able to separate chemicals that coelute (emerge from the chromatographic column at the same time). Also, GC–FID response is depressed by chlorine and other halogens, so it is not suitable for samples containing halogens. Mass spectrometry, by breaking chemicals into fragments and then identifying these fragments, is often capable of differentiating even among coeluting chemicals. However, since chemicals are identified by comparing these mass fragment spectra to existing libraries, and the libraries are incomplete, even GC–MS identifications are often tentative or mistaken. (One study using known mixtures of chemicals found about 75% accuracy of identification for several different GC–MS computerized spectral search systems.)

1.2.1.3 Olfactory analysis. Because of the complexity of most indoor and outdoor air samples, the cost and inaccuracy of measurement methods, and the almost complete lack of knowledge of the relationship of measured VOC levels within complex mixtures to resulting health effects, an alternative to chemical measurement methods has arisen in recent years—use of a trained panel to judge the possible health or comfort effects of an air sample directly. This method, pioneered by Ole Fänger of Denmark (Fänger, 1990), employs panels of six to ten persons who have been previously trained by sampling unknown mixtures of odorous compounds. When exposed to a test atmosphere, the judges provide an instantaneous estimate of its pollution potential, measured in units called decipols. One decipol is equivalent to the amount of pollution (body odor) produced by one person in a room ventilated at one air change per hour. The method is capable of predicting how persons will react to air of a given quality, and also of estimating the relative contribution of various sources (e.g. ventilation system, office machines, employees) to indoor air quality.

1.2.2 Water

The method of choice for sampling VOCs in water has been the purge-and-trap method developed by Bellar and Lichtenburg (1974). In this method, the VOCs are purged from the water sample into the headspace of a collection vessel, concentrated on a sorbent such as Tenax, and analyzed by GC–ECD, GC–FID or GC–MS. For drinking water samples collected from distribution systems that employ chlorination as a treatment process, the most important VOCs to sample are the trihalomethanes (particularly chloroform); detection will often be by ECD, a sensitive method for halogens.

1.2.3 Body fluids

Similar techniques to those above are used to sample breath, blood and urine. Exhaled breath can be collected in Tedlar bags, pumped across Tenax cartridges, and analyzed by GC–MS (Wallace and O'Neill, 1987); alternatively, the breath may be collected in electropolished canisters (Raymer *et al.*, 1990). Urine may be analyzed by a purge-and-trap approach similar to the Bellar and Lichtenburg method for water (Michael *et al.*, 1980). Mothers' milk has been analyzed by a similar purge-and-trap method (Pellizzari *et al.*, 1982). Blood may be analyzed by an isotope dilution technique (CDC, 1990).

1.3 Human exposure

1.3.1 Air

1.3.1.1 Personal air. Between 1979 and 1987, the US EPA carried out the TEAM (Total Exposure Assessment Methodology) studies to measure personal exposures of the general public to VOCs in several geographic areas in the United States (Pellizzari *et al.*, 1987a, b; Wallace, 1987). About 20 target

VOCs were included in the studies, which involved about 800 persons, representing 800 000 residents of the areas. Each participant carried a personal air quality monitor containing 1.5 g Tenax. A small battery-powered pump pulled about 20 l of air across the sorbent over a 12 h period. Two consecutive 12 h personal air samples were collected for each person. Concurrent outdoor air samples were also collected in the participants' back yards. In the studies of 1987, fixed indoor air samplers were also installed in the living rooms of the homes.

The initial TEAM pilot study (Wallace *et al.*, 1982) in Beaumont, TX and Chapel Hill, NC indicated that personal exposures to about a dozen VOCs exceeded outdoor air levels, even though Beaumont TX has major oil producing, refining and storage facilities. These findings were supported by a second pilot study in Bayonne-Elizabeth, NJ (another major chemical manufacturing and petroleum refining area) and Research Triangle Park, NC (Wallace *et al.*, 1984a). A succeeding major study of 350 persons in Bayonne-Elizabeth (Wallace *et al.*, 1984b) and an additional 50 persons in a nonindustrial city and a rural area (Wallace *et al.*, 1987a) reinforced these findings. A second major study in Los Angeles, CA and in Antioch-Pittsburg, CA (Wallace *et al.*, 1988), with a follow-up study to Los Angeles in 1987 (Wallace *et al.*, 1991a), added a number of VOCs to the list of target chemicals with similar results.

Major findings of these TEAM Studies included the following:

a) Personal exposures exceeded median outdoor air concentrations by factors of 2–5 for nearly all prevalent VOCs. The difference was even larger (factors of 10 or 20) when the maximum values were compared (Figure 1.1). This is despite the fact that most of the outdoor samples were collected in areas with heavy industry (New Jersey) or heavy traffic (Los Angeles).
b) Major sources (Table 1.1) are consumer products (bathroom deodorizers, moth repellents); personal activities (smoking, driving); and building materials (paints and adhesives). In the US, one chemical (carbon tetrachloride) has been banned from consumer products and exposure is thus limited to the global background of about 0.7 $\mu g/m^3$.
c) Traditional sources (automobiles, industry, petrochemical plants) contributed only 20–25% of total exposure to most of the target VOCs (Wallace *et al.*, 1991b). No difference in exposure was noted for persons living close to chemical manufacturing plants or petroleum refineries.

1.3.1.2 Indoor air

Concentrations. Three large studies of VOCs, involving 300–800 homes, have been carried out in The Netherlands (Lebret *et al.*, 1986), Germany (Krause *et al.*, 1987) and the US (Wallace, 1987). A smaller study of 15 homes was carried out in Northern Italy (De Bortoli *et al.*, 1986). Observed concen-

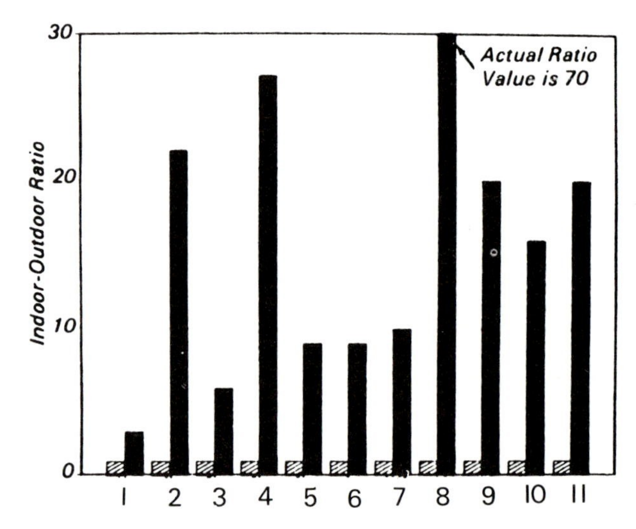

Figure 1.1 Comparison of unweighted 99th percentile concentrations of 11 prevalent chemicals in overnight outdoor (▨) and overnight personal (■) air in New Jersey (Fall, 1981). Even in an area with concentrated chemical plants and petroleum refineries, the highest exposures to 11 different toxic VOCs far outweigh the highest outdoor concentrations. VOCs are: 1, chloroform; 2, 1,1,1-trichloroethane; 3, benzene; 4, carbon tetrachloride; 5, trichloroethylene; 6, tetrachloroethylene; 7, styrene; 8, *m*-plus *p*-dichlorobenzene; 9, ethylbenzene; 10, *o*-xylene; and 11, *m*-plus *p*-xylene. From Wallace (1987).

Table 1.1 Common volatile organic chemicals and their sources

Chemicals	Major sources of exposure
Acetone	Cosmetics
Alcohols (ethanol, isopropanol)	Spirits, cleansers
Aromatic hydrocarbons (toluene, xylenes, ethylbenzene, trimethylbenzenes)	Paints, adhesives, gasoline, combustion sources
Aliphatic hydrocarbons (octane, decane, undecane)	Paints, adhesives, gasoline, combustion sources
Benzene	Smoking, auto exhaust, passive smoking, driving, pumping gas
Carbon tetrachloride	Fungicides, global background
Chloroform	Showering (10 min average), washing clothes, dishes
p-Dichlorobenzene	Room deodorizers, moth cakes
Formaldehyde	Pressed wood products
Methylene chloride	Paint stripping, solvent use
Styrene	Smoking
Tetrachloroethylene	Wearing or storing dry-cleaned clothes, visiting dry cleaners
1,1,1-Trichloroethane	Wearing or storing dry-cleaned clothes, aerosol sprays
Trichloroethylene	Unknown (cosmetics, electronic parts, correction fluid)
Terpenes (limonene, α-pinene)	Scented deodorizers, polishes, fabrics, fabric softeners, cigarettes, food, beverages

trations were remarkably similar for most chemicals, indicating similar sources in these countries. One exception is chloroform, present at typical levels of 1–4 $\mu g/m^3$ in the US but not found in European homes. This is to be expected, since the likely source is volatilization from chlorinated water (Wallace *et al.*, 1982; Andelman, 1985a,b); the two European countries do not chlorinate their water.

Major findings of these indoor air studies include:

a) Indoor levels in homes and older buildings (>1 year) are typically several times higher than outdoor levels. Sources include dry-cleaned clothes, cosmetics, air fresheners, cleaning materials.
b) New buildings (<1 month) have levels of some VOCs (aliphatics and aromatics) 100 times higher than outdoor, falling to 10 times outdoor about 2–3 months later. Major sources include paints and adhesives.
c) About half of 750 homes in the USA had total VOC levels (obtained by integrating the total ion current response curve of the mass spectrometer) greater than 1 mg/m^3, compared to only 10% of outdoor samples (Wallace *et al.*, 1992).
d) More than 500 volatile organic compounds (VOCs) were identified in four buildings in Washington, DC and Research Triangle Park, NC. (Sheldon *et al.*, 1988a).

One study (Wallace *et al.*, 1989) involved seven volunteers undertaking about 25 activities suspected of causing increased VOC exposures; a number of these activities (using bathroom deodorizers, washing dishes, cleaning an auto carburetor) resulted in 10- to 1000-fold increases in 8h exposures to specific VOCs.

Sources. Early studies of organics indoors were carried out in the 1970s in the Scandinavian countries (Berglund *et al.*, 1982a,b; Johansson, 1978; Mølhave and Møller, 1979). Mølhave (1982) showed that many common building materials used in Scandinavian buildings emitted organic gases. Seifert and Abraham (1982) found higher levels of benzene and toluene associated with storage of magazines and newspapers in German homes. Early US measurements were made in nine Love Canal residences (Pellizzari *et al.*, 1979); 34 Chicago homes (Jarke and Gordon, 1981); and in several buildings (Hollowell and Miksch, 1981; Miksch *et al.*, 1982).

Hundreds of VOCs have been identified in environmental tobacco smoke (Jermini *et al.*, 1976; Guerin *et al.*, 1987; Higgins, 1987; Löfroth *et al.*, 1989), which contaminates about 60% of all US homes and workplaces (Repace and Lowrey, 1980, 1985). Among these are several human carcinogens, including benzene. Benzene was elevated in the breath of smokers by a factor of ten above that in the breath of nonsmokers (Wallace *et al.*, 1987b.) The amount of benzene in mainstream smoke appears to be directly related to the amount of tar and nicotine in the cigarette (Higgins *et al.*, 1983). In the

US, it is calculated that the 50 million smokers are exposed to about half of the total nationwide 'exposure budget' for benzene (Wallace, 1990). Other indoor combustion sources such as kerosene heaters (Traynor *et al.*, 1990) and woodstoves (Highsmith *et al.*, 1988) may emit both volatile and semivolatile organic compounds (SVOCs).

Later studies also investigated building materials (Sheldon *et al.*, 1988a, b) but added cleaning materials and activities such as scrubbing with chlorine bleach or spraying insecticides (Wallace *et al.*, 1987c), and using adhesives (Girman *et al.*, 1986) or paint removers (Girman *et al.*, 1987). Knöppel and Schauenburg (1987) studied VOC emissions from ten household products (waxes, polishes, detergents); 19 alkanes, alkenes, alcohols, esters and terpenes were among the chemicals emitted at the highest rates from the ten products. All of these studies employed either headspace analysis or chambers to measure emission rates.

Other studies estimated emission rates from measurements in homes or buildings. For example, Wallace (1987) estimated emissions from a number of personal activities (visiting dry cleaners, pumping gas) by regressing measurements of exposure or breath levels against the specified activities. Girman and Hodgson (1987) extended their chamber studies of paint removers to a residence, finding similar (and very high—ppm) concentrations of methylene chloride in this more realistic situation.

The US National Aeronautics and Space Agency (NASA) has measured organic emissions from about 5000 materials used in space missions (Nuchia, 1986). Perhaps 3000 of these materials are in use in general commerce (Ozkaynak *et al.*, 1987). The chemicals emitted from the largest number of materials included toluene (1896 materials), methyl ethyl ketone (1261 materials) and xylenes (1111 materials).

A 41 day chamber study (Berglund *et al.*, 1987) of aged building materials taken from a 'sick' preschool indicated that the materials had absorbed about 30 VOCs, which they re-emitted to the chamber during the first 30 days of the study. Only 13 of the VOCs originally present in the first days of the study continued to be emitted in the final days, indicating that these 13 were the only true components of the materials. This finding has significant implications for remediating 'sick buildings'. Even if the source material is identified and removed, weeks may be needed before re-emission of organics from sinks in the building ceases.

Another study (Seifert and Schmahl, 1987) of sorption of VOCs and SVOCs on materials such as plywood and textiles concluded that sorption was small for the VOCs studied.

Emission rates of most chemicals in most materials are greatest when the materials are new. For 'wet' materials such as paints and adhesives, most of the total volatile mass may be emitted in the first few hours or days following application (Tichenor and Mason, 1987; Tichenor *et al.*, 1990). EPA studies of new buildings indicated that eight of 32 target chemicals measured within days after completion of the building were elevated 100-fold compared to outdoor levels: xylenes, ethylbenzene, ethyltoluene, trimethylbenzenes, decane and undecane (Sheldon *et al.*, 1988b). The half-lives of these chemicals varied from two to six weeks; presumably some other non-target chemicals, such as toluene, would have shown similar behavior. The main sources were likely to be paints and adhesives. Thus, new buildings would be expected to require about six months to a year to decline to the VOC levels of older buildings.

For dry building materials such as carpets and pressed wood products, emissions are likely to continue at low levels for longer periods. Formaldehyde from pressed wood products may be slowly emitted with a half-life of several years (Breysse, 1984). According to several recent studies, 4-phenylcyclohexene (4-PC), a reaction product occurring in the styrene–butadiene backing of carpets, is the main VOC emitted from carpets after the first few days. 4-PC is likely to be largely responsible for the new carpet odor.

A major category of human exposure to toxic and carcinogenic VOCs is room air fresheners and bathroom deodorants. Since the function of these products is to maintain an elevated indoor air concentration in the home or the office over periods of weeks (years with regular replacement), extended exposures to the associated VOCs are often the highest likely to be encountered by most (nonsmoking) persons. The main VOCs used in these products are *p*-dichlorobenzene (widely used in public restrooms), limonene and α-pinene. The first is carcinogenic to two species (NTP, 1986), the second to one (NTP, 1988), and the third is mutagenic. Limonene (lemon scent) and α-pinene (pine scent) are also used in many cleaning and polishing products, which would cause short-term peak exposures during use, but which might not provide as much total exposure as the air freshener.

Awareness is growing that most exposure comes from these small nearby sources. In California, Proposition 65 focuses on consumer products, requiring makers to list carcinogenic ingredients. Bills focusing on indoor air were introduced in both the House and the Senate in 1989 and 1990. Environmental tobacco smoke (ETS), was declared a known human carcinogen by the EPA in 1991; smoking has been banned from many public places and many private workplaces during the last few years.

1.3.1.3 Outdoor air. Outdoor air levels of many of the most common VOCs, even in heavily industrialized areas or areas with high densities of vehicles, are usually considerably lower than indoor levels (Wallace, 1987). This fact has not been fully recognized or incorporated into regulations. For example, in the US, the 1990 reauthorization of the Clean Air Act continues to deal only with outdoor air ("air external to buildings"), while adding 189 toxic chemicals to the list of those to be regulated. Many of these chemicals, which include common solvents and household pesticides, have been shown to be far more prevalent and at higher concentrations in homes than outdoors (Figure 1.2).

For many hydrocarbons, the major source of outdoor air levels is gasoline

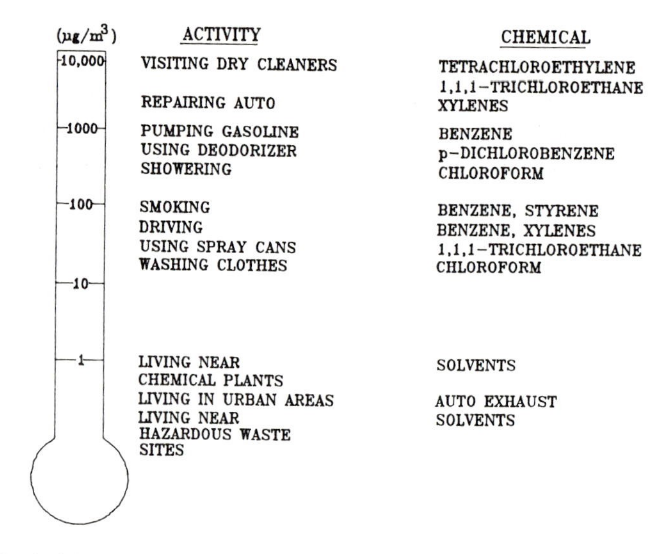

Figure 1.2 Activity toximeter. Common activities can lead to exposures that are far greater than living in urban industrial areas or near hazardous waste sites. All data based on measurements.

vapor or auto exhaust (Sigsby *et al.*, 1987; Zweidinger *et al.*, 1988). For example, about 85% of outdoor air benzene levels in the US is from mobile sources and only about 15% from stationary sources. Exposure to certain aromatics (benzene, toluene, xylenes, ethylbenzene) while inside the automobile can exceed ambient levels by a factor of six or so (SCAQMD, 1989).

1.3.2 Drinking water

In areas that chlorinate their drinking water, chlorination by-products such as chloroform and other trihalomethanes (THMs) contaminate the finished water (Bellar and Lichtenberg, 1974; Rook, 1974; IARC, 1991). The discovery of chloroform in the blood of New Orleans residents (Dowty *et al.*, 1975) led to the Safe Drinking Water Act (SDWA) of 1974, which set a limit of 100 μg/l for total THMs in finished water supplies.

A series of nationwide surveys of THM levels in water supplies have been carried out in the US since the passage of the SDWA. The National Organics Reconnaissance Survey of 80 treatment plants (Symons *et al.*, 1975) indicated that about 20% exceeded the THM standard of 100 μg/l. A more recent survey of 727 utilities (McGuire and Meadow, 1988) representing more than half of consumers indicated that only about 3.6% of water supplies surveyed continued to exceed the standard.

In succeeding years, it has become more evident that THM exposure occurs in ways other than drinking chlorinated water. Since treated water is normally used for all other household purposes, volatilization from showers, baths, and washing clothes and dishes is an important route of exposure. Studies of experimental laboratory showers by Andelman (1985a, b, 1990; Andelman *et al.*, 1986) resulted in the estimate that chloroform exposure from volatilization during a typical 8 min shower could range from 0.1 to 6 times the exposure from drinking water from the same supply, depending on the amount of water ingested per day. Other studies of full-scale showers corroborate these conclusions, with estimates of exposure equivalent to ingestion of 1.3–3.7 l/day of tap water at the same concentrations as the water in the shower (McKone, 1987; Hodgson *et al.*, 1988; Jo *et al.*, 1990a, b; McKone and Knezovich, 1991; Wilkes *et al.*, 1992). Inhalation of airborne chloroform during the rest of the day was also found to be comparable to ingestion, based on measurements of 800 persons and homes in the TEAM Study.

1.3.3 Food and beverages

Food and beverages made from treated water (ice cream, soft drinks) have been found to contain chloroform and other THMs (Entz and Diachenko, 1982). Some natural components of fruits and vegetables, such as limonene, have been found to be mutagenic and/or carcinogenic in animal testing (NTP, 1988). Although benzene has been reported at high levels in certain foods (Pearson and McConnell, 1975), more recent investigations (API, 1992) have failed to find benzene in any of a large number of foods tested from various food groups. Margarine and butter in supermarkets have been found to absorb tetrachloroethylene from nearby dry cleaners (Uhler and Diachenko, 1987), and trichloroethylene from the glue in the containers (Entz and Diachenko, 1988). Some chlorinated chemicals were found in food in Germany (Bauer, 1981).

1.3.4 Dust and soil

A few recent studies (Hodgson *et al.*, 1988; Kliest *et al.*, 1989) indicate that migration of VOCs through soil from hazardous waste sites or municipal landfills could contribute to exposure of nearby residents. Organics contained in soil could be absorbed through the skin; McKone (1990) estimates that for benzene in soil on the skin, uptake would be less than 3% over a 12 h period.

1.3.5 Dermal absorption

Several studies in Italy and the US indicate that dermal absorption during showers, baths, or swimming at indoor pools or spas may be an important route of exposure to VOCs in treated water. Early studies by Stewart and co-workers (Stewart and Dodd, 1964; Hake *et al.*, 1976) provided absorption coefficients for human volunteers who immersed their hands in pure solvents such as tetrachloroethylene. Experimental studies on animals (Bogen *et al.*, 1992; Shatkin and Brown, 1991) provided absorption coefficients for more realistic situations involving dilute solutions. One study of swimmers (Aggazzotti *et al.*, 1990) measured high concentrations of chloroform in blood, but could not disentangle the contributions of inhalation and dermal absorption. A study of chloroform in exhaled breath of volunteers showering both with and without rubber suits (Jo *et al.*, 1990a,b; Weisel *et al.*, 1990) indicated that dermal absorption during the shower was comparable to inhalation.

1.4 Body burden

Measurement of body burden provides a direct indication of the total dose through all important environmental pathways. Coupled with adequate pharmacokinetic models (Ramsey and Anderson, 1984), including knowledge of how the VOC partitions between blood, air and fat (Sato and Nakajima, 1979), measurement of a single body fluid such as breath or blood can give an indication of the total body burden and sometimes even an indication of the mode of exposure. For example, the breath measurements of smokers in the TEAM study indicated that they were exposed to 6–10 times as much benzene as nonsmokers. The personal air quality monitors had been unable to document this increase, since most of the exposure came from direct smoking with no external pathway.

1.4.1 Breath

For many VOCs, the most sensitive method of determining body burden is measurement of exhaled breath. Detection limits using Tenax sorbents or evacuated canister samples are usually well under 1 $\mu g/m^3$. For comparable sensitivity, blood measurements would have to detect levels as low as 10 ng/l. Early breath measurements at environmental (ppb) levels were made by Krotoczynski *et al.* (1977, 1979), who measured levels of more than 50 VOCs in the breath of 54 healthy nurses. Additional measurements were made of 800 participants in the TEAM study (Wallace, 1987; Wallace *et al.*, 1991a). These measurements first documented the overwhelming importance of smoking as a source of exposure to benzene and styrene.

Although it is true that breath measurements reflect only the most recent exposure, it is also true that if the time of taking the breath measurement is carefully chosen, it can reflect long-term exposure. That is, persons who have been in their normal environment for a few hours are not likely to be far out of equilibrium with their surroundings (provided the exposure of their previous environment was not vastly different). Also, some experimental studies using room-size chambers (Gordon *et al.*, 1988, 1990) or specially designed spirometers (Raymer *et al.*, 1991) have determined the residence times of more than a dozen VOCs in various 'compartments' of the body: blood (2–10 min), tissues rich in blood vessels (1–2 h) and 'vessel-poor' tissues (4–8 h).

1.4.2 Blood

Blood measurements at occupational levels have been made for many years (Monster and Houthooper, 1979; Monster and Smolders, 1984). Blood measurements at environmental levels have been made recently for chloroform (Pfaffenberger and Peoples, 1982; Aggazzotti *et al.*, 1990) and for benzene (Brugnone *et al.*, 1987). The 1989–93 National Health and Nutrition Examination Survey (NHANES III) will include 500–1000 persons whose blood will be analyzed for a dozen of the most prevalent VOCs.

1.4.3 Mothers' milk

Scores of VOCs were detected in broad-spectrum GC–MS analysis of samples collected from 40 nursing mothers (Pellizzari *et al.*, 1982). Tetrachloroethylene and *p*-dichlorobenzene were detected in the milk of 17 nursing mothers (Sheldon *et al.*, 1983).

1.5 Human health effects

1.5.1 Acute effects (sick building syndrome)

Acute effects of VOCs at high (ppm) levels include eye, nose and throat irritation, and central nervous system responses such as dizziness, headaches and loss of short-term memory. These symptoms are often seen in new buildings or in buildings undergoing renovation (Berglund *et al.*, 1984). Although the cause or causes of sick building syndrome (SBS) remain unknown, VOCs are suspected. Indoor levels of VOCs are often elevated, as indicated in section 1.3.1.2. Levels in new buildings or renovated buildings are even higher. Field studies indicate that total VOC levels exceeding 1–2 mg/m^3 are often found in 'sick buildings' (Mølhave, 1987). Laboratory studies have shown that certain symptoms (headache, eye irritation) can be produced in both sensitive and healthy subjects at TVOC levels of 5 and 25 mg/m^3 (Mølhave *et al.*, 1986; Otto *et al.*, 1990).

1.5.1.1 Large building studies. Four European studies and one American study have administered questionnaires to several thousand office workers in an attempt to learn more about building-related health complaints. The

European studies took place in the UK (Wilson and Hedge, 1987; Raw *et al.*, 1990), Denmark (Skov *et al.*, 1989, 1990; Gravesen *et al.*, 1990), Sweden (Stenberg *et al.*, 1990; Sundell *et al.*, 1990) and The Netherlands (Preller, *et al.*, 1990; Zweers *et al.*, 1990). The US study was carried out in Washington, DC (EPA, 1989, 1990). All four of the European studies involved a large number of buildings (28 to 61), whereas the US study included only four. Thus the European investigators were able to compare, for example, buildings with natural ventilation to those with mechanical ventilation, and buildings with spray humidification to those with steam humidification. On the other hand, since the US study involved a census rather than a sample of employees, it was possible to carry out a spatial analysis, testing the effect of all air handling systems employed in the buildings as well as the effect of new carpet installed in two of the buildings. (The Danish investigators also carried out a census of 4369 employees in 28 buildings, but did not analyze spatial variation within the buildings.)

1.5.1.2 Comparison of findings. In view of the different definitions of health symptoms (e.g. the minimum criterion for a 'positive' response varied from 2 days per year in the British study to about 100 days per year in the Danish study), a direct comparison of symptom frequencies in these studies appears to be impossible. A comparison with an earlier study of a healthy urban population (Lebowitz *et al.*, 1972a,b) showed that EPA employees reported higher frequencies of all nine of the symptoms common to both studies.

The only finding that was common to all studies was the increased symptom rates reported by females. It may be that the many factors that interact with gender (education, type of work, pay grade, type of workstation, number of persons sharing workspace, etc.) account for much or all of the observed gender difference.

In the US study, two of the four buildings studied had had carpet installed a year earlier, following which over 100 persons became ill. Because of the sharp onset of symptoms, it seems probable that VOC emissions from the new carpet or the adhesive were responsible in some way for the outbreak of symptoms. (However, in view of the findings regarding dust, it is also possible that ripping up the old carpet could have liberated spores and molds that could cause allergic symptoms to increase.) The US study investigated the effect on health of persons in the areas with the new carpet, and found only one set of throat-related symptoms (dry throat, sore throat, hoarseness) to be unequivocally associated with the newly carpeted areas. A second symptom (dizziness) was associated with men only. These effects are not limited to a small sensitive group, since those workers who had reacted strongly and were working at alternative workplaces were not included in this statistical analysis.

Although SBS symptoms are often relatively minor, they may take a heavy toll in terms of reduced productivity. For example, in the US study, half of the 4000 respondents reported experiencing a headache in the previous week, for an average of two days per week. This works out to about 200 000 headaches per year. If each headache causes ten minutes of lost work time, the lost work time would amount to about four man-years. A report by Hall *et al.* (1992) values headaches at about $8 if infrequent, or about $1.50 if frequent and accompanied by other symptoms. These values, assuming an average income of $30 000, would correspond to 10–50 man-years lost from a total workforce of 4000. When asked which symptom accounted for loss of work time, either through leaving work early or not coming in at all, headache was mentioned by 16% of the workers in the US study. Flu, colds, chills and fever, and other symptoms were mentioned by only 2–3% of workers.

1.5.1.3 Sensitivity to chemical odors. Nearly a third of people at all four buildings in the US study reported that they were unusually sensitive to a variety of chemical vapors. Such persons were significantly more likely to report health symptoms and odors. Women were more likely than men to report sensitivity to chemical fumes (37% vs. 25%), but the number of associations of this variable with health symptoms and odor problems was about the same in both sexes. Persons reporting sensitivity to chemicals were more likely to report being allergic to dust or molds. They were not, however, more likely to report any of the six psychological/social factors such as conflict at work or job dissatisfaction. These two findings are consistent with a possible immunological response as opposed to a purely psychological genesis. The sensitivity to chemical fumes in this study of sick building syndrome may bear some relation to the concept of *multiple chemical sensitivity* (Ashford and Miller, 1991).

1.5.2 Chronic effects (cancer)

Some VOCs are considered to be human carcinogens (benzene, vinyl chloride). Others are animal carcinogens and may be human carcinogens (methylene chloride, trichloroethylene, tetrachloroethylene, chloroform, *p*-dichlorobenzene). Others are mutagens (α-pinene) or weak animal carcinogens (limonene).

At present, methods of assessing carcinogenic risk are widely viewed as unsatisfactory. One method is to use the results of long-term animal studies to establish the carcinogenic potency of a given chemical (Gold, 1984), and then to extrapolate from animals to man and from high doses to low doses, assuming that there is no threshold. These twin extrapolations are the weak points of the method. For example, one species may have a completely different metabolic pathway to another, or may have a better mechanism of

DNA repair, which could lead to an effective threshold. These uncertainties are such that a given animal carcinogen may not cause human cancer at all—the actual cancer risk may be exactly zero.

A second method is to use short-term animal toxicity studies to estimate carcinogenic potential. This is based on the assumption that toxicity is correlated with carcinogenicity (Zeise *et al.*, 1986). If the assumption holds, this greatly increases the number of chemicals that can be considered, since many more toxicity studies have been done than carcinogenicity studies. However, the same uncertainties apply—plus the additional uncertainty of the basic assumption of toxicity as correlated with carcinogenicity.

In the US, the first method has been employed in a standardized way by the Environmental Protection Agency (EPA) to estimate upper limits of the carcinogenic potencies of more than 50 chemicals, including a number of VOCs. Since the method chooses the most conservative assumption at each of several points (only the results from the most sensitive animal species are used to extrapolate to humans; the dose–response curve giving the highest measure of risk at environmental concentrations is employed), the resulting potency is considered to be an upper-bound estimate—the true potency is unlikely to be greater, but could very well be smaller. A failing of this method is that it provides no best estimate—only an upper limit.

Several attempts to estimate risk from certain VOCs have been based on the first method above (Wallace, 1986), on the second (Tancrede *et al.*, 1987), or on both (McCann *et al.*, 1987). Since all these assessments depended on results from the TEAM study for estimates of population exposure, they tend to agree fairly well in the VOCs with the highest population risk: benzene, chloroform and *p*-dichlorobenzene.

In a more recent estimate of risk (Wallace, 1991), six VOCs exceeded the *de minimus* or negligible lifetime risk level of 10^{-6} (one chance in a million of contracting cancer) by a factor of ten or more: benzene, vinylidene chloride, *p*-dichlorobenzene, chloroform, methylene chloride and carbon tetrachloride.

Indoor sources accounted for the great majority (80–100%) of the total airborne risk associated with most of these chemicals. Carbon tetrachloride is the only one of the target chemicals for which outdoor sources account for a majority of the airborne risk, apparently due to its having been banned from consumer products. However, its long life in the atmosphere has led to a global background that is sufficiently high to result in a non-negligible risk.

Only one of these chemicals is considered a human carcinogen: benzene. Therefore, the risk estimate associated with benzene is on more solid ground than any of the others. Benzene is also the only one of the chemicals with human epidemiological studies showing a possible influence of environmental levels of exposure on cancer risk: two studies show that children of smokers die of leukemia at two or more times the rate of children of nonsmokers (Sandler *et al.*, 1985; Stjernfeldt *et al.*, 1986). The higher mortality rate is consistent with the measured elevated levels of benzene in the breath of smokers (suggesting exposure of the fetus in the womb of the pregnant smoker). Elevated levels of benzene in the air of homes have also been documented by both the TEAM study and the study in Germany (Krause *et al.*, 1987); however, the increase (on the order of 50% in both studies) does not seem enough to explain the increase in the mortality rate unless children are more susceptible to benzene-induced leukemia at some point in the first 8–9 years of life.

1.6 Summary

The main sources of exposure to VOCs are consumer products (e.g. room air deodorants), building materials (e.g. paints, adhesives) and personal activities (smoking, driving, taking showers). Exposure from traditional outdoor sources such as chemical plants, petroleum refineries, automobiles, and hazardous waste sites appears to be low, accounting generally for no more than 25% of personal exposure.

Short-term health effects (headache, eye irritation) appear to occur in buildings, particularly new or renovated ones. Although VOCs have not been conclusively shown as the major cause, they are thought to be important. Because of the loss of productivity involved in even minor ailments such as headaches, these effects may have a large economic impact.

Long-term health effects may include cancer. However, with the exception of benzene, a human carcinogen, most other prevalent VOCs are not known to cause human cancer, and risk estimates are extremely uncertain.

References

Aggazzotti, G., Fantuzzi, G., Tartoni, P.L. and Predieri, G. (1990) Plasma chloroform concentrations in swimmers using indoor swimming pools. *Arch. Environ. Health* **45** 175–179.

American Petroleum Institute (API) (1992) Personal communication from Will Ollison, American Petroleum Institute, Washington, DC.

Andelman, J.B. (1985a) Human exposures to volatile halogenated organic chemicals in indoor and outdoor air. *Environ. Health Perspect.* **62** 313–318.

Andelman, J.B. (1985b) Inhalation exposure in the home to volatile organic contaminants of drinking water. *Sci. Total Environ.* **47** 443–460.

Andelman, J.B. (1990) Exposure to volatile chemicals from indoor uses of water. *Total Exposure Assessment Methodology—A New Horizon*, pp. 300–311. Air and Waste Management Association, Pittsburgh, PA, VIP-16.

Andelman, J.B., Meyers, S.M. and Wilder, L.C. (1986) Volatilization of organic chemicals from indoor uses of water. In Lester, J.N., Perry, R. and Sterritt, R.M. (eds) *Chemicals in the Environment*, pp. 323–330. Selper Ltd., London.

Ashford, N. and Miller, C. (1991) *Chemical Exposures: Low Levels and High Stakes.* Van Nostrand Reinhold, New York.

Barkley, J., Bunch, J., Bursey, J.T., Castillo, N., Cooper, S.D., Davis, J.M., Erickson, M.D., Harris, B.S.H. III, Kirkpatrick, M., Michael, L., Parks, S.P., Pellizzari, E., Ray, M., Smith,

D., Tomer, K., Wagner, R. and Zweidinger, R.A. (1980) Gas chromatography mass spectrometry computer analysis of volatile halogenated hydrocarbons in man and his environment—a multimedia environmental study. *Biomed. Mass Spectrometry* **7**(4) 139–147.

Bauer, U. (1981) Human exposure to environmental chemicals—Investigations of volatile organic halogenated compounds in water, air, food, and human tissue (text in German). *Zbl. Bakt. Hyg., I. Abt. Orig. B.* **174** 200–237.

Bellar, T.A. and Lichtenberg, J.J. (1974) Determining volatile organics at microgram-per-liter levels by gas chromatograph. *J. Amer. Water Works Assoc.* **66** 739–744.

Berglund, B., Johansson, I. and Lindvall, T. (1982a) A longitudinal study of air contaminants in a newly built preschool. *Environ. Int.* **8** 111–115.

Berglund, B., Johansson, I. and Lindvall, T. (1982b) The influence of ventilation on indoor/outdoor air contaminants in an office building. *Environ. Int.* **8** 395–399.

Berglund, B., Berglund, U. and Lindvall, T. (1984) Characterization of indoor air quality and 'sick buildings'. *ASHRAE Trans.* **90**(1) 1045–1055.

Berglund, B., Johansson, I. and Lindvall, T. (1987) Volatile organic compounds from building materials in a simulated chamber study. In *Proceedings of the 4th International Conference on Indoor Air Quality and Climate*, Vol. 1, pp. 16–21. Institute for Soil, Water, and Air Hygiene, Berlin (West).

Bogen, K.T., Colston, B.W. and Machicao, L.K. (1992) Dermal absorption of dilute aqueous chloroform, trichloroethylene, and tetrachloroethylene in hairless guinea pigs. *Fund. Applied Toxicol.* **18** 30–39.

Breysse, P. A. (1984) Formaldehyde levels and accompanying symptoms associated with individuals residing in over 1000 conventional and mobile homes in the state of Washington. In Berglund, B., Lindvall, T. and Sundell, J. (eds) *Indoor Air: Sensory and Hyperreactivity Reactions to Sick Buildings, Volume 3*, pp. 403–408. Swedish Council for Building Research, Stockholm, Sweden.

Brugnone, F., Perbellini, L., Faccini, G.B. and Pasini, F. (1987) Benzene in the breath and blood of general public. In *Proceedings of the 4th International Conference on Indoor Air Quality and Climate*, Vol. 1, pp. 133–138. Institute for Soil, Water, and Air Hygiene, Berlin (West).

Centers for Disease Control (CDC) (1990) *Protocol for Measurement of Volatile Organic Compounds in Human Blood using Purge/Trap Gas Chromatography Mass Spectrometry.* Centers for Disease Control, Public Health Service, US Department of Health and Human Services, Atlanta, GA.

Coutant, R.W., Lewis, R.G. and Mulik, J.D. (1986) Modification and evaluation of a thermally desorbable passive sampler for volatile organic compounds in air. *Anal. Chem.* **58** 445–448.

De Bortoli, M. *et al.* (1986) Concentrations of selected organic pollutants in indoor and outdoor air in northern Italy. *Environ. Int.* **12** 343–350.

Dowty, B., Carlisle, D., Laseter, J.L. and Storer, J. (1975) Halogenated hydrocarbons in New Orleans drinking water and blood plasma. *Science* **187** 75–77.

Entz, R.C. and Diachenko, G.W. (1988) Residues of volatile halocarbons in margarines. *Food. Add. Contd.* **5** 267–276.

Entz, R. Thomas, K. and Diachenko, G. (1982) Residues of volatile halocarbons in food using headspace gas chromatography. *J. Agric. Food Chem.* **30** 846–849.

Environmental Protection Agency (EPA) (1989) *Indoor Air Quality and Work Environment Survey: EPA Headquarters Buildings. Volume I: Employee Survey.* US Environmental Protection Agency, Washington, DC.

Environmental Protection Agency (EPA) (1990) *Indoor Air Quality and Work Environment Survey: EPA Headquarters Buildings. Volume IV: Multivariate Statistical Analysis of Health, Comfort and Odor Perception as Related to Personal and Workplace Characteristics.* US Environmental Protection Agency, Washington, DC.

Fänger, O. (1987) A solution to the sick-building mystery. In Seifert, B. (ed.) *Indoor Air. 87. Proceedings of the 4th International Conference on Indoor Air*, Vol. 4, pp. 49–57. Institute for Soil, Water, and Air Hygiene, Berlin (West).

Girman, J.R. and Hodgson, A.T. (1987) Exposure to methylene chloride from controlled use of a paint remover in a residence. Paper presented at the 80th annual meeting of the Air Pollution Control Association in New York, NY, June 21–26, 1987. Lawrence Berkeley Laboratory, Berkeley, CA, *Report LBL 23078.*

Girman, J.R., Hodgson, A.T., Newton, A.W. and Winkes, A.W. (1986) Volatile organic emissions from adhesives with indoor applications. *Environ. Int.* **12** 317–321.

Girman, J.R., Hodgson, A.T. and Wind, M.L. (1987) Considerations in evaluating emissions from consumer products. *Atmos. Environ.* **21** 315–320.

Gold, L.S. *et al.* (1984) A carcinogenic potency data base of the standardized results of animal bioassays. *Environ. Health Perspect.* **58** 9–319.

Gordon, S., Wallace, L., Pellizzari, E. and O'Neill, H. (1988) Breath measurements in a clean air chamber to determine washout times for volatile organic compounds at normal environmental concentrations. *Atmos. Environ.* **22** 2165–2170.

Gordon, S.M., Wallace, L.A., Pellizzari, E.D. and Moschandreas, D.J. (1990) Residence times of volatile organic compounds in human breath following exposure to air pollutants at near-normal environmental concentrations. In *Total Exposure Assessment Methodology—A New Horizon*, pp. 247–256. Air and Waste Management Association, Pittsburgh, PA, VIP-16.

Gravesen, S., Skov, P., Valbjorn, O. and Lowenstein, H. (1990) The role of potential immunogenic components of dust (MOD) in the sick building syndrome. In Walkinshaw, D. (ed.) *Indoor Air '90: Proceedings of the 5th International Conference on Indoor Air Quality and Climate*, Toronto, Canada, 29 July to 3 August 1990, Vol. 1, pp. 9–13. Canada Mortgage and Housing Association, Ottawa, Canada.

Guerin, M.R., Higgins, C.E. and Jenkins, R.A. (1987) Measuring environmental emissions from tobacco combustion: sidestream cigarette smoke literature review. *Atmos. Environ.* **21** 291–297.

Hake, C.L., Stewart, R.D., Wu, A. and Graff, S.A. (1976) Experimental human exposure to perchloroethylene. *Toxicol. Appl. Pharmacol.* **37** 175.

Hall, J.F., Winer, A.M., Kleinman, M.T., Lurmann, F.W., Brajer, V. and Colome, S.D. (1992) Valuing the health benefits of clean air. *Science* **255** 812–817.

Higgins, C. *et al.* (1983) Applications of Tenax trapping to cigarette smoking. *J. Assoc. Off. Anal. Chem.* **66** 1074–1083.

Higgins, C.E. (1987) Organic vapor phase composition of sidestream and environmental tobacco smoke from cigarettes. In *Proceedings of the 1987 EPA/APCA Symposium on Measurement of Toxic and Related Air Pollutants*, pp. 140–151, Air and Waste Management Association, Pittsburgh, PA.

Highsmith, V.R., Zweidinger, R.B. and Merrill, R.G. (1988) Characterization of indoor and outdoor air associated with residences using woodstoves: a pilot study. *Environ. Int.* **14** 213–219.

Hodgson, A. T., Girman, J.R. and Binenboym, J. (1986) A multisorbent sampler for volatile organic compounds in indoor air. Presented at the 79th annual meeting of the Air Pollution Control Association in Minneapolis, MN, June 1986. Air Pollution Control Association, Pittsburgh, PA, paper no. 86-37.1. Lawrence Berkeley Laboratory Berkeley, CA, *Report LBL-21378.*

Hodgson, A.T., Garbesi, K., Sextro, R.G. and Daisey, J.M. (1988) *Evaluation of Soil-Gas Transport of Organic Chemicals into Residential Buildings*, Lawrence Berkeley Laboratory, Berkeley, CA, *Report LBL-25465.*

Hollowell, C.D. and Miksch, R.R. (1981) Sources and concentrations of organic compounds in indoor environments. *Bull. N.Y. Acad. Med.* **57** 962–977.

International Agency for Research on Cancer (IARC) (1991) Monographs on the evaluation of carcinogenic risks to humans. Vol. 52: *Chlorinated drinking-water; chlorination by-products; some other halogenated compounds; cobalt and cobalt compounds.* International Agency for Research on Cancer, Lyon, France.

Jarke, F.H. and Gordon, S.M. (1981) Recent investigations of volatile organics in indoor air at sub-ppb levels. Presented at the 74th annual meeting of the Air Pollution Control Association, PO Box 2861, Pittsburgh, PA 15230, paper no. 81-57.2, (16 pages).

Jermini, C., Weber, A. and Grandjean, E. (1976) Quantitative determination of various gas-phase components of the sidestream smoke of cigarettes in room air (in German), *Int. Arch. Occup. Environ. Health* **36** 169–181.

Jo, W.K., Weisel, C.P. and Lioy, P.J. (1990a) Routes of chloroform exposure and body burden from showering with contaminated tap water. *Risk Anal.* **10** 575–580.

Jo, W.K., Weisel, C.P. and Lioy, P.J. (1990b) Chloroform exposure and the health risk associated with multiple uses of chlorinated tap water. *Risk Anal.* **10** 581–586.

Johansson, I. (1978) Determination of organic compounds in indoor air with potential reference to air quality. *Atmos. Environ.* **12** 1371–1377.

Kliest, J., Fast, T., Boleij, J.S.M., van de Wiel, H. and Bloemen, H. (1989) The relationship between soil contaminated with volatile organic compounds and indoor air pollution. *Environ. Inter.* **15** 419–425.

Knöppel, H. and Schauenburg, H. (1987) Screening of household products for the emission of volatile organic compounds. *Environ. Inter.* **15** 443–447.

Krause, C., Mailahn, W., Nagel, R., Schulz, C., Seifert, B. and Ullrich, D. (1987) Occurrence of volatile organic compounds in the air of 500 Homes in the Federal Republic of Germany. In *Proceedings of the 4th International Conference on Indoor Air Quality and Climate*, Vol. 1, pp. 102–106. Institute for Soil, Water, and Air Hygiene, Berlin (West).

Krost, K.J., Pellizzari, E.D., Walburn, S.G. and Hubbard, S.A. (1982) Collection and analysis of hazardous organic emissions. *Anal. Chem.* **54** 810–817.

Krotoszynski, B.K., Gabriel, G. and O'Neill, H. (1977) Characterization of human expired air: A promising investigation and diagnostic technique. *J. Chromatogr. Sci.* **15** 239–244.

Krotoszynski, B.K., Bruneau, G.M. and O'Neill, H.J. (1979) Measurement of chemical inhalation exposure in urban populations in the presence of endogenous effluents. *J. Anal. Toxicol.* **3** 225–234.

Lebowitz, M.D., Cassell, E. and McCarroll, J. (1972a) Health and the urban environment. XI. The incidence and burden of minor illness in a healthy population: Methods, symptoms, incidence. *Amer. Rev. Resp. Dis.* **105** 824–834.

Lebowitz, M.D., Cassell, E. and McCarroll, J. (1972b) Health and the urban environment. XII. The incidence and burden of minor illness in a healthy population: Duration, severity, and Burden. *Amer. Rev. Resp. Dis.* **105** 835–841.

Lebret, E., van de Weil, H.J., Noij, D. and Boleij, J.S.M. (1986) Volatile hydrocarbons in Dutch homes. *Environ. Int.* **12** 323–332.

Löfroth, G., Burton, B., Forehand, L., Hammond, S.K., Seila, R., Zweidinger, R. and Lewtas, J. (1989) Characterization of environmental tobacco smoke. *Env. Sci. Technol.* **23** 610–614.

Mailahn, W., Seifert, B. and Ullrich, D. (1987) The use of a passive sampler for the simultaneous determination of long-term ventilation rates and VOC concentrations. In *Proceedings of the 4th International Conference on Indoor Air Quality and Climate*, Vol. 1, pp. 149–153. Institute for Soil, Water, and Air Hygiene, Berlin (West).

McCann, J., Horn, L., Girman, J. and Nero, A.V. (1987) Potential risks from exposure to organic carcinogens in indoor air. Lawrence Berkeley Laboratory, Berkeley, CA, *Report LBL-22473.*

McGuire, M.J. and Meadow, R.G. (1988) AWWARF trihalomethane survey. *J. Amer. Water Works Assoc.* **80** 61.

McKone, T.E. (1987) Human exposure to VOCs in household tap water: The indoor inhalation pathway. *Env. Sci. Technol.* **21** 1194–1201.

McKone, T.E. (1990) Dermal uptake of organic chemicals from a soil matrix. *Risk Anal.* **10** 407–419.

McKone, T.E. and Knezovich, J. (1991) The transfer of trichloroethylene (TCE) from a shower to indoor air: experimental measurements and their implications. *J. Air Waste Manage. Assoc.* **40** 282–286.

Michael, L.C., Erickson, M.D., Parks, S.P. and Pellizzari, E.D. (1980) Volatile environmental pollutants in biological matrices with a headspace purge technique. *Anal. Chem.* **52** 1836–1841.

Miksch, R.R., Hollowell, C.D. and Schmidt, H.E. (1982) Trace organic chemical contaminants in office spaces. *Environ. Int.* **8** 129–137.

Mølhave, L. (1987) The sick buildings—a sub-population among the problem buildings? In *Indoor Air '87: Proceedings of the 4th International Conference on Indoor Air Quality and Climate*, August 17–21, 1987, pp. 469–474. Institute for Water, Soil, and Air Hygiene, Berlin (West).

Mølhave, L. (1982) Indoor air pollution due to organic gases and vapours of solvents in building materials. *Environ. Int.* **8** 117–127.

Mølhave, L. and Møller, J., (1979) The atmospheric environment in modern Danish dwellings: measurements in 39 flats. In Fänger, O. and Valbjørn, O. (eds) *Indoor Climate*, pp. 171–186, SBI, Hørsholm, Denmark.

Mølhave, L., Bach., B. and Pedersen, O.F. (1986) Human reactions to low concentrations of volatile organic compounds. *Environ. Int.* **12** 167–175.

Monster, A.C. and Houthooper, J.M. (1979) Estimation of individual uptake of trichloroethylene, 1,1,1-trichloroethane, and tetrachloroethylene from biological parameters. *Int. Arch. Occup. Environ. Health* **42** 319–323.

Monster, A.C. and Smolders, J.F.J. (1984) Tetrachloroethylene in exhaled air of persons living near pollution sources. *Int. Arch. Occup. Environ. Health* **53** 331–336.

National Toxicology Program (NTP) (1988) *Technical Report on the Toxicity and Carcinogenesis of 1,4-Dichlorobenzene (CAS #106-46-7) in F344/N Rates and B6C3F1 Mice (gavage study)*. NTP Technical Report #319, Board Draft, March 1986.

National Toxicology Program (NTP) (1986) *Technical Report on the Toxicity and Carcinogenesis of d-Limonene (CAS #5989-27-5) in F344/N Rates and B6C3F1 Mice (gavage study)*. NTP Technical Report #347, NIH Publication #88-2802.

Nuchia, E. (1986) *MDAC—Houston Materials Testing Database Users' Guide*. McDonnell Douglas Corp. Software Technology Development Laboratory, Houston, TX.

Oliver, K.D., Pleil, J.D. and McCleanny, W.A. (1986) Sample integrity of trace level volatile organic compounds in ambient air stored in summa polished canisters. *Atmos. Environ.* **20** 1403.

Otto, D., Mølhave, L., Rose, G., Hudnell, H.K. and House, D. (1990) Neurobehavioral and sensory irritant effects of controlled exposure to a complex mixture of volatile organic compounds. *Neurotox. Teratol.* **12** 649–652.

Ozkaynak, H., Ryan, P.B., Wallace, L.A., Nelson, W.C. and Behar, J.V. (1987) Sources and emission rates of organic chemical vapors in homes and buildings. In *Proceedings of the 4th International Conference on Indoor Air Quality and Climate*, Vol. 1, pp. 3–7. Institute for Soil, Water, and Air Hygiene, Berlin (West).

Pearson, C.R. and McConnell, G. (1975) Chlorinated C_1 and C_2 hydrocarbons in the environment. *Proc. Royal Soc. London Ser. B* **189** 305.

Pellizzari, E.D., Erickson, M.D. and Zweidinger, R. (1979) *Formulation of a Preliminary Assessment of Halogenated Organic Compounds in Man and Environmental Media*. US Environmental Protection Agency, Washington, DC.

Pellizzari, E.D., Hartwell, T.D., Harris, B.S.H. III, Waddell, R.D., Whitaker, D.A. and Erickson, M.D. (1982) Purgable organic compounds in mothers' milk. *Bull. Environ. Cont. Toxicol.* **28** 322–328.

Pellizzari, E.D., Perritt, K., Hartwell, T.D., Michael, L.C., Whitmore, R., Handy, R.W., Smith, D. and Zelon, H. (1987a) *Total exposure assessment methodology (TEAM) sutdy: Elizabeth and Bayonne, New Jersey; Devils Lake, North Dakota: and Greensboro, North Carolina, Vol. II*. US Environmental Protection Agency, Washington, DC.

Pellizzari, E.D., Perritt, K., Hartwell, T.D., Michael, L.C., Whitmore, R., Handy, R.W., Smith, D. and Zelon, H. (1987b) *Total exposure assessment methodology (TEAM) sutdy: selected communities in northern and southern California, Vol. III*. US Environmental Protection Agency, Washington, DC.

Pffaffenberger, C.D. and Peoples, A.J. (1982) Long-term variation study of blood plasma levels of chloroform and related purgable compounds. *J. Chromatog.* **239** 217–226.

Preller, L., Zweers, T., Brunekreef, B. and Boleij, J. (1990) Sick leave due to work-related health complaints among office workers in the Netherlands. In Walkinshaw, D. (ed. *Indoor Air '90: Proceedings of the 5th International Conference on Indoor Air Quality and Climate*, Toronto, Canada, 29 July to 3 August 1990, Vol. 1, pp. 227–230. Canada Mortgage and Housing Association, Ottawa, Canada.

Ramsey, J.C. and Andersen, M.E. (1984) A physiologically based description of the inhalation pharmacokinetics of styrene in rats and humans. *Toxicol. Appl. Pharm.* **73** 159–175.

Raw, G., Roys, M. and Leaman, A. (1990) Further findings from the office environment survey: Productivity. In Walkinshaw, D. (ed.) *Indoor Air '90: Proceedings of the 5th International Conference on Indoor Air Quality and Climate*, Toronto, Canada 29 July to 3 August 1990, Vol. 1, pp. 231–236. Canada Mortgage and Housing Association, Ottawa, Canada.

Raymer, J.H., Thomas, K.W., Cooper, S.D. and Pellizzari, E.D. (1990) A device for sampling human alveolar breath for the measurement of expired volatile organic compounds. *J. Anal. Toxicol.* **14** 337–344.

Raymer, J.H., Pellizzari, E.D., Thomas, K.W. and Cooper, S.D. (1991) Elimination of volatile organic compounds in breath after exposure to occupational and environmental microenvironments. *J. Expos. Anal. Environ. Epidem.* **1** 439–451.

Repace, J.L. and Lowrey, A.H. (1980) Indoor air pollution, tobacco smoke, and public health. *Science* **208** 464–472.

Repace, J.L. and Lowrey, A.H. (1985) A quantitative estimate of non-smokers' lung cancer risk from passive smoking. *Environ. Internat.* **11** 3–22.

Rook, J.J. (1974) Formation of haloforms during chlorination of natural water. *J. Water Treat. Exam.* **23** 234.

Sandler, D.P., Everson, R.B., Wilcox, A.J. and Browder, J.P. (1985) Cancer risk in adulthood from early life exposure to parents' smoking. *Am. J. Public Health* **75** 467.

Sato, A. and Nakajima, T. (1979) Partition coefficients of some aromatic hydrocarbons and ketones in water, blood and oil. *Br. J. Ind. Med.* **36** 231–234.

SCAQMD (1989) *In-vehicle Characterization Study in the South Coast Air Basin.* South Coast Air Quality Management District, Los Angeles, CA.

Seifert, B. and Abraham, H.J. (1982) Indoor air concentrations of benzene and some other aromatic hydrocarbons. *Ecotoxicol. Environ. Safety* **61** 190–192.

Seifert, B. and Abraham, H.J. (1983) Use of passive samplers for the determination of gaseous organic substances in indoor air at low concentration levels. *Int. J. Environ. Anal. Chem.* **13** 237–253.

Seifert, B. and Schmahl, H.-J. (1987) Quantification of sorption effects for selected organic substances present in indoor air. In *Proceedings of the 4th International Conference on Indoor Air Quality and Climate*, Vol. 1, pp. 252–256. Institute for Soil, Water, and Air Hygiene, Berlin (West).

Shatkin, J.A. and Brown, H.S. (1991) Pharmacokinetics of the dermal route of exposure to volatile organic chemicals in water: a computer simulation model. *Environ. Res.* **56** 90–108.

Sheldon, L.S., Handy, R.W., Hartwell, T.D., Leininger, C. and Zelon, H. (1983) *Human Exposure Assessment to Environmental Chemicals: Nursing Mothers Study*. Final report. EPA, Washington, DC.

Sheldon, L.S., Handy, R.W., Hartwell, T.D., Whitmore, R.W., Zelon, H.S. and Pellizzari, E.D. (1988a) *Indoor Air Quality in Public Buildings*, US Environmental Protection Agency, Washington, DC. EPA 600/6-88/009a.

Sheldon, L.S., Easton, C., Hartwell, T.D., Zelon, H.S. and Pellizzari, E.D. (1988b) *Indoor Air Quality in Public Buildings. Vol. II.* US Environmental Protection Agency, Research Triangle Park, NC. EPA 600/6-88/009b.

Sigsby, J.E., Tejada, S. and Ray, W. (1987) Volatile organic compound emissions from 46 in-use passenger cars. *Environ. Sci. Technol.* **21** 466–475.

Skov, P., Valbjorn, O., Pedersen, B.V. and the Danish Indoor Climate Study Group (1989) Influence of personal characteristics, job-related factors, and stress factors on the sick building syndrome. *Scand. J. Work Environ. Health* **15** 286–295.

Skov, P., Valbjorn, O., Pedersen, B.V. and the Danish Indoor Climate Study Group (1990) Influence of indoor climate on the sick building syndrome in an office environment. *Scand. J. Work Environ. Health* **16** 363–371.

Spicer, C.W. *et al.* (1986) Intercomparison of sampling techniques for toxic organic compounds in indoor air. In Hochheiser, S. and Jayanti, R.K.M. (eds) *Proceedings of the 1986 EPA/APCA Symposium on the Measurement of Toxic Air Pollutants*, pp. 45–60. Air Pollution Control Association, PO Box 2861, Pittsburgh, PA 15230.

Stenberg, B., Hanson-Mild, K.H., Sandstrom, M. Lonnberg, G., Wall, S., Sundell, J. and Zingmark, P.A. (1990) The office illness project in northern Sweden. Part I: A prevalence study of sick building syndrome (SBS) related to demographic data, work characteristics and ventilation. In Walkinshaw, D. (ed.) *Indoor Air '90: Proceedings of the 5th International Conference on Indoor Air Quality*, Vol. 4, pp. 627–632. Canada Mortgage and Housing Association, Ottawa, Canada.

Stewart, R.D. and Dodd, H.C. (1964) Absorption of carbon tetrachloride, trichloroethylene, tetrachloroethylene, methylene chloride, and 1,1,1-trichloroethane through human skin. *J. Am. Ind. Hyg. Assoc.* **25** 439–446.

Stjernfeldt, M., Berglund, K., Lindsten, J. and Ludvigsson, J. (1986) Maternal smoking during pregnancy and risk of childhood cancer. *Lancet* June 14 1350–1352.

Sundell, J., Lonnberg, G., Wall, S., Stenberg, B. and Zingmark, P.-A. (1990) The office illness project in northern Sweden. Part III: A case-referent study of sick building syndrome (SBS) in relation to building characteristics and ventilation. In Walkinshaw, D. (ed.) *Indoor Air '90: Proceedings of the 5th International Conference on Indoor Air Quality and Climate*, Vol. 4, pp. 633–638. Canada Mortgage and Housing Association, Ottawa, Canada.

Symons, J.M., Bellar, T.A., Carsell, J.K., DeMarco, J., Kropp, K.L., Robeck, G.G., Seeger, D.R., Slocum, C.J., Smith, B.L. and Stevens, A.A. (1975) National organics reconnaissance survey for halogenated organics. *J. Amer. Water Works Assoc.* **67** 708–729.

Tancrede, M., Wilson, R., Zeise, L. and Crouch, E.A. (1987) The carcinogenic risk of some organic vapors indoors: a theoretical survey. *Atmos. Environ.* **21** 2187–2205.

Tichenor, B.A. and Mason, M.A. (1987) Organic emissions from consumer products and building materials to the indoor environment. *J. Air Pollution Control Assoc.* **38** 264–268.

Tichenor, B.A., Sparks, L.E., White, J.B. and Jackson, M.D. (1990) Evaluating sources of indoor air pollution. *J. Air Waste Management Assoc.* **41** 487–492.

Traynor, G.W., Apte, M.G., Sokol, H.A., Chuang, J.C., Tucker, W.G. and Mumford, J.L. (1990) Selected organic pollutant emissions from unvented kerosene space heaters. *Environ. Sci. Technol.* **24** 1265–1270.

Uhler, A.D. and Diachenko, G.W. (1987) Volatile halocarbon compounds in process water and processed foods. *Bull. Environ. Contam. Toxicol.* **39** 601–607.

Wallace, L.A. (1986) Cancer risks from organic chemicals in the home. In *Environmental Risk Assessment: Is Analysis Useful?* APCA Specialty Conference Proceedings, Air Pollution Control Association, Pittsburgh, PA. SP-55.

Wallace, L.A. (1987) *The TEAM Study: Summary and Analysis: Volume I*, US Environmental Protection Agency, Washington, DC. EPA 600/6-87/002a. NTIS PB 88-100060.

Wallace, L.A. (1990) Major sources of exposure to benzene and other volatile organic compounds. *Risk Anal.* **10** 59–64.

Wallace, L.A. (1991) Comparison of risks from outdoor and indoor exposure to toxic chemicals. *Environ. Health Perspect.* **95** 7–13.

Wallace, L.A. and O'Neill, I.K. (1987) Personal air and biological monitoring of individuals for exposure to environmental tobacco smoke. In *Environmental Carcinogenesis: Selected Methods of Analysis: Volume 9—Passive Smoking*, Chapter 7. International Agency for Research on Cancer (IARC), Lyon, France.

Wallace, L.A., Zweidinger, R., Erickson, M., Cooper, S., Whitaker, D. and Pellizzari, E.D. (1982) Monitoring individual exposure: Measurement of volatile organic compounds in breathing-zone air, drinking water, and exhaled breath. *Environ. Int.* **8** 269–282.

Wallace, L.A., Pellizzari, E., Hartwell, T., Rosenzweig, R., Erickson, M., Sparacino, C. and Zelon, H. (1984a) Personal exposure to volatile organic compounds: I. Direct measurement in breathing-zone air, drinking water, food, and exhaled breath. *Environ. Res.* **35** 293–319.

Wallace, L.A., Pellizzari, E., Hartwell, T., Sparacino, C., Sheldon, L. and Zelon, H. (1984b) Personal exposures, indoor–outdoor relationships and breath levels of toxic air pollutants measured for 355 persons in New Jersey. *Atmos. Env.* **19** 1651–1661.

Wallace, L.A., Pellizzari, E.D., Hartwell, T.D., Sparacino, C., Whitmore, R., Sheldon, L., Zelon, H. and Perritt, R. (1987a) The TEAM study: Personal exposures to toxic substances in air, drinking water, and breath of 400 residents of New Jersey, North Carolina, and North Dakota. *Environ. Res.* **43** 290–307.

Wallace, L.A., Pellizzari, E., Hartwell, T., Perritt, K. and Ziegenfus, R. (1987b) Exposures to benzene and other volatile organic compounds from active and passive smoking. *Arch. Environ. Health* **42** 272–279.

Wallace, L.A., Pellizzari, E., Leaderer, B., Hartwell, T., Perritt, R., Zelon, H. and Sheldon, L. (1987c) Emissions of volatile organic compounds from building materials and consumer products. *Atmos. Envir.* **21** 385–393.

Wallace, L.A., Pellizzari, E.D., Hartwell, T.D., Whitmore, R., Perritt, R. and Sheldon, L. (1988) The California TEAM study: Breath concentrations and personal exposures to 26 volatile compounds in air and drinking water of 188 residents of Los Angeles, Antioch, and Pittsburgh, CA. *Atmos. Environ.* **22** 2141–2163.

Wallace, L.A., Pellizzari, E.D., Hartwell, T.D., Davis, V., Michael, L.C. and Whitmore, R.W. (1989) The influence of personal activities on exposure to volatile organic compounds. *Environ. Res.* **50** 37–55.

Wallace, L.A., Nelson, W.C., Ziegenfus, R. and Pellizzari, E. (1991a) The Los Angeles TEAM study: Personal exposures, indoor–outdoor air concentrations, and breath concentrations of 25 volatile organic compounds. *J. Expos. Anal. Environ. Epidem.* **1**(2) 37–72.

Wallace, L.A., Nelson, C.J. and Dunteman, G. (1991b) Workplace characteristics associated with health and comfort concerns in three office buildings in Washington, DC. In *IAQ '91:*

Healthy Buildings. Proceedings of a conference in Washington, DC, September 1991. American Society of Heating, Refrigerating, and Air-Conditioning Engineers, Inc., Atlanta, GA.

Wallace, L.A., Pellizzari, E. and Wendel, C. (1992) Total volatile organic concentrations in 2700 personal, indoor, and outdoor air samples collected in the US EPA TEAM Studies. *Indoor Air* **1**(4) 465–477.

Weisel, C.P., Jo, W.K. and Lioy, P.J. (1990) Routes of chloroform exposure from showering with chlorinated water. In *Measurement of Toxic and Related Air Pollutants: Proceedings of the 1990 EPA/Air Waste Management Association International Symposium*. Air and Waste Management Association, Pittsburgh, PA, VIP-17. EPA 600/9-90/026.

Wilkes, C.R., Small, M.J., Andelman, J.B., Giardino, N.J. and Marshall, J. (1992) Inhalation exposure model for volatile chemicals from indoor uses of water. *Atmos. Environ.* **26A**(12) 2227–2236.

Wilson, S. and Hedge, A. (1987) *The Office Environmental Survey: A Study of Building Sickness*. Building Use Studies, Ltd., London, England.

Zeise, L., Crouch, E.A.C. and Wilson, R. (1986) A possible relation between toxicity and carcinogenicity. *J. Am. Coll. Toxicol.* **5** 137–151.

Zweers, T., Preller, L., Brunekreef, B. and Boleij, J. (1990) Relationships between health and indoor climate complaints and building workplace, job and personal characteristics. In Walkinshaw, D. (ed.) *Indoor Air '90: Proceedings of the 5th International Conference on Indoor Air Quality and Climate*, Toronto, Canada, 29 July to 3 August 1990, Vol. 1, pp. 495–500. Canada Mortgage and Housing Association, Ottawa, Canada.

Zweidinger, R.B., Sigsby, J.E., Tejada, S.B., Stump, F.D. Dropkins, D.L. and Ray, W.D. (1988) Detailed hydrocarbon and aldehyde mobile source emissions from roadway studies. *Environ. Sci. Technol.* **22** 956–962.

2 VOCs and the environment and public health—health effects

G.J.A. SPEIJERS

2.1 General introduction

In this chapter the behaviour of volatile organic compounds (VOCs) in the body after exposure by inhalation or after oral and dermal exposure, together with the toxic effects caused by exposure to VOCs via the different routes are discussed. Due to the number of VOCs it is impossible to cover all compounds and their toxicity. The aim is to include the most important representatives of the different classes of VOCs with respect both to toxic properties and to occurrence. Exposure to VOCs is discussed in chapter 1, while toxicity and exposure in the occupational health situation, with special emphasis on dermal toxicity, are discussed in chapter 7. This chapter will focus on the toxicokinetics, biotransformation and toxicity after oral exposure and exposure by inhalation, especially in experimental animals. Towards the end of the chapter new developments, such as the use of physiologically based models to determine toxicokinetic and dynamic properties of VOCs in risk assessment and risk evaluation, and attention to groups at risk and consideration of the combined effects of different VOCs are reviewed.

2.2 Toxicokinetics and biotransformation

2.2.1 Introduction

Inhalation of VOCs seems to be the most important route of exposure, although for several VOCs oral and dermal exposure are also important. In many cases toxicological data are only adequately available for one exposure route, usually oral exposure but also, to a lesser extent, inhalation. Consequently, knowledge about the toxicokinetic behaviour of certain VOCs after inhalation, ingestion or dermal exposure gives more insight into the toxicological profile of other related VOCs. It is important to see: (1) whether absorption of a VOC, which may differ for the three exposure routes, leads ultimately to comparable distribution of the VOC and its metabolites in the body of animals and man; and (2) which physical and chemical properties play an important role in the distribution. Among these properties, lipophilicity and solubility in water are extremely important for the behaviour of a compound in the body. Besides distribution, biotransformation is often important for the detoxification or activation (toxification) of VOCs. In many cases the liver plays the most important role in modifying VOCs in such a way that the metabolites become directly water soluble or able to undergo conjugation with other compounds produced in the body. The metabolites can also undergo reduction in order to become water soluble. Detoxification/toxification reactions, in which the enzymes located in the liver microsomes are often involved, are classified as phase I reactions. Conjugations to, for example, glucuronic acid, are classified as phase II reactions. Once the metabolites of a VOC are water soluble they can be eliminated via the urinary excretion route or by exhalation. The lipophilic compounds and metabolites may be eliminated via the entero-hepatic pathway. Renal elimination greatly depends on kidney function. If this function is impaired by VOC or its metabolites elimination may be impaired. Gaseous exchange in the lungs and thus lung function, are important for elimination by exhalation. If a VOC has a direct toxic effect on the lungs this has consequences for its elimination by exhalation and complicates the toxicokinetics of the VOC at toxic exposure levels.

In this chapter the toxicokinetic and biotransformation of VOCs will be discussed using important representatives as examples. Some examples of conditions which influence toxicokinetics or biotransformation will be reviewed.

The occurrence and onset of toxic properties of VOCs are greatly governed by the toxicokinetic behaviour in terms of both quantity and quality. The concentration of a VOC or its metabolites, and thus the ultimate toxic effects in target organs may differ for different exposure routes. Toxicokinetics give information on the internal exposure. In the case of a direct effect on the respective organ or tissue this internal exposure level can be applied in physiologically based models for risk evaluations, provided that information on the mechanism of toxicity of the VOC in the tissue is available. This aspect will be discussed later in this chapter. However, some toxic effects are secondary to other effects of the compound.

The toxicokinetic properties of VOCs are characterized and summarized by parameters such as absorption rate, distribution half-life ($t_{1/2\ \mathrm{dis}}$), elimination half-life ($t_{1/2\ \mathrm{el}}$), distribution volume, blood and tissue levels, and the main metabolites and amounts of the VOC or its metabolites (in absolute quantities or relative to the total exposure). These parameters are described in the literature, for example in IPCS/WHO (1986).

2.2.2 *Conditions influencing the toxicokinetic behaviour*

In the case of exposure by inhalation the conditions influencing the toxicokinetic behaviour and the final tissue concentrations of a VOC or its metabolites in humans and animals include: (1) exposure level and duration; (2) pulmonary ventilation; (3) diffusion of the solvent through the alveolo-capillary membrane; (4) solubility of the VOC in blood; (5) circulation of the blood through the lungs and other organs; (6) diffusion of the VOC through the tissue membrane; and (7) solubility of the VOC in tissues (Astrand, 1975; Engström *et al.*, 1978a; Withey and Collins, 1979; McKenna *et al.*, 1982; Carlsson, 1983; Stott and McKenna, 1984; Anon, 1988; Paterson and Mackay, 1989). These conditions can be greatly modified by smoking, physical exercise, stress and other factors such as diseases affecting the respiratory tract. Astrand (1975) examined the uptake of solvents in the blood and tissues of humans exposed to different VOCs (toluene, methylchloroform, styrene, white spirit, methylene chloride and trichloroethylene) both at rest and after different levels of exercise. In healthy people the magnitude of pulmonary ventilation and blood circulation is mainly governed by metabolic rate. Under normal conditions the magnitude of the metabolic rate depends entirely upon the degree of physical exercise. The uptake of a gas in the lungs differs greatly at rest when compared to uptake during exercise. Astrand (1975) showed that when alveolar ventilation increased in conjunction with physical exercise, the concentration in alveolar air rose. During exercise the arterial blood concentrations of VOCs were higher than at rest. The rate of increase for the VOCs tested correlated with blood solubility. For some VOCs (methyl-chloroform, aliphatic components of white spirit, methylchloride, trichloro-ethylene and toluene) the blood/air partition coefficients were hardly or not changed, whereas for the aromatic components of white spirit the partition coefficient rose considerably during exercise. The lower the solubility of a VOC in blood, the less the time needed for the equilibrium to settle. Blood flow also plays a role in uptake in various organs. Blood flow through adipose tissue is slight at rest, and probably even less during physical exercise. Blood flow through the kidneys, the liver and the gastro-intestinal tract may be reduced during physical exercise. However, blood flow supplying the lungs and the nervous system, including the brain, increases along with coronary blood flow. Thus physical exercise produces conditions favouring large uptake in the lungs, slight delivery to certain organs and extensive delivery to others. This effect of physical exercise contributed to the relatively high blood concentrations of VOCs found by Astrand (1975). Although arterio-venous differences declined for all VOCs tested during heavy exercise, the total amount released from arterial blood per unit of time was greater. In other studies with methylene chloride humans also showed an increase in pulmonary absorption during physical exercise (Astrand and Gamberale, 1978). Carlsson (1983) concluded that the total uptake of toluene by inhalation increased during physical exercise.

The biotransformation of VOCs, which mainly takes place in liver and lungs after exposure by inhalation, might be modulated by factors influencing the metabolic capacity of these organs. Such factors include induction time, other inducing compounds in food or the environment, level of exposure, and time of exposure (Andersen *et al.*, 1984; Anon, 1988). Using inhalation experiments Andersen *et al.* (1984) showed that pretreatment with pyrazole essentially abolished *in vivo* styrene metabolism, while pretreatment with phenobarbital increased the maximum metabolic velocity (V_{max}) of the metabolism about six-fold. This shows that once VOCs are absorbed they are subjected to the same factors that interfere with the biotransformation of orally ingested xenobiotics. These facts are well described in the literature.

The previously reported influences on toxicokinetics and biotransformation indicate the importance for toxicological research to incorporate these interfering factors (e.g. physical exercise and stress) in inhalation studies.

If the metabolites of a VOC are volatile exhalation also plays a role in the elimination of the VOC and its metabolites (Falk *et al.*, 1991). Thus the same conditions that influence the uptake of a VOC may also modulate the excretion of the VOC and its metabolites. Once VOCs are metabolised to water-soluble metabolites the main route of elimination is urinary excretion, and thus conditions influencing kidney function may interfere with the elimination of the metabolites. Lipophilic VOCs or metabolites may be excreted by the entero-hepatic pathway, although marked species differences are observed. (In humans this excretion route is less important.) Thus factors influencing the production of bile acids may modulate this elimination route.

In the case of oral exposure to VOCs, conditions influencing uptake include: water solubility, fat solubility, the vehicle of the VOC, the composition of the food, and the composition of the gut microflora (IPCS/WHO, 1986; Lam *et al.*, 1990). The distribution of a VOC and its metabolites largely depends on the route of transportation in the body. In the case of lipophilic compounds this may mainly be the lymphatic system, whereby the VOC reaches the liver for biotransformation only after it has been circulated in the blood. The VOC first passes the respiratory system in which it can be partly exhaled depending on its volatility (Figure 2.1(a)). By contrast, water-soluble VOCs enter the circulation by the venal portal system of the liver and are directly subjected to biotransformation processes (Figure 2.1(b)). Biotransformation will mainly take place in the liver. Conditions modulating the metabolic function of the liver will therefore enhance or impair the biotransformation, and thus the elimination of VOCs. The elimination of VOCs and their metabolites after oral intake will follow the same routes of elimination as after inhalatory exposure, although there may be considerable difference in the quantities eliminated via exhalation, urine and faeces.

In the case of dermal exposure, conditions influencing the uptake of VOCs through the skin include water solubility, fat solubility, dehydrating properties, the vehicle in which the VOC is dissolved, the skin condition, and occlused or non-occlused exposure to the VOC (Dutkiewicz and Tyras, 1967; Tsuruta, 1975, 1977; Baranowska-Dutkiewicz, 1982; Szejnwald-Brown *et al.*, 1984; Susten *et al.*, 1986).

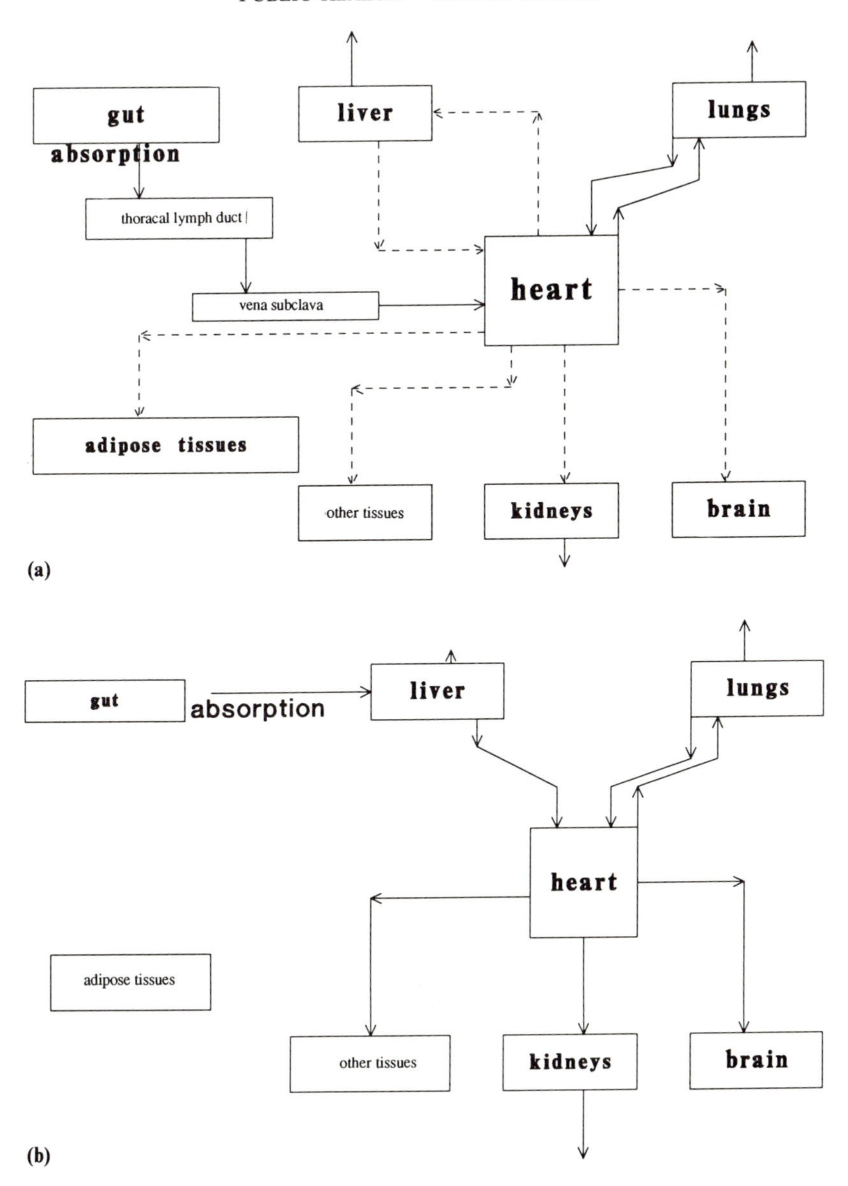

Figure 2.1 Schematic presentation of the metabolic routes of (a) lipophilic and (b) hydrophilic VOCs after oral intake.

After dermal absorption, the VOC is first circulated through the lungs, therefore elimination by exhalation may occur to a certain extent depending on the volatility of the VOC (Tsuruta, 1975; Susten *et al.*, 1990). On reaching the liver biotransformation takes place.

The distribution of VOCs after dermal exposure also depends on conditions such as lipophility or hydrophility, exposure route, and transportation route in the body (Lam *et al.*, 1990). Although there may be considerable differences in the distribution pattern, both quantitatively and qualitatively, for the different routes of exposure, there are several examples of studies with VOCs which show that distribution can be rather similar. Timchalk *et al.* (1991) concluded from studies with ^{14}C-1,2-dichloropropane in the rat that the pharmacokinetics and metabolism of ^{14}C-dichloropropane were similar, regardless of the route of exposure, or sex.

2.2.3 Toxicokinetic behaviour of styrene

After inhalatory exposure the styrene concentration in the blood of animals and humans is directly detectable, indicating rapid uptake (Astrand *et al.*, 1974; Withey, 1976, 1978; Withey and Collins, 1979). According to the calculations of Andersen *et al.* (1984) the rate of uptake in rats exposed by inhalation to 430–1150 mg/m^3 ranged from 6.13 to 49.71 mg/kg/hr. In mice an absorption of 29.8% was recorded after inhalation of styrene (25 600 mg/m^3, 10 min) (Bergman, 1979). Studies in humans revealed that 45–68% of inhaled styrene was absorbed (Oedkvist *et al.*, 1982; Mahrlein *et al.*, 1983; Wigeaus *et al.*, 1983), although Fernandez and Caperos (1977) recorded an absorption of 82–93% by the lungs after inhalatory exposure (294–865 mg/m^3, 4 h). Ramsey *et al.* (1986) calculated an absorption rate of 100 mg/h in humans exposed by inhalation (336 mg/m^3, 6 h). Ramsey and Young (1978) found that a 15-fold increase in the styrene concentration in inhaled air led to an 80-fold increase in the blood concentration, indicating saturation of the biotransformation mechanisms in the case of increasing intensity and/or duration of styrene exposure.

After oral administration of styrene dissolved in corn oil to mice and rats, peak concentrations in blood and tissues were recorded within a few hours (Plotnick and Weigel, 1979; Sbrana *et al.*, 1983). This absorption is even more rapid in aqueous solutions. The peak concentration in blood appears almost as fast as after intravenous administration (Withey, 1976, 1978).

In rats dermally exposed by putting their tail in undiluted styrene for 1 h concentrations in the liver and brain were calculated at 50–70% of the concentration normally measured after a 4 h inhalatory exposure to the LC_{50} (lethal concentration causing 50% mortality) (11 800 mg/m^3) (Shugaev and Yaroslavl, 1969). In humans the absorption rate through the skin after exposure to aqueous solutions of styrene was 0.04–0.18; after exposure to undiluted styrene it was 9–15 mg/cm^2/h (Dutkiewicz and Tyras, 1968). The absorption of styrene vapour through the skin was 1.9% of the estimated amount absorbed after inhalation of the same concentration (Riihimaki and Pfaffi, 1978).

Animal experiments showed that styrene is rapidly distributed over the whole body, regardless of the route of exposure; it mainly heads for the adipose tissues. Elimination from most tissues is rapid and concentrations were below the detection limit within 48 h. Adipose tissues retain styrene for much longer (Savolainen and Vaino, 1977; Plotnick and Weigel, 1979; Löf *et al.*, 1983, 1984). The distribution pattern largely depends on exposure levels because these modify the organ/blood ratio (Withey, 1976, 1978; Withey and Collins, 1977, 1979). At exposure levels saturating the metabolic processes the distribution ranking is adipose tissue > liver > brain > kidneys > blood > heart and spleen (Withey, 1976, 1978; Withey and Collins, 1977, 1979; Teramato and Hariguchi, 1981). Styrene is eliminated from blood and tissues following a dose-dependent two-compartment model, in which the elimination rate decreases with increasing dose. Adipose tissue is the second compartment (Ramsey and Young, 1978). The initial phase of elimination is rapid and is followed by a slow phase if the exposure level does not cause saturation (Withey and Collins, 1977, 1979; Teramoto and Hariguchi, 1981).

In humans styrene is also eliminated from the body according to a linear two-compartment model (Ramsey *et al.*, 1980). The elimination half-life ($t_{1/2\,el}$) ranges from 30 to 40 minutes for the initial phase and from 3 to 13 hours for the second phase (Ramsey and Young, 1978; Teramoto and Hariguchi, 1981). The $t_{1/2\,el}$ for adipose tissue is estimated to be 2.2–5.2 days (Engström, 1978; Engström *et al.*, 1978b).

The biotransformation of styrene in experimental animals mainly takes place in the liver (Harkonen, 1978), but also occurs in other tissues such as the lungs, intestines, kidneys, spleen and heart (James *et al.*, 1976; Ryan *et al.*, 1976; Salmona *et al.*, 1976; Cantoni *et al.*, 1978). The major degradation route is thought to be oxidation of styrene to styrene-7,8-epoxide (James and White, 1967; Bardodej and Bardodejova, 1970; Ohtsuji and Ikeda, 1971) catalysed by the cytochrome P450-dependent monooxygenase in the liver microsomes or other tissues (Leibman and Ortiz, 1969; Leibman, 1975; Salmona *et al.*, 1976; Belvedere *et al.*, 1977; Cantoni *et al.*, 1978; Duverger-Van Bogaert *et al.*, 1978; Van Bogaert *et al.*, 1978; Vainio and Elovaara, 1979). The epoxide is subsequently hydrolysed to styrene-glycol, a reaction catalysed by the epoxide hydrolase system present in the liver microsomes or other tissues (Belvedere *et al.*, 1977; Dansette *et al.*, 1978; Duverger-Van Bogaert *et al.*, 1978; Watabe *et al.*, 1981a, 1983; Pacifici and Rane, 1982; Pacifici *et al.*, 1983a, b; Smith *et al.*, 1983a). Styrene-7,8-oxide can also be conjugated with glutathione, catalysed by glutathione-*S*-transferase which leads to the formation of mercapturic acid derivatives (Delbressine, 1981; Pacifici and Rane, 1982; Steele *et al.*, 1981; Warholm *et al.*, 1983). In rabbits styrene is partly converted in a glucuronide but the major metabolic pathway is oxidation to amandelic acid and phenylglyoxylic acid. In rats and rabbits styrene-glycol is transformed to benzoic acid, which is excreted in the urine as hippuric acid (James and White, 1967; Ohtsuji and Ikeda, 1971; Bardodej, 1978). A less important metabolic pathway is the formation of arene oxide (styrene-1,2-oxide, styrene-3,4-oxide) from which, by hydration, phenols

(2-vinylphenol and 4-vinylphenol) are formed in rats (Bakke and Scheline, 1970; Pantarotto *et al.*, 1982; Watabe *et al.*, 1981c).

Both in experimental animals and in humans, a small percentage of unchanged styrene is exhaled, regardless of the route of exposure (Bardodej and Bardodejova, 1970; Fernandez and Caperos, 1977; Engström *et al.*, 1978a; Ramsey and Young, 1978; Bergman, 1979; Caperos *et al.*, 1979; Ramsey *et al.*, 1980). The main part of the absorbed styrene (up to 97%) is metabolised and rapidly excreted via the urine (Bardodej and Bardodejova, 1970; Astrand *et al.*, 1974; Caperos *et al.*, 1979; Guillemin and Bauer, 1979; Plotnick and Weigel, 1979; Ramsey *et al.*, 1980). Only a small proportion of the absorbed styrene is excreted in the faeces (Plotnick and Weigel, 1979). In rats and mice exposed to styrene via inhalation or subcutaneous injection, excretion by exhalation of CO_2 has been recorded.

In general the metabolites of styrene detected in the urine of humans and experimental animals are the same, but differ in quantity. In man, amandelic acid and phenylglyoxyl acid are the most important metabolites, whereas hippuric acid is only a minor metabolite (Stewart *et al.*, 1968; Bardodej and Bardodejova, 1970; Astrand *et al.*, 1974; Ikeda and Imamura, 1974; Guillemin and Bauer, 1979). The $t_{1/2\,el}$ for exhalation varies from 1 to 5 hours (Stewart *et al.*, 1968). For humans $t_{1/2}$ values for the excretion ($t_{1/2\,ex}$) of amandelic acid as measured in urine were reported to be 3.9–24.7 h; for phenylglyoxylic acid the values were 9–10.5 h (Ikeda and Imamura, 1974; Engström *et al.*, 1976; Teramoto, 1978; Guillemin and Bauer, 1979).

2.2.4 *Toxicokinetic behaviour of toluene*

Absorption through the respiratory tract after inhalatory exposure to toluene has been studied less extensively in experimental animals than in humans. Toluene was readily absorbed (90%) from both the upper and lower respiratory tract by dogs (Egle and Gochberg, 1976). Hobara *et al.* (1984), in their study with dogs, found that after 1 h of inhalatory exposure to toluene, 27% was absorbed. In humans the absorption of toluene was about 50% after inhalation of 375 mg toluene/m^3 for several hours (Nomiyama and Nomiyama, 1974a, b; Carlsson and Lindqvist, 1977; Veulemans and Masschelein, 1978; Carlsson, 1983).

Toluene is completely absorbed from the gastro-intestinal tract after oral dosing both in experimental animals and humans (El Masri *et al.*, 1956; DGMK, 1985). If toluene is dissolved in vegetable oil absorption is less efficient (Pyykko *et al.*, 1977). Although it appears to penetrate excised rat skin *in vitro*, toluene does not penetrate the skin *in vivo* (Tsuruta, 1982).

In humans the dermal resorption velocity was found to be 0.17 mg/cm^2/h after exposure to 35 and 230 mg toluene (DGMK, 1985). Absorption after dermal exposure resulted in a blood concentration of 15–25% of the concentration after inhalation of toluene.

In the case of inhalation toluene is rapidly distributed to the more vasculated tissues such as the liver and brain (Peterson and Bruckner, 1978; Benignus *et al.*, 1981, 1984; Bruckner and Peterson, 1981a; Fishbein, 1985a). Studies carried out by Carlsson and Lindqvist (1977) and Pyykko *et al.* (1977) with rats, and by Bergman (1979) with mice showed that tissue distributions of toluene after oral and inhalatory exposure were almost comparable. The highest concentration was found in adipose tissue, followed by the adrenals, kidneys, liver, cerebrum, cerebellum, stomach, blood and bone marrow. However, distribution appeared to be delayed by absorption from the digestive tract.

The biotransformation of toluene in animals and humans is thought to be similar. The initial step in the metabolism of toluene to benzoic acid appears to be side-chain hydroxylation to benzyl alcohol by the microsomal mixed-function oxidase system. Ikeda and Ohtsuji (1971) demonstrated that induction of hepatic mixed-function oxidase by phenobarbital increased the metabolism of toluene and resulted in high levels of benzoic acid in the blood, while in non-induced rats benzoic acid was not found in the blood. The biotransformation of toluene seems to occur primarily in the liver (EPA, 1983). Toluene metabolism occurs rapidly, as shown by the appearance of metabolites in the liver, kidneys and blood of mice within minutes of exposure to toluene (Koga, 1978; Bergman, 1979) and by increased urinary excretion of hippuric acid in rabbits within 0.5 h of inhalation (Nomiyama and Nomiyama, 1974b).

Benzoic acid can also be conjugated to glucuronide acid to form benzoyl glucuronide (EPA, 1983). Ring hydroxylation by mixed-function oxidases occurs only to a minor degree, and results in the formation of *o*-cresol and *p*-cresol. This has been demonstrated after oral administration (Bakke and Scheline, 1970). In humans, 15–20% of the absorbed toluene is exhaled unchanged in the lungs (Cohr and Stokholm, 1979); 60–70% is excreted by the kidneys as hippuric acid (Veulemans and Masschelein, 1979); 10–20% is excreted as glucuronide conjugate; and at the most 1% is excreted as *o*- or *p*-cresol, either as glucuronide or as sulphates (EPA, 1983). The levels of nine metabolites of toluene returned to background levels within 24 h inhalatory exposure to 187–375 mg/m^3 for several hours (Veulemans and Masschelein, 1979).

2.2.5 Toxicokinetic behaviour of phenol

Phenol and conjugated phenols are normally found in animal and human tissues, urine, faeces, saliva and sweat. The concentration of phenol in the blood depends on the composition of the food. Phenol is easily absorbed regardless of the route of exposure (Deichmann and Keplinger, 1963; Van Koten-Vermeulen *et al.*, 1986). The route of absorption through the skin depends largely on the exposed surface (Liao and Oehme, 1980), whereas damage to the skin depends on the concentration of the phenol solution

(Deichmann *et al.*, 1952). The horn layer inhibits absorption (Behl *et al.*, 1983). In humans phenol vapour and solutions are mostly absorbed through the skin. High rates of absorption of phenol vapour (Ruedemann and Deichmann, 1953) are also illustrated by sometimes fatal effects (Hinkel and Kintzel, 1968; Piotrowsky, 1971; Baranowska-Dutkiewicz, 1981; Lewin and Cleary, 1982). Absorption of phenol in the lungs of volunteers exposed to 5–25 mg/m^3 decreased from 80 to 70% during exposure periods of 6–8 h (Piotrowsky, 1971; Ohtsuji and Ikeda, 1972).

Fifteen minutes after administration of 0.5 g phenol/kg body weight the highest concentration is found in the liver, followed by the central nervous system, the lungs and the blood. After 82 min the phenol is equally distributed over all tissues. The ratio of conjugated phenol:phenol gradually increases with time (Deichmann, 1949). After oral administration of ^{14}C-phenol to rats the highest value relative to plasma was found in the liver (42%), followed by the spleen, kidney, adrenals, thyroid and lungs, with peak levels after 30 min (Liao and Oehme, 1980, 1981). After intravenous administration to mice and rats the highest levels in tissues were recorded after 2 h in the kidneys and the liver (Gbodi and Oehme, 1978; Wheldrake *et al.*, 1978; Greenlee *et al.*, 1981).

Phenol is metabolised in the liver to carbon dioxide (23%) and small quantities of catechols and hydrochinones; 72% of the compound is excreted in the urine, both as unchanged and as conjugated phenols, 1% is excreted in the faeces, and only minimal quantities of unchanged phenol are exhaled. Only 4% is thought to be retained in the body (Deichmann and Keplinger, 1963; Cassidy and Houston, 1980; Houston and Cassidy, 1982). In the urine of different animal species several phenol metabolites have been found, namely phenylsulphate, phenylglucuronide, chinolsulphate and chinolglucuronide. The relative proportions of the different metabolites excreted depend on the animal species, and on the amount of phenol administered (Oehme and Davies, 1970; Capel *et al.*, 1972a, b; Miller *et al.*, 1976; Mehta *et al.*, 1978; Kao *et al.*, 1979; Weitering *et al.*, 1979; Baudinette *et al.*, 1980; Koster *et al.*, 1981; Ramli and Wheldrake, 1981). A restriction in the conjugation of phenol with sulphate or glucuronic acid was noticed in pigs and cats (Capel *et al.*, 1979). In some cases (cat, sheep) phenyl and/or chinolphosphates were also detected (Capel *et al.*, 1974; Kao *et al.*, 1979).

Excretion via urine is, both for experimental animals and humans, the main route of elimination. The excretion rate depends on the exposure level, exposure route and animal species. Excretion is most rapid in rats, in which 95% of orally administered phenol of 25 mg/kg body weight was excreted via urine within 24 h. Excretion (91%) was slowest in the rhesus monkey (Capel *et al.*, 1972a, b). The amount of unmetabolised phenol exhaled or excreted by the faeces was minimal (Deichmann and Keplinger, 1963).

There are cases where phenol or its conjugates are excreted via bile, probably where excretion by urine is blocked (Abdou-el-Makarem *et al.*, 1967; Weitering *et al.*, 1979).

If a one-compartment model for the elimination of phenol after inhalation or absorption through the skin is applied, the $t_{1/2\,el}$ is about 3.5 h (Piotrowski, 1971).

2.2.6 *Toxicokinetic behaviour of chlorophenols*

The majority of data on the kinetics and biotransformation of chlorophenols refers to pentachlorophenol (PCP). However, the volatility of PCP is restricted and problems of inhalation are mainly confined to aerosols. For this reason this section will focus on the kinetics and biotransformation of mono-, di- and trichlorophenols, although data for these compounds are very limited. Based on the available data and the physicochemical properties of chlorophenols it is assumed that all chlorophenols are readily absorbed and excreted, with urine being the major route of elimination (Janus *et al.*, 1990). No specific absorption studies with mono- and dichlorophenols after dermal or inhalatory exposure are available. However, toxicity studies with respective exposure routes, both in experimental animals and humans, indicate that chlorophenols are readily absorbed via dermal and inhalatory exposure routes (Janus *et al.*, 1990).

After a single parenteral administration of 2,4-dichlorophenol (2,4-DCP) or 2,4,6-trichlorophenol (2,4,6-TCP), the highest concentrations were found in the liver. In both cattle and sheep fed 2,4-dichlorophenoxyacetic acid for 28 days, the highest and second highest 2,4-DCP concentrations were found in the kidneys and liver, respectively (WHO, 1989a).

Studies on biotransformation using different laboratory animals, different routes of exposure and different compounds revealed that the lower chlorinated chlorophenols (monochlorophenol (MCP), dichlorophenol (DCP) and trichlorophenol (TCP)) are present in tissues and body fluids mainly as glucuronide and sulphate conjugates (WHO, 1989a). No quantitative data on the excretion of MCP, DCP, TCP and their metabolites are available (Janus *et al.*, 1990).

2.2.7 *Toxicokinetic behaviour of benzene*

After exposure of rabbits to 10–16 g/m^3 benzene by inhalation absorption ranged from 37 to 55% (Sandmeyer, 1982). Absorption in humans after exposure to 30–350 mg benzene/m^3 for 1–7 h ranged from 75–20% (Srbova *et al.*, 1950; Hunter and Blair, 1972; Sherwood, 1972a, b; Nomiyama and Nomiyama, 1974a; Saita, 1976).

No human data on the absorption of benzene from the gastro-intestinal tract are available (Slooff, 1987).

Although benzene seems to increase the permeability of the skin, absorption in experimental animals and humans is low—not higher than 0.2% (Conca and Maltagliati, 1955; Slooff, 1987).

After inhalatory exposure to 1650–4200 mg/m^3, benzene was readily distributed in the body of dogs (<2h), with the highest concentrations measured in the bone marrow and adipose tissue (Rickert *et al.*, 1979). In rats exposed to benzene by inhalation, distribution was achieved within 2–6h, with the highest concentrations in the blood, adipose tissue and bone marrow. The half-life for elimination ($t_{1/2\,el}$) from these tissues ranged from 0.4 to 0.8h, except for adipose tissue in which the $t_{1/2}$ was 1.6h (Rickert *et al.*, 1979). In mice the highest concentrations of benzene after 3h dermal exposure were recorded in the liver, spleen, bone marrow and blood. The concentration of CO_2 reached a maximum one hour after dermal administration (Gill and Ahmed, 1981). Exhalation of unchanged benzene is the major excretion route in dogs, rabbits, mice and rats (IARC, 1982a). Up to 50% of benzene is excreted unchanged by exhalation, whereas the remainder is transformed to phenol by the mixed-function oxidase system in the liver (Snyder *et al.*, 1981).

Intermediate metabolites of benzene are benzene-epoxide, small proportions of the dihydroxy derivative catechol, hydroquinone, and very small proportions of the dihydroxy derivate hydroxyquinone. These hydroxy derivatives are mainly excreted into urine as glucuronide or sulphate esters. The relative quantities depend on the animal species. In rabbits and mice glucuronides are the major metabolites (Garton and Williams, 1949; Snyder, 1974) whereas in rats sulphate esters are the major metabolites (Cornish and Ryan, 1965; Gerarde and Ahlstrom, 1966).

In rabbits oral administration of benzene also leads to the formation of *trans-trans* muconic acid (1–3%) and phenylmercapturic acid (0–5%) (Fielder, 1982). These metabolites are also formed after dermal and inhalatory exposure (Synder, 1974). In urine high amounts of phenol (up to 80%), phenolsulphate (20–50%), catechol (3%), quinone (1%) and traces of hydroxyquinone, phenylglucuronide, phenylmercapturic acid, *trans* muconic acids and conjugated compounds have been detected (Teisinger *et al.*, 1952; Williams, 1963; Sherwood and Carter, 1970; Hunter and Blair, 1972; Sherwood, 1976).

Almost 80% of the administered amount of benzene is excreted within 2–3 days (Parke and Williams, 1953). In humans exposed to benzene (169–192 mg/m^3) ·34% of the absorbed dose is exhaled unchanged. This exhalation continues for about 15h. The exhalation proceeds in three phases, with different elimination rates reflected in $t_{1/2\,el}$ values of a few minutes, 20 min and 20–30 h, respectively (Nomiyama and Nomiyama, 1974a,b). Other investigators have observed at least two elimination phases: (i) an initially fast phase; and (ii) a slow phase (Berlin *et al.*, 1979, 1980). Täuber (1983) concluded that the excretion of benzene can be described by a three-compartment model: (1) blood and well-vasculated tissues; (2) less-vasculated tissues; and (3) adipose tissues. Only a small proportion of benzene was excreted by exhalation of CO_2 (Täuber, 1983). In humans and experimental animals benzene metabolites are excreted by the urine. The $t_{1/2\,el}$ values of phenol derivatives (sulphates and glucuronides) have been estimated to be 4–8h (Teisinger *et al.*, 1952; Docter and Zielhuis, 1967; Sherwood and Carter, 1970).

2.2.8 *Toxicokinetic behaviour of chloroform*

After inhalatory exposure of humans, 60–80% of the inhaled quantity of chloroform is absorbed. After oral administration chloroform is readily and almost completely absorbed in animals and humans. Blood peak levels of ^{13}C- or ^{14}C-labelled chloroform are reached within 1–1.5 h post dosing. Chloroform is also absorbed by the skin.

Chloroform distributes throughout the whole body. Maximal tissue concentrations are reached in fat, blood, liver, lungs, kidneys and nervous system. Binding to tissue components occurs in the liver and kidneys (especially in the renal cortex of male mice). Chloroform will also pass the placenta.

Chloroform is thought to be metabolised by a cytochrome P450-dependent enzyme system to phosgene, which mainly reacts with cysteine or glutathione. Phosgene may also become bound to tissue macromolecules. However, the formation of phosgene from chloroform has not been proven irrefutably. A minor biotransformation pathway of chloroform is the cytochrome P450- and glutathione-dependent formation of carbon monoxide. At low exposure levels, phosgene seems to be the actual toxic principle, whereas at higher exposure levels, chloroform itself contributes to the toxic effects (anaesthesia). In contrast with other halomethanes, chloroform does not seem to give rise to trihalomethyl radicals. The main metabolising organ is the liver. Monkeys and humans metabolise chloroform to a lesser extent than mice and rats. There is also a sex difference in metabolic activity; male mice show considerable metabolising activity in the kidneys.

Chloroform and its main metabolite, carbon dioxide, are excreted by the lungs. Excretion via other routes is negligible. Elimination of chloroform by the pulmonary route appears to be very slow and the excretion rate amounts to only a few percent of the body burden. In obese persons retention is higher. The quantity of exhaled chloroform reflects the rate of metabolism. Slow metabolisers expire relatively large amounts of chloroform. In humans a terminal plasma $t_{1/2\,el}$ of chloroform of about 90 min has been reported. However, in exhaled air the $t_{1/2}$ may be as long as two days. These findings are indicative of a deep fatty compartment, from which chloroform is only slowly removed. In humans only 55% of an oral dose is available, due to hepatic pulmonary first-pass effects (Van der Heijden *et al.*, 1987).

2.2.9 *Toxicokinetic behaviour of methylene chloride*

The absorption of methylene chloride has been studied after inhalatory exposure in several mammalian species. Pulmonary absorption in sedentary humans ranged from 55 to 72% of the total exposure (Astrand and Gamberale, 1978; Divincenzo and Kaplan, 1981). In rats absorption of methylene chloride as studied after oral administration in different vehicles was relatively rapid in aqueous solutions. Percutaneous absorption rates were 1291 and 1140 nmol/min/cm^2 skin in mice and rats, respectively (Van Apeldoorn *et al.*, 1987).

Inhalatory exposure of rats to ^{14}C-methylene chloride for 6h resulted in overall tissue concentrations ranging from 7 to 23% of the total body burden. Blood concentrations of the radioactive label reached steady state after 2h of exposure and declined exponentially after termination of exposure. The highest concentrations were found in the liver, kidneys and lungs (McKenna *et al.*, 1982). One hour after inhalatory exposure to a high concentration of methylene chloride, high concentrations were observed in white adipose tissue and the adrenal glands, as well as in the liver and kidneys (Carlsson and Hultengren, 1978). Elimination of 25–90% of the accumulated radioactivity from these tissues was observed within 2h of exposure. In cases of prolonged exposure methylene chloride was detected in perirenal fat and brain tissue. The distribution was similar in mice. Data from distribution studies with radioactively labelled methylene chloride in rats and mice indicated greater accumulation of radioactivity in the liver and kidneys, and more rapid elimination from body fat and brain in comparison with other tissues (Van Apeldoorn *et al.*, 1978). After inhalatory exposure of humans to methylene chloride the concentration in the blood gradually increased and eventually reached a plateau. Methylene chloride also increased in the adipose tissues (Engström and Bjurström, 1977; Divicenzo and Kaplan, 1981).

The biotransformation of methylene chloride is enzymatically catalysed and therefore shows capacity-limited phenomena at high concentrations (Anderson, 1981). First-order kinetics are observed at low concentrations, and zero-order kinetics are observed at high concentrations. The overall time course for biotransformation is complex due to contributions from at least two pathways with varying kinetic parameters. The carbon monoxide-producing pathway (P450 system) appears to be a high-affinity, low-capacity pathway. The glutathione-mediated pathway (glutathione transferase) is a low-affinity, high-capacity first-order process at all accessible exposure concentrations. The maximum metabolic velocity (V_{max}) is a composite of two pathways. Because of the saturable character of methylene chloride metabolism, the discrepancy between the administered dose and the effective dose becomes greater with increasing exposure (Anon, 1988).

After inhalatory exposure of rats to ^{14}C-methylene chloride, 5–55% of the body burden is excreted unchanged by exhalation (McKenna *et al.*, 1982). The non-volatile radioactivity is excreted by urine in amounts of 1–8% (Divincenzo *et al.*, 1972). After intraperitoneal injection 98% of the radioactivity was eliminated in the breath as methylene chloride (91.5%), CO (2%), CO_2 (3%) and an uncharacterised metabolite (1.5%), (McKenna *et al.*, 1982). Although urinary excretion of non-volatile metabolites was the primary elimination mode for inhaled methylene chloride in rats and mice, a small proportion was also detected in faeces (McKenna *et al.*, 1982). After termination of inhalatory exposure less than 5% of the 70% methylene chloride absorbed was excreted by exhalation, indicating that biotransformation plays a major role in the elimination. Urinary excretion levels of the metabolites were related to inhalatory exposure levels (Divincenzo *et al.*, 1972).

2.2.10 Toxicokinetic behaviour of volatile chlorobenzenes

Monochlorobenzene (MCB) is well absorbed after oral and inhalatory exposure. Absorption after dermal exposure seems to be minimal. Shortly after oral administration, MCB was found in all tissues examined, with adipose tissue containing highest amounts. A rapid tissue clearance was observed. The principal metabolites of MCB were 4-chlorophenol, 4-chlorocatechol and 4-chlorophenylmercapturic acid. The majority of metabolites were excreted in the urine, with only small amounts present in the faeces. Urinary metabolites of MCB were ethereal sulphates (33.9%), glucuronides (33.6%), mercapturic acids (23.6%), diphenols (4.2%), monophenols (2.8%) and 3,4-dihydro-3,4-dihydroxychlorobenzene (0.6%) (Smith and Carlson, 1980).

Experimental studies have indicated that dichlorobenzenes (DCBs) are absorbed after different routes of exposure. After both oral inhalatory and subcutaneous exposure, tissue distribution appeared to be similar. Highest levels occurred in adipose tissue, liver, kidneys and lungs. After inhalatory exposure to 1,4-DCB a difference in organ distribution was observed between male and female rats; females had significantly higher levels of 1,4-DCB in their livers, whereas the kidneys of the males contained significantly higher levels compared to those of the female rats. The 1,2-DCB and 1,4-DCB were mainly oxidised to 3,4- and 2,5-dichlorophenol, respectively, which were further conjugated. Within 24 h of exposure 50–90% of the administered dose was excreted in the faeces as metabolites. The main excretion products of 1,2-DCB were glucuronides (48% of the administered dose), ethereal sulphate (21%), mercapturic acids (5%), 3,4-dichlorophenol (30%) and 2,3-dichlorophenol (9%). Only a small amount of the DCBs was excreted in the faeces. The DCBs were absorbed via the entero-hepatic circulation.

Trichlorobenzenes (TCBs) are absorbed after oral, inhalatory and dermal exposure. After oral exposure of rats highest amounts were found in the liver, kidneys, adipose tissues, bladder and gastro-intestinal tract. Increased contents were also reported for skin and muscle tissues. Tissue levels were generally highest after exposure to 1,3,5-TCB. TCBs are metabolised to trichlorophenols, which, in rats, are further conjugated with glutathione. In contrast, monkeys do not use glutathione in the metabolism of TCBs, although intermediate metabolite arene oxides are hydrolysed. The resulting dihydrodiol is excreted as a glucuronide. Due to the additional metabolic steps excretion in monkeys is lower than that in rats. Within 24 h of oral exposure rats excreted approximately 70% in the urine and about 15% in the faeces, whereas monkeys excreted 40% in the urine and less than 1% in the faeces. TCBs were also reabsorbed via the entero-hepatic cycles (Hesse *et al.*, 1991).

2.2.11 *Toxicokinetic behaviour of ethylene and propylene oxide*

Ethylene oxide is very soluble in blood, therefore pulmonary uptake after inhalatory exposure is fast and total (Altman and Dittmer, 1974; Ehrenberg *et al.*, 1974). No specific data on skin absorption are available (Vermeire *et al.*, 1985). Accidental exposure of the skin of industrial workers to 1% aqueous ethylene oxide resulted in systemic effects on the central nervous system (Sexton and Henson, 1949), which indicates that ethylene oxide is absorbed through the skin. Ethylene oxide is rapidly distributed throughout the body. In mice dosed by intravenous injection the concentration of ethylene oxide first increased in the liver, kidneys and pancreas, and later appeared in the intestinal mucosa, stomach contents, lungs, epididymus, testes and cerebellum (Appelgren *et al.*, 1977). Directly after inhalation by mice the highest concentrations of ethylene oxide or its metabolites were found in the liver, kidneys and lung. The radioactivity in the liver and kidneys dropped exponentially and approached the level in the lungs, testes, spleen and brain within 4 h indicating rapid metabolism and excretion (Ehrenberg *et al.*, 1974). A $t_{1/2}$ value for the first-order elimination of ethylene oxide from rat and mouse tissue is estimated to be 9–10 min, based on alkylation data in different tissues (Ehrenberg *et al.*, 1974) or haemoglobin alkylation data (Osterman-Golkar, 1983). For humans similar values were estimated (Calleman *et al.*, 1978). The elimination of the metabolite 1,2-methanediol was slower ($t_{1/2\,el}$ value of 3–4 h) in dogs, rats and monkeys than in mice (McChesney *et al.*, 1971).

The limited data on biotransformation of ethylene oxide indicate two possible pathways: (i) hydrolysis; and (ii) glutathione conjugation (Gessner *et al.*, 1961; McChesney *et al.*, 1971; Jones and Wells, 1981). Metabolites indentified in humans and experimental animals were 1,2-ethanediol, hydroxy-acetaldehyde, hydroxyacetic acid, oxoacetic acid, ethanediocic acid, formic acid, *s*-2-hydroxyethyl glutathione, *s*-2-hydroxyethylcysteine and *N*-acetyl-*s*-(2-hydroxyethyl) cysteine. A major fraction of 1,2-ethanediol was excreted unchanged in the urine of rats, rabbits and monkeys (Gessner *et al.*, 1961; McChesney *et al.*, 1971).

Ehrenberg *et al.* (1974) found that an average of 74% of labelled ethylene oxide inhaled by mice was excreted in the urine within 24 h as unidentified metabolites, and only 4% within the next 24 h. The available data show that a significant proportion of ethylene oxide or its metabolites is not excreted immediately but is probably bound to macromolecules or to a part of the carbon pool of the body (Vermeire *et al.*, 1985). It is known that ethylene oxide alkylates proteins and DNA (Ehrenberg *et al.*, 1974; Osterman-Golkar, 1983; Segerbäck, 1983).

There are no experimental data on the absorption of propylene oxide. Like the chemically related ethylene oxide inhaled propylene oxide is probably completely absorbed and distributed throughout the body and rapidly metabolised. The $t_{1/2\ el}$ value in rat tissues has been estimated to be 40 min (Vermeire *et al.*, 1988).

2.3 Introduction to toxicological studies

In this part of the chapter the general toxicology for representative VOCs is summarised. This toxicology includes acute, short-term (subacute and subchronic) and long-term toxicity, carcinogenicity, mutagenicity and teratogenicity studies for the different exposure routes. These data are normally required for health risk evaluation according to several guidelines. In addition to general toxicology the effects of VOCs on several organ systems and the existence of probable risk groups are discussed. Attention is paid to differences between humans and experimental animals. Finally new approaches to the risk evaluation of toxic compounds, including VOCs, are reviewed. Emphasis is also placed on the physiologically based kinetic models (PBKs).

2.3.1 Acute toxicity in animals

In general acute oral toxicity of VOCs is low to moderate. The oral LD_{50} (lethal dose causing 50% mortality) is 200 mg/kg body weight (bw) or higher. In some cases toxicity is higher after inhalatory or dermal exposure. Toxic effects at (sub)lethal dose levels can differ according to the exposure route. This difference in toxicity may be explained by the distribution pattern and metabolism of VOCs as discussed in section 2.2. In cases of inhalatory exposure the lungs are involved. Although there are differences in the toxicological profiles of different VOCs, hepatotoxicity and renal toxicity are often involved at these (sub)lethal dose levels. Several VOCs also have anaesthetic effects, while a few cause effects on the central nervous system. In the following sections this is illustrated by summarising the acute toxicity of a number of representative VOCs.

2.3.1.1 Benzene. The oral LD_{50} in rats for benzene has been shown to range from 3900 to 5970 mg/kg bw and the inhalatory LC_{50} from 15 000 to 60 000 mg/m^3 (Fielder, 1982; IARC, 1982a). Although no dermal LD_{50} is available, toxicity after subcutaneous exposure (leucocytopenia) has been shown in mice and rabbits at 5000 mg/kg bw or more (Gerarde and Ahlstrom, 1966; Kissling and Speck, 1972; Gill *et al.*, 1979). The (sub)lethal toxic effects included an anaesthetic effect, tremors and, in the case of inhalatory exposure, pathological changes in the lungs and liver. In the case of oral exposure stomach lesions were also observed (Slooff, 1987).

2.3.1.2 Styrene. The oral LD_{50} for styrene in rats ranges from 2150 to 8057 mg/kg bw (Hinz *et al.*, 1980; Lewis and Tatken, 1982). This is higher than that for mice of 316 to 660 mg/kg (Lewis and Tatken, 1982). This difference in toxicity of styrene was not seen for the inhalatory route; the LC_{50} in rats varied from 11 340 to 19 396 mg/m^3, and in mice from 10 202 to 21 000 mg/m^3 (Shugaev and Yaroslavl, 1969; Bonnet *et al.*, 1979; Jaeger *et al.*, 1979; Lewis and Tatken, 1982). No data on acute dermal toxicity are available. The toxic effects of styrene at (sub)lethal dose levels include liver and kidney damage; in the case of inhalatory exposure pneumonia and effects on the central nervous system have also been observed (Parkki, 1978; Agrawal *et al.*, 1982; Van Apeldoorn *et al.*, 1986).

2.3.1.3 Toluene. The oral LD_{50} for toluene in rat varies from 5000 to 7500 mg/kg bw (Smyth *et al.*, 1969; Kimura *et al.*, 1971; Van der Heijden *et al.*, 1988). The inhalatory LC_{50} for toluene in mice and rats (exposure time 4–7 h) ranges from 26 000 to 72 000 mg/m^3 (Van der Heijden *et al.*, 1988). The lowest published lethal dose after subcutaneous exposure to toluene ranged from 1100 to 8700 mg/kg bw (EPA, 1983). The toxic effects at these (sub)lethal dose levels included tremors, elevated levels of neurotransmitters and loss of coordination (Hobara *et al.*, 1984). The cause of death was reported to be respiration failure (EPA, 1983).

2.3.1.4 Xylenes. The oral LD_{50} for mixed xylenes in rats ranged from 4300 to 8000 mg/kg bw (Wolff *et al.*, 1956; Hine and Zuidema, 1970). When exposure is by inhalation, the xylene isomers have different acute toxicities (Low *et al.*, 1989). In rats the LC_{50} for *o*-xylene was 4595 mg/m^3, for *m*-xylene 5267 mg/m^3 and for *p*-xylene 3907 mg/m^3 (Bonnet *et al.*, 1979). The LC_{50} for mixtures of xylenes and ethylbenzene ranged from 6350 to 6700 mg/m^3 (Hine and Zuidema, 1970; Carpenter *et al.*, 1975). Cats exposed to 9430 mg xylene/m^3 exhibited classical central nervous system symptoms characterised by salivation, ataxia, tonic and clonic spasms and anaesthesia, and died within two hours exposure (Carpenter *et al.*, 1975). The lethal subcutaneous dose in rats for *m*- and *p*-xylene ranged from 4000 to 8000 mg/kg bw and for *o*-xylene from 2000 to 4000 mg/kg bw (Cameron *et al.*, 1938).

2.3.1.5 Phenol. The oral LD_{50} for phenol in mice, rats, rabbits and dogs ranged from 300 to 600 mg/kg bw, whereas the inhalatory LC_{50} in rats was >900 mg/m^3 (8 h exposure). The dermal LD_{50} values in rats and rabbits were 670 and 850 mg/kg bw, respectively (Conning and Hayes, 1970; Brown *et al.*, 1975; Flickinger, 1976). The toxic effects at these (sub)lethal dose levels were tremors, convulsion and elevated heart frequency, followed by a low and non-regular pulse, increased salivary production and decline in body temperature (Pullin *et al.*, 1978; Liao and Oehme, 1980; Van Koten-Vermeulen *et al.*, 1986). After oral administration mucous tissues of the throat and oesophagus showed erythema, corrosion and necrosis. After inhalatory exposure lung lesions, infarcts, broncho-pneumonia, purulent bronchitis and hyperplasia of peribronchial tissues were observed (Van Oettingen, 1949). Other effects observed at this high dose level of phenol were demyelinisation of nerve tissues, degeneration and necrosis of the myocardium (Deichmann and Keplinger, 1963; Liao and Oehme, 1980), and liver and kidney damage (Oehme and Davis, 1970; Coan *et al.*, 1982).

2.3.1.6 Chlorophenols. The oral LD_{50} for monochlorophenols (MCPs) in rats ranged from 260 to 2400 mg/kg bw; for dichlorophenols (DCPs) the values were 465 to 4000 mg/kg bw; for trichlorophenols (T3CPs) 455 to 2960 mg/kg bw; and for tetrachlorophenols (T4CPs) and pentachlorophenols (PCPs) 25 to 295 mg/kg bw. The inhalatory LC_{50} values for MCP and DCP in rats were 11 mg/m^3 and 225–355 mg/m^3, respectively. No LC_{50} data are available for phenols with a higher degree of chlorination. The dermal LD_{50} in rats, rabbits and other mammals ranged from 1000 to 1500 mg/kg bw; for 4-MCP; for 2,4-DCP the value was 790 mg/kg bw; for 2,4,6-T3CP 700 mg/kg bw; for 2,3,5,6-T4CP 280–485 mg/kg bw; for 2,3,5,6-T4CP 300–500 mg/kg bw; and for PCP 40–330 mg/kg bw. These data indicate that acute toxicity increases with chlorination of phenols after oral and dermal exposure (RTECS, 1989; WHO, 1987a, 1989a; Janus *et al.*, 1990).

Most signs and symptoms at (sub)lethal exposure levels of different chlorophenols are similar, and include motor weakness, an increase in respiration rate and body temperature, tremors, depression of the central nervous system, convulsions, dyspnea and coma. Tremor activity was high for phenols with a low degree of chlorination, whereas uncoupling of phosphorylation was higher for phenols with a higher degree of chlorination (Ahlborg and Thunberg, 1978; Borzelleca *et al.*, 1985; WHO, 1987a; Janus *et al.*, 1990).

2.3.1.7 Chloroform. The oral LD_{50} for chloroform dissolved in sesame oil ranged from 36 to 463 mg/kg bw in rats and from 353 to 1366 mg/kg bw in mice; when dissolved in aqueous solution the oral LD_{50} values were 1120 and 1400 mg/kg bw, respectively (Kimura *et al.*, 1971; Bowman *et al.*, 1978; Pericin and Thomann, 1979; Chu *et al.*, 1980). No inhalatory LC_{50} and dermal LD_{50} data are available. The lethal inhalatory concentration in rabbits and guinea pigs ranged from 10 000 to 100 000 mg/m^3. The toxic effects after inhalatory exposure at these (sub)lethal dose levels included anaesthesia, cardiotoxicity, respiratory acidosis and hepatotoxicity (Brondeau *et al.*, 1983; Van der Heijden *et al.*, 1987). After oral exposure to chloroform the major toxic effects were hepatoxicity and renal toxicity. After dermal exposure to 75–1500 mg/kg bw toxic effects included hepatic and renal damage (Smith *et al.*, 1983; Van der Heijden *et al.*, 1987).

2.3.1.8 Methylene chloride. The oral LD_{50} for methylene chloride in rats was 3000 mg/kg bw (Kimura *et al.*, 1971); the inhalatory LC_{50} in rats and mice varied from 49 300 to 79 000 mg/m^3 (Bonnet *et al.*, 1980; Kashin *et al.*, 1980; Van Apeldoorn *et al.*, 1987). No dermal LD_{50} data for methylene chloride are available. The toxic effects observed at (sub)lethal dose levels of methylene chloride included depression of the central nervous system resulting in reversible narcosis, reversible changes in motor activity, and pathological changes in the liver, lungs and heart.

2.3.1.9 Chlorobenzenes. The oral LD_{50} for monochlorobenzene (MCB) in rats, rabbits, mice and guinea pigs ranged from 430 to 2830 mg/kg bw; for 1,2-dichlorobenzene (1,2-DCB) the value was 2138–3375 mg/kg bw; for 1,4-dichlorobenzene (1,4-DCB) 500–2812 mg/kg bw; and for 1,2,4-trichlorobenzene (1,2,4-TCB) 756–766 mg/kg bw (EPA, 1984; Hesse *et al.*, 1991). The inhalatory LC_{50} for MCB in rats and mice ranged from 1886 to 2965 mg/m^3 bw, whereas the dermal LD_{50} in rabbits was $>10\,000$ mg/kg bw. At (sub)lethal dose levels MCB, DCB and TCB all caused narcotic signs, and depression and toxic effects on the liver, kidneys and central nervous system; in the case of inhalation effects on the lungs were also observed (EPA, 1984; Hesse *et al.*, 1991).

2.3.1.10 Ethylene oxide. The oral LD_{50} for ethylene oxide in rats and mice ranged from 261 to 383 mg/kg bw; the inhalatory LC_{50} values (4 h) in mice, dogs and rats were 1500, 1730 and 2630 mg/m^3 bw, respectively (Vermeire *et al.*, 1985). No dermal LD_{50} data are available. The toxic effects of (sub)lethal levels of ethylene oxide were seen in the lungs and nervous system (ataxia, prostation, laboured breathing and tonic convulsions) (Woodard and Woodard, 1971; Vermeire *et al.*, 1985).

2.3.1.11 Propylene oxide. The oral LD_{50} for propylene oxide in rats ranged from 540 to 1140 mg/kg bw; the inhalatory LC_{50} values (4 h exposure) in rats and mice were 9500 and 4100 mg/m^3 bw, respectively. After oral exposure the main toxic effects of (sub)lethal exposure were necrosis of the stomach and hepatotoxicity. After inhalatory exposure the lungs were affected and neuropathological effects were seen (Vermeire *et al.*, 1988).

2.3.1.12 Nitropropanes. The oral LD_{50} for 1-nitropropane (1-NP) in rats was 455 mg/kg bw, whereas for 2-nitropropane (2-NP) the lowest lethal dose was 500 mg/kg bw. The inhalatory LC_{50} for 1-NP in rats was 3100 ppm and for 2-NP 400 ppm (George *et al.*, 1989).

2.3.1.13 Formaldehyde. The oral LD_{50} of a 2% formaldehyde solution in rats ranged from 500 to 800 mg/kg bw, while in guinea pigs it was 260 mg/kg bw. For a 37% solution of formaldehyde a dose of 523 mg/kg bw was lethal to man (Restani and Galli, 1991). The inhalatory LC_{50} in rats (4 h exposure) ranged from 310 to 720 mg/m^3. The LC_{50} values in mice and cats were in the same range: 620 and 810 mg/m^3 bw, respectively. The percutaneous LD_{50} in rabbits was 270 mg/kg bw (Appelman, 1985).

2.3.2 *Short-term toxicity*

2.3.2.1 Benzene. Short-term experiments with benzene after repeated oral administration, dermal or inhalatory exposure produced as major toxic effects changes in the blood and blood-forming tissues and organs. The most common effect was leucocytopenia. For inhalation experiments, which represent the bulk of short-term studies with benzene, no clinical signs of toxicity were observed below 1600 mg/m^3. The effects on the number of white blood cells reflect a decrease in the number of lymphocytes at benzene concentrations $\geqslant$150 mg/m^3 (Fielder, 1982; Slooff, 1987). The effect level in the case of oral administration (187 days) was 10 mg benzene/kg bw (Wolff *et al.*, 1956), whereas after subcutaneous application (10 days) the effect level was 1.0 ml benzene/kg bw (Irons *et al.*, 1979).

2.3.2.2 Styrene. In short-term oral toxicity experiments with styrene physiological, biochemical and histopathological examinations showed neurotoxic effects. In a semichronic experiment these effects were seen at styrene dose levels of $\geqslant$200 mg/kg bw (Agrawal *et al.*, 1982). The neurotoxic effects included impaired neurotransmitter function, significant increases in the concentrations of serotonine and noradrenaline in the brain, decreased monoaminooxidase activity in the brain, changes in nerve conduction velocity, and diffuse loss of Purkinje cells in the brain (Wolff *et al.*, 1956; Husain *et al.*, 1980).

At styrene dose levels of >400 mg/kg bw in rats the detoxification system of the liver is affected (see section 2.2.3); at these dose levels alanine aminotransferase (ALAT) and aspartate aminotransferase (ASAT) activities are also elevated, indicating hepatotoxicity. This was confirmed by histopathological examinations which revealed focal necrosis of the parenchyma cells, cellular inflammation, disappearance of glycogen and anisokaryosis. Oral doses of $\geqslant$500 mg styrene/kg bw in rats caused decreased growth and irritation of the oesophagus and stomach, and hyperkeratosis in the forestomach (Wolff *et al.*, 1956; Parkki, 1978; Hinz *et al.*, 1980; Van Apeldoorn *et al.*, 1986).

In short-term inhalation experiments with rats styrene, at exposure levels of $\geqslant$420 mg/m^3, also caused glutathione depletion in the liver. Induction of mixed-function oxidases in the liver and the kidneys were observed in rats and mice at an exposure level of 1260 mg/m^3 (Sandell *et al.*, 1978; Van Apeldoom *et al.*, 1986). At higher concentrations (6000–9500 mg/m^3) irritation of the eyes and nose occurred and, in rats and guinea pigs, inhibition of growth (Van Apeldoorn *et al.*, 1986). After inhalatory exposure (1260–6300 mg/m^3) neurotoxic effects similar to those noted after oral administration were observed (Mutti *et al.*, 1984). Mice exposed by inhalation to 1260 mg styrene/m^3 showed microscopical changes in the lungs (Morisset *et al.*, 1979). Rats exposed to 630 mg/m^3 showed ultramicroscopic changes in the mucosa of the nose and trachea, and a decrease in motility of the ciliary epithelium (Ohashi *et al.*, 1982). Throughout the inhalatory studies all doses had an effect (Van Apeldoorn *et al.*, 1986).

2.3.2.3 Toluene. Most short-term toxicity studies with toluene have been carried out to investigate the effects of repeated inhalatory exposure; only a few studies with oral or dermal administration of toluene have been performed. Rats and mice have most frequently been used as test animals. Several experiments revealed an increase in liver weight, but no histopathological liver changes (Van der Heijden *et al.*, 1988). Induction of mixed-function oxidase enzymes in the liver were seen at inhalatory exposure levels of 9400 mg/m^3. Changes in the collecting tubules of rat kidneys were seen with dose levels of 2280 mg/m^3 (EPA, 1983). Although no histopathological changes have been observed in the liver, the enzymes ALAT and ASAT, indicative for hepatotoxicity were found at inhalatory dose levels ranging from 12 000 to 45 000 mg/m^3 (Khinkova, 1974; Bruckner and Peterson, 1981a,b). Both in mice and rats slight changes in the blood and blood-forming tissues (bone marrow) were reported by Horiguchi and Inoue (1977) when animals were exposed to 3800 mg/m^3. The no-observed-toxic-effect level after inhalatory exposure in short-term experiments with rats, guinea pigs, dogs and monkeys appears to be 400 mg/m^3. The neurophysiological effects of toluene are reported in section 2.3.6.3.

2.3.2.4 Xylene. Most short-term toxicity studies with xylene comprise inhalation studies, in which slight effects on both red and white blood cells of rats have been observed at exposure levels of $\geqslant$1150 mg/m^3. In guinea pigs slight liver degeneration and lung inflammation were observed after inhalation of 300 mg/m^3 (64 days) (Low *et al.*, 1989). In subcutaneous experiments only slight reductions in red blood cells were noted in rats exposed to 800–1600 mg/kg bw/day (Low *et al.*, 1989). In general, xylene does not appear to cause adverse haematological effects in animals and is not myelotoxic. However, at high exposure levels some subtle haematological changes (e.g. reduction of red blood cells, leucocytes, lymphocytes and thrombocytes) have been noted, although such effects appear to be reversible when exposure is terminated (Low *et al.*, 1989).

In spite of extensive studies carried out in animals, there is little evidence that xylene damages the liver, kidneys or lungs. Reports of renal, pulmonary and hepatic damage have been few. In recent studies by Carpenter *et al.* (1975), blood chemistry and histopathological examination showed no evidence of liver, kidney or lung injury in rats exposed to 860 mg xylene/m^3 for up to 13 weeks. Enzyme induction of the cytochrome P450 system and related mono-oxygenases in the liver and kidneys were reported in rats exposed to *m*-xylene vapours, but no increase in liver or kidney weight, and no biochemical or histopathological signs of hepatotoxicity were observed (Elovaara *et al.*, 1984; Toftgard *et al.*, 1986a). Such inductive effects are typical of the alkylbenzene and other xenobiotics and may be viewed as adaptive rather than toxic (Low *et al.*, 1989; Simmons *et al.*, 1991); however, depending on the toxicity of the formed metabolites, they may influence the toxic end-point.

In a short-term oral toxicity experiment in rats, with 43, 131, 435 and 1305 mg xylene/kg bw, no changes in haematological parameters and no histopathological changes in the liver and kidneys were observed, whereas growth retardation and water retention by the kidneys were noted (De Vries *et al.*, 1983).

2.3.2.5 Phenol. Short-term toxicity studies with orally administered phenol in rats and mice caused hepatoxicity and renal toxicity (Van Koten-Vermeulen *et al.*, 1986). The toxic effects were seen from 50 mg phenol/kg bw onwards. After subcutaneous administration to rats effects on leucocytes were also observed at 50 mg/kg bw. Cats were extremely vulnerable to the toxic effects of phenol, especially on the nervous system (Ernst *et al.*, 1961). Dermal toxicity of phenol (dissolved in water) in rabbits revealed both systemic effects (tremors, mortality) at 130 mg/kg, and local skin irritation (hyperaemia, necrosis) at 100 mg/kg bw dose levels (Van Koten-Vermeulen *et al.*, 1986). Short-term toxicity experiments with rats, rabbits and guinea pigs revealed effects on the lungs and heart and, to a lesser extent, on the liver and kidneys after inhalatory exposure to high concentrations of phenol (100–200 mg/m^3). Histopathological examination showed acute lobular pneumonia, lesion of the arterial vascular system, and liver and kidney lesions (Deichmann, 1944). Guinea pigs exposed to 3.8–152 mg/m^3 by inhalation showed severe changes in the respiratory tract. The myocardium, liver and kidneys were also damaged (Kurnatowski, 1960; Van Koten-Vermeulen *et al.*, 1986). Minimal effects were seen in mice, monkeys and rats exposed to 19 mg/m^3 by inhalation (Sandage, 1961).

2.3.2.6 Chlorophenols. Oral short-term toxicity experiments with chlorophenols comprise mainly subacute studies designed as range-finding experiments for teratology, reproduction or (semi)chronic toxicity studies. Effects on survival, body weight, organ weights, and also gross pathology and histopathology are considered to be most relevant effects in these subacute studies. A few semichronic studies are available for 2,4-DCP, 2,4,5-T3CP and, especially, different grades of PCP (Janus *et al.*, 1990). For MCP an oral 2-week experiment in mice was reported in which high dose levels (175 mg/kg bw/day) led to mortality, while lower dose levels (69 mg/kg bw/day) led to reduced body weights (Borzelleca *et al.*, 1985). In a similar experiment in mice 2,4-DCP at a dose level of 638 mg/kg bw did not cause any effect. A

2-week oral toxicity experiment in rats revealed reduced food intake and body weight gain at 2000 mg 2,4-DCP/kg bw/day (NTP, 1986a). In a 3-week oral toxicity experiment with 2,4,5-T3CP, increases in kidney weight were recorded at dose levels $\geqslant$750 mg/kg bw/day. In a similar experiment very slight kidney and liver changes were reported at levels of 70 and 350 mg/kg bw/day. For a 7-week 2,4,6-T3CP oral toxicity experiment in rats and mice reduced body weight gain was observed; for the rats the spleen and liver were the target organs (NCI, 1979). In the case of oral exposure of rats to 2,3,4,6-T4CP (50 mg/kg bw/day) liver necrosis was observed (Hattula *et al.*, 1981). Oral exposure to PCP (70 mg/kg bw/day) for two weeks revealed liver lesions in rats (NTP, 1989b). An oral semichronic toxicity experiment in mice with 2,4-DCP revealed, at dose levels higher than 700 mg/kg bw/day, hepatocellular necrosis as the most sensitive effect. For rats orally dosed with 500 mg/kg bw/day, bone marrow atrophy was observed (NTP, 1989a). Oral exposure of rats to 2,4,5-T3CP for three months at doses of 150 mg/kg bw/day resulted in pathological changes to the liver and kidneys (McCollister *et al.*, 1961). Oral exposure of mice to PCP at doses of 7 and 70 mg/kg bw/day in semichronic experiments resulted in liver lesions (mild to moderate swelling of hepatocytes, nuclear swelling of hepatocytes, nuclear vacuolisation and eosinophilic inclusion bodies within nuclear vacuoles). Oral exposure of rats to PCP (technical grade) at dose levels of 10 and 30 mg/kg bw/day resulted in increased liver and kidney weights, pathological liver changes, induction of microsomal enzymes in the liver, and changes in haematological parameters (Johnson *et al.*, 1973; Knudsen *et al.*, 1974). Subacute experiments in animals involving inhalatory exposure to PCP and other chlorophenols have not been conducted (Janus *et al.*, 1990). Semichronic inhalation experiments performed with PCP in rats and rabbits revealed effects on blood parameters at 29.4 mg/m^3 (anaemia, leucocytosis, eosinophilia), dystrophic processes in the liver, and increases in weight of the lungs, kidneys, liver and adrenals (WHO, 1987a; Janus *et al.*, 1990). The toxicity observed for PCP might be partly due to impurities in technical grade PCP.

Short-term experiments with different chlorophenols show a trend of increasing toxicity with chlorination, although the experimental data are limited (Janus *et al.*, 1990).

2.3.2.7 Chloroform. Repeated dosing of mice and rats with chloroform leads to liver and kidney damage (Van der Heijden *et al.*, 1987). In mice the effects (including centrolobular cytoplasmic pallar, marked cell proliferation and focal inflammation in the liver, and intratubular mineralisation, epithelial hyperplasia and cytomegalia in the kidneys) were observed after a 14-day oral dosing of $\geqslant$37 mg chloroform/kg bw/day (Condie *et al.*, 1983). In addition, in mice exposed to 50 mg chloroform/kg bw/day a reduction in the number of antibody-forming cells in the blood, and a mild lymphoid atrophy of the spleen were observed (Munson *et al.*, 1982). In mice dosed with $\geqslant$50 mg chloroform/kg bw/day by gavage for 90 days liver and kidney damage and depression of the antibody-forming cells were observed (Munson *et al.*, 1982). If chloroform was administered orally in drinking water the effects on the liver were less pronounced at dose levels of $\geqslant$66 mg/kg bw/day. At higher dose levels mild lymphoid atrophy of the spleen was seen, whereas no kidney lesions were observed. The depression of the central nervous system by chloroform was dose related (Jorgenson and Rushbrook, 1980). In Sprague-Dawley rats intubated with chloroform dissolved in peanut oil, liver effects were noted at dose levels of 0.5 mg/kg bw/day (Groger and Grey, 1979). Administration of chloroform in drinking water resulted in a decrease in relative kidney weight (0.6 mg/kg bw/day) and a reduction in the number of neutrophilic leucocytes (63 mg/kg bw/day). These effects were not observed after oral exposure (in drinking water) to doses of up to 40 mg/kg bw/day for 90 days. At this dose level liver and thyroid lesions were found (Chu *et al.*, 1982a,b). In Osborne-Mendel rats, the uptake of doses of $\geqslant$38 mg/kg bw/day for 30 days resulted in an increase in the serum cholesterol content, while the uptake of 160 mg/kg bw/day led to a decrease in the serum triglyceride concentration (Jorgenson and Rushbrook, 1980). Inhalation of chloroform vapours (110–410 mg/m^3) in a 6-month study with rats elicited renal and hepatic damage (Torkelson *et al.*, 1976).

2.3.2.8 Methylene chloride. Oral subacute experiments with methylene chloride in mice revealed both kidney (113 mg/kg bw/day) and liver (>333 mg/kg bw/day) damage (Condie *et al.*, 1983). In a 90-day experiment with drinking water mice revealed liver damage at dose levels ranging from 166 to 1669 mg/kg bw/day (Van Apeldoorn *et al.*, 1987). In a semichronic inhalation experiment with rats liver and kidney lesions were noted at dose levels of 87 and 347 mg/m^3, respectively (Haun *et al.*, 1972). Similar changes in the liver were found in dogs and monkeys exposed to 3470 mg/m^3. The dogs further showed vacuolar changes in their renal tubules (Haun *et al.*, 1972). Mice exposed to methylene chloride (347–17 350 mg/m^3) by inhalation revealed similar toxic effects (Weinstein and Diamond, 1972; Weinstein *et al.*, 1972; Norpoth *et al.*, 1974; Kurppa and Vainio, 1981; Toftguard *et al.*, 1982). Other investigators reported effects in the lungs after inhalatory exposure to methylene chloride (Sagai and Tappel, 1979; Sahu *et al.*, 1980; Sahu and Lowther, 1981). Except for the lung effects, similar toxicity levels were found for both oral and inhalatory exposure to methylene chloride (Van Apeldoorn *et al.*, 1987).

2.3.2.9 Chlorobenzenes. In short-term toxicity experiments with MCB effects on the liver, kidneys, spleen, thymus, gastro-intestinal mucosa, haemopoietic tissues and porphyrin metabolism were observed (NTP, 1985a; Hesse *et al.*, 1991). These effects were seen in rats at dose levels of 60 mg MCB/kg bw/day for oral administration and 2000 mg/m^3 for inhalatory exposure; in dogs the effects were recorded with dose levels of 750 mg/m^3. In one inhalatory experiment with chlorobenzene, neurotoxic effects were reported at a concentration of 100 mg/m^3 (EPA, 1984). Similar toxic effects to those observed with MCB were recorded with DCBs in rats and mice (NTP, 1985b, 1987). The no-observed-adverse-effect levels (NOAEL) in semichronic toxicity studies in rats and mice (using hepatotoxicity as the most sensitive effect) were 125 mg/kg bw/day for 1,2-DCB, and 150 mg/kg bw/day for 1,4-DCB (NTP, 1987). For rabbits and mice the toxic effect levels were 500 and 600 mg 1,4-DCB/kg bw/day, respectively. In short-term inhalatory toxicity experiments with DCB in rats, rabbits and guinea pigs eye irritation and lung lesions were also observed. The toxic effects were recovered after inhalatory exposure to dose levels of 1038 mg DCB/m^3 (Hesse *et al.*, 1991). For T3CBs the toxic effects in short-term experiments were similar to those seen with MCB and DCB after oral, dermal and inhalatory exposure (Carlson and Tardiff, 1976; Smith and Carlson, 1980; Kociba *et al.*, 1981; Robinson *et al.*, 1981; Rao *et al.*, 1982). Semichronic oral toxicity studies in rats revealed toxic effects at dose levels of 10 mg 1,2,4-T3CB/kg bw/day (Carlson and Tardiff, 1976; Carlson, 1977). In semichronic inhalatory experiments in rats, rabbits and monkeys marginal effect levels were observed at dose levels of 742 mg 1,2,4-T3CB/m^3 (Coate *et al.*, 77; Sasmore *et al.*, 1983).

There is a clear similarity between the effects of the different chlorobenzenes. The liver and kidneys are target organs of all chlorobenzenes; exposure resulted in increased organ weights, enzyme induction and histopathological changes. All chlorobenzenes disturbed porphyrin metabolism. The lower chlorinated chlorobenzenes suppressed the blood-forming activities of bone marrow, spleen and thymus (Hesse *et al.*, 1991).

2.3.2.10 Ethylene and propylene oxide. Oral short-term toxicity experiments with ethylene oxide (EO) revealed marked loss of body weight, gastric irritation and liver damage (Hollingworth *et al.*, 1956a). In short-term inhalatory toxicity experiments rats, mice, dogs and monkeys tolerated 90–200 mg/m^3 ethylene oxide over a period of about 30 weeks without adverse effects. Higher doses cause lung injury, delayed effects on the neuromuscular system and, in rats, histopathological changes in the liver, kidneys, and testes. Concentrations of approximately 1500 mg/m^3 caused one hundred percent mortality. Oral doses of up to 30 mg/kg bw over a period of 30 days had no adverse effect in rats, whereas 100 mg over a period of 21 days caused gastric irritation and slight liver damage (Hollingworth *et al.*, 1956a; Jacobson *et al.*, 1956; Vermeire *et al.*, 1985). According to Vermeire *et al.* (1985) the experimental data available were restricted and relatively out of date.

No oral short-term toxicity experiments with propylene oxide have been conducted. In short-term inhalatory studies with rodents irritation of the respiratory tract, lung oedema and effects on the central nervous system became apparent at levels above 1000 mg/m^3 (Rowe *et al.*, 1956; NTP, 1984; Young *et al.*, 1985; Vermeire *et al.*, 1988). Slight neuropathological effects, the significance of which is uncertain, were reported in monkeys exposed by inhalation to propylene oxide for two years (Young *et al.*, 1985).

2.3.2.11 Nitropropane. After inhalatory exposure of rats to high concentrations of 2-nitropropane (200 ppm), both hepatotoxicity and hepatocarcinogenicity were observed (Lewis *et al.*, 1979; Denk *et al.*, 1990).

2.3.2.12 Formaldehyde. In short-term oral toxicity studies with formaldehyde in rodents, at a dose level of 125 mg/kg bw, the prevalent effects were hyperkeratosis in the forestomach and focal gastritis in the glandular stomach (Til *et al.*, 1989; WHO, 1989; Restani and Galli, 1991). In several short-term oral experiments no-observed-adverse-effect levels were claimed by the authors; these levels ranged from 10 to 100 mg/kg bw/day (Restani and Galli, 1991). Formaldehyde is also characterized by its cutaneous effects after dermal exposure. These effects include dermatitis and allergic contact dermatitis or contact urticaria, and were reported to be induced by 3–10% solutions of formaldehyde. Formaldehyde can easily be combined with a variety of proteins to induce immunological sensitivity. Subacute and subchronic inhalatory exposure of animals resulted in effects on the respiratory system, liver, eyes, plasma, locomotion, antibody inhibition and body weight at dose levels varying from 120 to 15 000 μg/m^3 (WHO, 1989b).

2.3.3 Long-term toxicity and carcinogenicity

2.3.3.1 Benzene. Chronic toxicity/carcinogenicity studies with rats orally administered benzene revealed carcinogenic properties. Chronic inhalation studies with benzene showed carcinogenicity both in rats and mice. The carcinogenicity consisted of dose-related increased incidences of cymbal kidney carcinoma, and slight increased incidences of leukaemia, carcinoma of the mouth cavity, hepatocarcinoma, lymphoreticular neoplasia and, in the case of oral administration, possible mammae carcinoma (Slooff, 1987).

2.3.3.2 Styrene. In oral chronic toxicity and carcinogenicity studies with rats and mice styrene produced hepatotoxicity and renal toxicity. Carcinogenic effects (lung tumours) occurred in mice after oral exposure to extremely high dose levels, whereas in rats no carcinogenicity was observed. Subcutaneous and intraperitoneal administration of styrene to rats also produced no carcinogenic effect. In inhalatory chronic toxicity/carcinogenicity studies some weak indications for leukaemia and lymphosarcoma were noted; nevertheless, the results did not lead to the conclusion that styrene was carcinogenic. The inhalatory studies revealed toxic effects in the liver and lungs. All inhalatory exposure levels tested revealed toxic effects, therefore a no-observed-adverse effect level could not be established (Van Apeldoorn *et al.*, 1986).

2.3.3.3 Toluene. Only a few long-term toxicity studies have been performed with toluene. Inhalatory exposure of rats to 1100 mg/m^3 for 24 months did not induce any histopathological changes. Except for one oral study, in which there was a slightly increased incidence of neoplasm in rats, no oral, dermal or inhalatory chronic studies with rats and mice have revealed any carcinogenicity property. Observed toxic effects include hepatotoxicity and effects on red and white blood cell parameters (Van der Heijden *et al.*, 1988).

2.3.3.4 Xylene. Xylene has not been shown to be carcinogenic (Fishbein, 1985b; Low *et al.*, 1989). In a 12-month inhalatory experiment with mice xylene caused haematological and immunological changes at a dose level of 11.5 ppm (Kashin *et al.*, 1968 cited in Low *et al.*, 1989).

2.3.3.5 Phenol. Chronic carcinogenicity experiments with mice and rats revealed no carcinogenic effects of phenol after oral exposure, whereas with dermal exposure in rats phenol showed tumour promoting activities. No other chronic toxicity experiments are available (Van Koten-Vermeulen *et al.*, 1986).

2.3.3.6 Chlorophenols. Long-term toxicity and carcinogenicity studies are available for 2-MCP, 2,4-DCP, 2,4,6-T3CP and PCP. For 2-MCP and 2,4-DCP no carcinogenicity was reported in rats and mice. Studies with 2,4,6-T3CP revealed increased incidences of both hepatocellular carcinomas and adenomas in mice, and leukaemias in male rats. The carcinogenicity of PCP was only studied in mice. Increased incidences of hepatocellular carcinomas and adenomas, and benign medullary phaeochromocytomas were noted in males; in females hemangiosarcomas in spleen and liver were also seen. Limited carcinogenicity studies in rats and other strains of mice did not reveal carcinogenic effects (Janus *et al.*, 1990). In long-term experiments DCP and T3CP revealed hepatotoxic effects in the bone marrow and kidneys. For 2,4-DCP the no-observed-toxic effect level (NOEL) was 120 mg/kg bw/day, while for 2,4,5-T3CP the NOEL was 50 mg/kg bw/day. 2,4,6-T3CP was reported to be carcinogenic (Janus *et al.*, 1990). PCP revealed the same hepatotoxicity as seen with the other chlorophenols, but the hepatic porphyria was more pronounced in technical grade PCP. For pure PCP the NOEL ranged from 3 to 5 mg/kg bw/day (Janus *et al.*, 1990).

2.3.3.7 Chloroform. Long-term inhalatory or dermal toxicity experiments with chloroform have not been made. Chloroform produces tumours in the liver and kidneys of some strains of mice and rats after oral exposure to dose levels that also produce toxic effects (Van der Heijden *et al.*, 1987).

Exposure of mice to chloroform ($\geqslant$30 mg/kg bw/day) in drinking water for at least 6 months resulted in an increased liver fat content (Jorgenson *et al.*, 1982). In rats all dose groups in a drinking water study using chloroform at $\geqslant$30 mg/kg bw/day revealed changes in several biochemical blood parameters; however, the most distinct effects were decreased concentrations of cholesterol and triglycerides in the serum (Van der Heijden *et al.*, 1987). Oral administration of chloroform to dogs (15–30 mg/kg bw, six times per week) for approximately 7 years resulted in elevated enzyme activities in the serum (ALAT, ASAT, alkaline phosphatase (alk.pase), γ-GT) and histopathological changes in the liver (Heywood *et al.*, 1979).

2.3.3.8 Methylene chloride. Although carcinogenicity studies of both oral and inhalatory exposure to methylene chloride have revealed only very weak evidence for carcinogenic properties in animals, methylene chloride has been considered to be carcinogenic (EPA, 1982; IARC, 1986). If this evidence is combined with the conflicting results of mutagenicity studies (see section 2.3.4.8) and the absence of DNA alkylation by methylene chloride, it seems appropriate to establish a NOEL based on long-term toxicity studies (Van Apeldoorn *et al.*, 1987). The most sensitive criterion is hepatoxicity, which was clearly observed after oral administration of $\geqslant$58 mg methylene chloride/kg bw/day, and after inhalatory exposure to $\geqslant$694 mg/m^3 (Van Apeldoorn *et al.*, 1987). An oral NOEL of 6 mg methylene chloride/kg bw has been derived from a long-term drinking water study with rats. The NOEL for inhalatory exposure (173 mg/m^3) was derived from a long-term inhalation study with rats (Nitsche *et al.*, 1982).

2.3.3.9 Chlorobenzenes. There is no evidence that MCB is carcinogenic (Hesse *et al.*, 1991). In a chronic oral study with mice and rats, in which the liver was the most sensitive target organ (neoplastic noduli), the NOEL was 60 mg MCB/kg bw/day (NTP, 1985a). The IARC (1987) concluded that there was inadequate evidence for carcinogenicity of 1,2-DCB to animals. For 1,4-DCB there is only limited evidence of carcinogenicity in experimental animals (Hesse *et al.*, 1991). Long-term studies with rats revealed a NOEL of 19 mg/kg bw/day for both 1,2-DCB and 1,4-DCB, (Hollingworth *et al.*, 1956b, 1958), whereas a chronic oral study in mice revealed a NOEL of 60 mg/kg bw/day (tubular regeneration in kidneys) (Hesse *et al.*, 1991). At the lowest level tested 1,4-DCB produced harmful effects in the liver (hepatocellular degeneration); kidney lesions (nephropathy) were also seen (NTP, 1987). In inhalatory studies of 6 to 7 months duration in rats, mice, guinea pigs, rabbits and monkeys the NOEL for DCB was 577 mg/m^3 (Hollingworth *et al.*, 1956b). No data are available on the possible carcinogenicity of T3CB or on its long-term toxicity (Hesse *et al.*, 1991). Of the T3CBs, 1,2,4-T3CB is considered to be the most toxic isomer. In subchronic studies the marginal effect level after oral administration of 1,2,4-T3CB is 10 mg/kg bw/day (Hesse *et al.*, 1991), while for inhalatory exposure the NOEL is 100 mg/m^3 (Sasmore *et al.*, 1983).

2.3.3.10 Ethylene and propylene oxide. Ethylene oxide has been shown to induce tumours in mice and rats. After long-term inhalation by rats, leukaemia and primary tumours were induced in both sexes, with peritoneal mesotheliomas, and subcutaneous fibromas in male rats. On ingestion, female rats developed squamous cell carcinomas in the forestomach. No adequate dermal experimental data are available (Vermeire *et al.*, 1985). As ethylene oxide is clearly carcinogenic after long-term exposure, a NOEL is not relevant. Toxic effects are mainly seen in the spleen, red and white blood cells, lungs, kidneys and adrenals (Vermeire *et al.*, 1985).

Propylene oxide has also caused increased incidences of several types of tumours. Because of the carcinogenic properties of propylene oxide, other long-term toxicity data for the establishment of a NOEL are lacking. The long-term toxic effects are similar to those observed for ethylene oxide (Hayes *et al.*, 1988; Vermeire *et al.*, 1988).

2.3.3.11 Nitropropane. 2-Nitropropane (2-NP) induces liver tumours in rats following exposure by inhalation (Lewis *et al.*, 1979; Griffin *et al.*, 1980) or gavage (Fiola *et al.*, 1987a), whereas 1-NP is not carcinogenic (Hadidian *et al.*, 1968; Griffin *et al.*, 1982; Fiola *et al.*, 1987; Roscher *et al.*, 1990). Hepatocellular proliferation and effects on hepatocellular function (cholestasis and necrosis) were mild to moderate after exposure to 2-NP, whereas equivalent administration of 1-NP by gavage (0.5, 1 or 2 mmol/kg) revealed only minimal effects, probably due to slight dehydration (Cunningham and Mathews, 1991).

2.3.3.12 Hexane. The most characteristic toxic property of hexane is neurotoxicity which will be reviewed in section 2.3.6.1.

2.3.3.13 Formaldehyde. According to the IARC (1987) there is sufficient evidence that formaldehyde is carcinogenic to rats. Several skin initiation/promotion studies with formaldehyde did not produce evidence for skin carcinogenicity in mice, while the results with respect to promotion were either negative or inconclusive (WHO, 1989b). None of the epidemiological studies available provides conclusive evidence for carcinogenicity of formaldehyde in humans (WHO, 1989b). In rats and mice long-term toxic effects of formaldehyde after inhalatory exposure consisted of metaplasia of the nasal turbinates (1200 μg/m^3) and squamous metaplasia and basal cell hyperplasia of the respiratory epithelium of the nasal cavity (3700 μg/m^3). In a 3-year drinking water study with rats high dose levels of formaldehyde caused severe damage to the gastric mucosa, but no tumours. At a dose level of 25 mg/kg bw/day these effects were not observed (Til *et al.*, 1989).

2.3.4 *Genotoxicity and mutagenicity*

2.3.4.1 Benzene. Benzene does not cause gene mutation in the Ames test (either with or without activation) nor in the host-mediated assay or in *Drosophila* (Kale and Baume, 1983; De Flora *et al.*, 1984). It can induce *in vitro* and *in vivo* chromosome aberrations in plants and somatic mammalian cells (Gomez-Arroyo and Villalobos-Pietrini, 1981; Van Raalte and Grasso, 1982; Morimoto *et al.*, 1983; Gad-El-Karim *et al.*, 1984). Benzene did not affect DNA repair in bacterial cells (De Flora *et al.*, 1984). The clastogenic capacity of benzene is at least partly due to hydroxylated metabolites (Van Raalte and Grasso, 1982), although DNA binding is also reported in rat liver and in the bone marrow of mice (Snyder *et al.*, 1978; Gill and Ahmed, 1981; Slooff, 1987).

2.3.4.2 Styrene. Styrene is mutagenic in several test systems, but only in the presence of an activating system. It induces gene mutations in prokaryotic and eukaryotic cells, and in *Drosophila* and mammalian cells *in vitro*, and it causes chromosome aberrations in mammalian cells (De Flora *et al.*, 1984; Van Apeldoorn *et al.*, 1986). *In vivo* tests for the detection of chromosome aberrations revealed conflicting results; only high exposure levels were able to induce positive effects (Loprieno *et al.*, 1978; Bianchi *et al.*, 1982; Pero *et al.*, 1982; Van Apeldoorn *et al.*, 1986). Styrene induced no DNA repair synthesis. Metabolites of styrene, including styrene-7,8 oxide, bind both *in vitro* and *in vivo* in a covalent manner to macromolecules (including DNA) (Parkki, 1978; Zitting *et al.*, 1980; Van Apeldoorn *et al.*, 1986).

2.3.4.3 Toluene. Toluene is not genotoxic in conventional *in vitro* test systems. The results of Liang *et al.* (1983) indicate that toluene block mitosis *in vivo*. With the exception of experiments carried out in the former USSR, which have consistently indicated that toluene induces chromosome aberrations in the bone marrow of rats toluene has not been demonstrated to be genotoxic *in vivo* either to somatic or to germ cells; this discrepancy may be due to the presence of contaminants in the former samples (Van der Heijden *et al.*, 1988).

2.3.4.4 Xylene. Xylene has not been shown to be mutagenic (Bos *et al.*, 1981; Fishbein, 1985; Low *et al.*, 1989). Although 'technical grade xylene' has shown some weak mutagenic effects in the recessive lethal test in *Drosophila*, no mutagenicity was observed when pure *o*- and *m*-xylene were tested (Donner *et al.*, 1980). Xylene also has no clastogenic properties in rats (Donner *et al.*, 1980; Fishbein, 1985b; Low *et al.*, 1989).

2.3.4.5 Phenol. Phenol has been tested in several test systems, both *in vitro* and *in vivo*, to detect mutagenicity. The results of these tests revealed conflicting data, i.e. both positive and negative mutagenic reactions, therefore no definite conclusion on the possible mutagenic effect of phenol can be drawn (Van Koten-Vermeulen *et al.*, 1986). There are clear indications that phenol affects DNA metabolism and mitosis (Poirier *et al.*, 1975; Morimoto *et al.*, 1976; Painter and Howard, 1982; Amlacher and Rudolph, 1983).

2.3.4.6 Chlorophenols. Only six chlorophenols (2,4-DCP, 2,6-DCP, 2,4,5-T3CP, 2,4,6-T3CP, 2,3,4,6-T4CP and PCP) have been studied for gene mutation in at least one prokaryotic test system (bacterial) and one eukaryotic test system (mammalian cells and/or yeast). 2,4-DCP, 2,4,6-T3CP and PCP were also studied in tests for allied genotoxic effects, primarily chromosomal aberrations and sister chromatid exchanges in mammalian cells. The majority of tests with these compounds resulted in negative responses. The other chlorophenols were studied only for gene mutation in one test system; most results were negative (Janus *et al.*, 1990). *In vivo* mammalian studies with 2-MCP, 2,3-DCP, 2,4-DCP, 2,5-DCP, 2,6-DCP and PCP for bone marrow sister chromatid exchange, sperm morphology and/or testicular DNA synthesis consistently resulted in negative responses, with the exception of 2,5-DCP, which revealed equivocal results in one test. A mammalian spot test and host-mediated assays in mice with 2,4,6-T3CP and PCP were negative. A sex-linked lethal test with *Drosophila*, and a test for nondisjunction and loss of sex chromosomes were negative (Janus *et al.*, 1990). For the six chlorophenols studied there is insufficient evidence for mutagenicity, whereas for other chlorophenols adequate data on mutagenicity are lacking (Janus *et al.*, 1990).

2.3.4.7 Chloroform. Chloroform has been studied in a variety of mutagenicity and genotoxicity tests. From the available data only sparsely positive effects have been reported, therefore chloroform does not seem to be mutagenic. Chloroform did not induce DNA damage or DNA repair in a variety of bacterial and mammalian test systems, either *in vitro* or *in vitro*. In contrast, sister chromatid exchange was induced by chloroform in hamster and human cells *in vitro* (although only in the absence of metabolic activation) and in mice *in vivo* (Van der Heijden *et al.*, 1987).

2.3.4.8 Methylene chloride. Methylene chloride is mutagenic in bacterial tests and there is some evidence for mutagenic activity in fungi. The results for *Drosophila* are conflicting. No gene mutations were detected in mammalian cells *in vitro*. In some tests with mammalian cells *in vitro*, chromosomal aberrations were induced by methylene chloride, but *in vivo* mammalian studies for chromosomal aberrations and micronuclei revealed no mutagenic activity. No induction of unscheduled DNA repair synthesis by methylene chloride in mammalian cells *in vitro* was observed. The inhibition of DNA synthesis due to a metabolic block of the synthesis was reversible. Thus, methylene chloride had mutagenic properties in several *in vitro* tests, but was negative when tested *in vivo* (Van Apeldoorn *et al.*, 1987).

2.3.4.9 Chlorobenzenes. Data on genotoxicity are rather limited for chlorobenzenes. *In vitro* tests with MCB, 1,4-DCB and T3CBs were made in mammalian cells for gene mutations or chromosome aberrations and sister chromatid exchanges. These tests revealed only one equivocal result for DCB, or equivocal/positive for MCB, whereas all other test were negative. *In vivo* genotoxicity was only performed for 1,4-DCB (and HCB) and was negative. However, one other mouse micronucleus test with MCB, DCBs and T3CBs revealed positive results. The data on genotoxicity available for volatile chlorobenzene do not justify a definitive conclusion (Hesse *et al.*, 1991).

2.3.4.10 Ethylene and propylene oxide. Ethylene oxide directly alkylates proteins and DNA both *in vitro* and *in vivo*. It is mutagenic to plants, microorganisms, insects and mammalian cells *in vitro* and *in vivo*, and induces both gene mutation and chromosome aberrations (Vermeire *et al.*, 1985). DNA repair has also been induced, but at higher exposure levels (Cumming and Michaud, 1979). Propylene oxide breaks DNA and induces chromosomal abnormalities in mammalian cells *in vitro*. *In vivo* an increased frequency of micronuclei was induced in mice after intraperitoneal exposure, but not after oral exposure. No clear dominant lethal effects were found in rodents, and no sperm head abnormalities in mice and monkeys. No chromosome aberrations or sister chromatid exchanges were observed in monkeys exposed for two years (Vermeire *et al.*, 1988).

2.3.4.11 Hexane. No data on mutagenic or genotoxicity of hexane are available.

2.3.4.12 Nitropropane. Several mutagenicity tests with 1- and 2-nitropropane (NP) have been published. Among the *in vitro* tests, only the Ames test uniformly discriminated between the carcinogenic 2-NP and the non-carcinogenic 1-NP isomers. 1-NP was negative and 2-NP acted as a direct mutagen (Fiola *et al.*, 1987b; George *et al.*, 1989; Conaway *et al.*, 1991). *In vitro* tests with mammalian cells (gene mutations) and unscheduled DNA synthesis (UDS) assays were less discriminating (Cunningham and Matthews, 1991; Haas Jobelius *et al.*, 1991). A liver UDS test revealed a negative result for 1-NP and a positive result for 2-NP (Andrae *et al.*, 1988). Liver micronucleus tests revealed a clastogenic effect of 2-NP in the liver, indicating that 2-NP induces chromosome aberrations as well as DNA repair *in vivo*, but that it seems to act organospecifically. Micronucleus tests in bone marrow *in vivo* were reported to be negative for both 1-NP and 2-NP (Kliesch and Adler, 1987). For 1-NP a slightly increased incidence of micronuclei was found in the liver, and was accompanied by a markedly increased mitotic index (George *et al.*, 1989). 2-NP cause oxidative DNA and RNA damage in rats (Fiola *et al.*, 1987b; Guo *et al.*, 1990; Robbiano *et al.*, 1991).

2.3.4.13 Formaldehyde. Formaldehyde has been shown to be mutagenic in several *in vitro* systems and can be classified as a weak mutagen (Appelman, 1985; WHO, 1989b). It induced mutation, gene conversion, DNA strand breaks and DNA protein cross-links in fungi, and mutation and DNA damage in bacteria. In rodent cells, it induced transformation of mouse C3H10T1/2 cells, chromosomal aberrations, sister chromatid exchanges, DNA strand breaks and DNA protein cross-links (IARC, 1982b, 1987). *In vivo*, formaldehyde has been shown to be mutagenic only in male *Drosophila* larvae, but most studies were negative (IARC, 1982b, 1987; WHO, 1989b; Restani and Galli, 1991). Exposure to formaldehyde vapour was ineffective.

2.3.5 Reproductive toxicological effects and teratogenicity

2.3.5.1 Benzene. Benzene caused foetotoxic effects (decreased foetus weight and skeletal effects) after inhalatory exposure to $\geqslant 160\,mg/m^3$. Although maternal toxicity was seen at these dose levels, no teratogenic effects were observed. In one study slight skeletal malformations were noted, but in rats, mice and rabbits no teratogenic effects were observed at dose levels of benzene, which did not cause maternal toxicity (Slooff, 1987).

2.3.5.2 Styrene. In teratogenicity experiments with oral or inhalatory exposure to styrene no teratogenic effects were observed in mice and rats. Embryotoxic effects were observed at inhalatory exposure levels of $\geqslant 1050$ mg/m^3 (Murray *et al.*, 1978; Vergieva *et al.*, 1979; Kankaanpaa *et al.*, 1980; Van Apeldoorn *et al.*, 1986).

2.3.5.3 Toluene. There is no evidence for teratogenic properties of toluene, even at embryotoxic concentrations (Nawrot and Staples, 1979; Tatrai *et al.*, 1979; Ungvary, 1985; Courtney *et al.*, 1986; Van der Heijden *et al.*, 1988). Embryotoxicity was observed in mice after oral exposure to $\geqslant 430$ mg toluene/kg bw and in rats after inhalatory exposure to 1000 mg toluene/m^3 (Nawrot and Staples, 1979; Ungvary, 1985).

2.3.5.4 Xylene. In animals, the three xylene isomers have not been shown to be teratogenic and there are only limited data for the embryotoxic and foetotoxic effects of xylenes in rats (Barlow and Sullivan, 1982).

2.3.5.5 Phenol. The progeny of rats exposed to 7000 mg phenol/l drinking water during three generations showed retarded growth and decreased viability (Heller and Pursell, 1938). In one teratogenicity experiment with

Sprague-Dawley rats, phenol did not cause any teratogenic effect (Minor and Becker, 1971; Van Koten-Vermeulen *et al.*, 1986).

2.3.5.6 Chlorophenols. Not all chlorophenols have been tested in teratogenicity and reproduction studies. 2,4-DCP, 2,3,4,6-TCP and PCP were not teratogenic. Embryo-/foetotoxic effects of 2,4-DCP, 2,4,6-T3CP and PCP were observed at exposure levels at which maternal toxicity was not evident in rats. The NOELs for these effects were 5 mg/kg bw/day for 2-MCP, 3 mg/kg bw/day for 2,4-DCP, 3 mg/kg bw/day for 2,4,6-T3CP and 2.5–4 mg/kg bw/day for the different PCP formulations. Studies in which progeny were exposed both pre- and postnatally resulted in effects at lower dose levels than those affecting reproductive performance. In these studies immunocompetence, liver and spleen weights were found to be sensitive parameters (Janus *et al.*, 1990).

2.3.5.7 Chloroform. The effects of chloroform on the development of embryos and foetuses were investigated in several species, after oral administration. These developmental effects were observed with dose levels of $\geqslant$25 mg/kg bw/day in mice and 50 mg/kg bw/day in rats during foetal organogenesis. After inhalation of 146 mg chloroform/m^3, embryo- and foetotoxicity, and retarded ossification were found. In a three-generation study with mice, effects on fertility were only observed at dose levels (1000 mg/kg bw) which caused maternal toxicity. Chloroform did not cause teratogenic effects (Van der Heijden *et al.*, 1987).

2.3.5.8 Methylene chloride. Methylene chloride was not teratogenic in animal experiments. No reproductive impairment in rats was found, either in a one-generation study with 125 mg/l administered in drinking water 13 weeks before mating, nor in a two-generation reproduction study at exposure levels up to and including 5200 mg/m^3 (Van Apeldoorn *et al.*, 1987).

2.3.5.9 Chlorobenzenes. Inhalatory exposure to up to 450 ppm chlorobenzenes caused no adverse effects on reproduction or fertility of rats. Inhalatory exposure to up to 590 ppm chlorobenzenes did not result in teratogenic effects in rabbits, but did caused a slight delay in skeletal development of rats, accompanied by maternal toxicity.

Oral exposure of rats to 1,4-DCB (250–1000 mg/kg bw/day) provided no evidence for teratogenicity. An embryotoxicity effect occurred at 500 mg/kg bw/day. Inhalation of up to 800 ppm caused no teratogenic or embryotoxic effects in rabbits. In a teratogenicity study the highest dose of 1,2-DCB (400 ppm; 2400 mg/m^3) caused slight maternal toxicity in rats and rabbits, but no embryotoxic or teratogenic effects.

In a reproduction study with rats, exposed through drinking water to 1,2,4-TCB (25–400 mg/l), no effects on reproduction were found. Similarly, a teratogenicity study with rats did not result in any effects at doses up to 300 mg/kg bw (1,2,4-TCB) or 600 mg/kg bw (1,2,3- and 1,3,5-TCB). In another study with rats a dose of 360 mg/kg bw caused retarded embryonal growth, accompanied by maternal toxicity; this was not reported at 120 mg/kg bw (Hesse *et al.*, 1991).

2.3.5.10 Ethylene and propylene oxide. Embryotoxicity, foetotoxicity and effects on reproductive performance were detected in rats after inhalatory exposure to up to 173 mg ethylene oxide/m^3, and in mice after intravenous doses of 150 mg/kg bw. Teratogenic effects were observed after high intravenous doses of ethylene oxide, but not after inhalation by rats and rabbits (Laborde and Kimmel, 1980; Hackett *et al.*, 1982; Snellings *et al.*, 1982a,b; Vermeire *et al.*, 1985).

An inhalation study in rats revealed foetotoxicity and a few development effects at dose levels of 1190 mg propylene oxide/m^3. These changes were not observed in rabbits. Effects on spermatogenesis functions were observed in monkeys at dose levels of $\geqslant 240$ mg/m^3 (Vermeire *et al.*, 1988).

2.3.5.11 Hexane. The toxicity of hexane at different concentrations (500, 800, 1000 and 1500 ppm) was examined to determine its effect on pregnant rats and their offspring. After inhalatory exposure to *n*-hexane a concentration-dependent increase of intrauterine mortality was observed. In cases of only prenatal exposure reduced body growth was noted from 500 ppm upwards; in these cases, a delay in the maturation of cellular cortex was also observed. In cases of pre- and postnatal hexane exposure, the effects of malnutrition were added to the solvent-induced retardation. This resulted in an extreme delay in maturation, accompanied by retarded cell maturation. Hexane had no teratogenic effect in pregnant rodents and their offspring. Neurotoxic effects were restricted to the known structures of the adult animals, whose axons were damaged (Stoltenburg-Didinger *et al.*, 1990).

2.3.5.12 Formaldehyde. Several teratogenicity studies with formaldehyde after inhalatory exposure, oral exposure and dermal exposure did not show any evidence of the embryos being unusually sensitive to formaldehyde, and there is no information to show that formaldehyde is teratogenic or has reproductive effects in rodents at oral or dermal dose levels without maternal toxicity (WHO, 1987b). In one teratogenicity study with mice formaldehyde injected intraperitoneally on days 7–14 of pregnancy led to increased incidence of prenatal death and teratogenic effects (particularly cleft palates and malformation of the extremities) (IARC, 1982b). Although several studies with humans have indicated possible teratogenic and reproductive effects, the presence of potentially confounding factors which were not evaluated means that no definite conclusions can be drawn (WHO, 1987b). According to WHO (1989b), formaldehyde does not have any adverse effects on reproduction and is not teratogenic.

2.3.6 *Effects on the nervous system*

Some VOCs of different categories appear to cause hazardous effects to the nervous system. One of the best studied VOCs with respect to neurotoxicity is hexane, but neurotoxic effects of other VOCs have also been published.

2.3.6.1 Hexane. Animals exposed to hexane (700 ppm) for a duration of nine weeks developed motor weakness, beginning in the hind limbs and leading to total hind limbs paralysis. The grip strength was determined for the fore limb and was also reduced by hexane exposure. Light-microscopic examination of the peripheral nerves revealed the characteristic pattern of scattered multifocal giant axonal swelling (Altenkirch *et al.*, 1990). Chronic hexane intoxication in experimental animals results in symmetrical, distal degeneration of axons in the peripheral nervous system (PNS) and in some long tracts of the central nervous system (CNS) (Spencer and Schaumburg, 1977; Schaumburg and Spencer, 1978, 1979; Chang, 1987; Jörgenson and Cohr, 1981). As was reported in section 2.3.5.11, hexane caused a delay in the maturation of the cerebellar cortex of offspring exposed pre- and postnatally (Stoltenburg-Didinger *et al.*, 1990). Both in humans and in experimental animals (Spencer *et al.*, 1980) it has been found that hexane can induce axonal swelling, with myelin changes in the peripheral nerves. Similar pathological changes involving the terminal portion of axons in the CNS have been reported in experimental animals (Spencer and Schaumburg, 1977; Schaumburg and Spencer, 1978, 1979). Clinical evidence suggesting a chronic neurotoxic effect of *n*-hexane on the CNS has rarely been mentioned in patients with polyneuropathy (Spencer *et al.*, 1980). However, Chang (1987) reported somatosensory potential abnormalities in patients with *n*-hexane neuropathy, indicating a chronic toxic effect of *n*-hexane on the CNS. Chronic hexane intoxication in experimental animals results in symmetrical, distal degeneration of the axons in the PNS and in some long tracts of the CNS (Chang, 1990). Nevertheless, Chang (1990) concluded that neuropathy induced by *n*-hexane has a good prognosis, and that spasticity due to damage to the CNS is functionally reversible, whereas muscle cramps and dyschromatopsia persisted longer. Another study in which long-term exposure (8 h) to 86–109 ppm hexane caused polyneuropathy in man, showed that the patients completely recovered when exposure was stopped (Huang *et al.*, 1991). Chronic exposure of humans and experimental animals to *n*-hexane is associated with peripheral neuropathy, but a metabolite 2,5-hexanedione has been identified as the neurotoxic agent (Spencer *et al.*, 1980; Toftgard *et al.*, 1986b). Recently, a case of Parkinsonism in an Italian leather worker due to *n*-hexane exposure was reported (Pezzoli *et al.*, 1989). Following this finding mice and rats were exposed to both hexane and its neurotoxic metabolite (400 mg/kg/day intraperitoneally in mice, 25 mg/kg by automatic intravenous infusion in rats) for approximately 3–4 weeks. In mice dopamine and homovannilic acid levels in the striata were significantly higher compared with controls. Based on these experimental data the authors suggest that exposure to hexane and its metabolites may play a role in inducing Parkinsonism in humans (Pezzoli *et al.*, 1990).

2.3.6.2 Styrene. In acute and short-term experiments styrene caused increased sensitivity of dopamine receptors (Agrawal *et al.*, 1982). Local irritation by styrene primarily affects the CNS leading to incoordination, tremors, ataxia and unconsciousness (6000 mg/m^3) (Spencer *et al.*, 1942). Sensoric irritation and behavioural changes (>2562 mg/m^3) were noted in mice (Ceaurriz *et al.*, 1983), and changes in the spontaneous motoric activity and behaviour were seen in rats. In short-term experiment with styrene changes in nerve conduction velocities in the slow conductive nerves were observed (Wolff *et al.*, 1979). Styrene induced alterations in the neurotransmitter functions. Significant increases in the concentrations of serotonine and noradrenalin, and a decrease in mono-oxidase activity in the brain were found (Husain *et al.*, 1980). After inhalatory exposure to styrene for 11 weeks (1260 mg/m^3, 5 days per week, 6 h per day) slight changes in the axon proteins in the spinal cord and biochemical changes in the brain were observed (Savolainen and Vainio, 1977). In a later study these effects were not confirmed (Savolainen *et al.*, 1980).

2.3.6.3 Toluene. The neurophysiological effects of toluene have been extensively studied using a variety of behavioural and electrophysiological measures as parameters. Behavioural effects include disruption of operant behaviour, decreased locomotor activity and self administration of the solvent. These kind of tests may indicate early signs of toxicity to the CNS (Shigeta *et al.*, 1980; Ceaurriz *et al.*, 1983; Lorenzana Jimenez and Salvas, 1983; Robert *et al.*, 1983; Moser and Balster, 1985). Exposure of rats to 3800 mg toluene/m^3 (14 h per day for 2 weeks) resulted in ototoxicity, measurable by behavioural and electrophysiological methods. Lower concentrations (1500–2000 mg/m^3) were without effect, even after 16 weeks of exposure (Himnan, 1984; Pryor *et al.*, 1984). Ototoxicity of toluene was more pronounced in young prepubertal rats than in older rats (Pryor *et al.*, 1984).

2.3.6.4 Xylenes. Intraperitoneal injection of xylene has been shown to increase the excitation threshold in the cerebral motor cortex of guinea pigs, thereby leading to clonic muscle contractions and tremors (Mikiska, 1960; Mikiskova, 1960). Battig and Grandjean (1964) found that inhalatory exposure of rats to 550–750 ppm levels for 2.5 h does not affect avoidance conditioning. Desi *et al.* (1967) reported that subcutaneous injections (0.5 ml/kg bw) of xylene caused rats to run a maze much more slowly than controls, suggesting that learned behaviour might be impaired by xylene (Low *et al.*, 1989). Exposure to xylene isomers at a concentration of 3000 ppm showed toxic effects on the rotarod performance of rats. The toxic effects of *o*- and *m*-xylene were more pronounced than that of *p*-xylene (Korzak *et al.*, 1990).

In a human volunteer study some sensory effects were seen, such as a decreased latency in the pattern of the visual evoked potential (400 ppm) and some activation of the arousal level (auditorial evoked potential) of subjects after the most intensive exposure situations (Seppalainen *et al.*, 1989). Neurological disturbance in humans at acute exposure to high concentrations of xylene causes CNS depression or narcosis characterised by dizziness, excitement, loss of coordination and staggering gait; some subjects suffered from retrograde amnesia preceding their loss of consciousness (Johnstone and Miller, 1960; Morley *et al.*, 1970).

In a behavioural study in rats with 1600 ppm *p*-xylene (4h per/day for 5 days), in which effects on autoshaping, motor activity and reversal learning were examined, inhaled *p*-xylene at 1600 ppm suppressed response rates in an automaintained reversal learning paradigm without affecting reversal rate. Studies of motor activity showed that while vertically-directed activity was unaffected by *p*-xylene, horizontally-directed activity was increased by about 30% for the first period time of each daily 25 min test. Increased horizontally-directed activity was not correlated with enhancement of autoshaping. These results indicate that *p*-xylene inhalation at 16 times the threshold limit value induced hyperactivity and facilitated autoshaping by different means, and that the facilitation of autoshaping involved a change in response topography and not a direct effect on learning ability *per se* (Bushnell, 1989).

2.3.6.5 Methylene chloride. Methylene chloride caused depression of the CNS in dogs, monkeys, rats, rabbits and guinea pigs (34 700 mg/m^3, 7h per day, 5 days). The animals became inactive, sometimes after initial excitement (Van Apeldoorn *et al.*, 1987). In controlled studies with humans exposed to methylene chloride neurobehavioural changes were observed at low exposure levels (694 mg/m^3); these changes included vigilance disturbance impaired combined tracking monitoring performance, and (at 1040 mg/m^3) a decrease in the critical flicker fusion frequency, a measure of sensory function (Putz *et al.*, 1976). At higher exposure levels the psychomotor performance (2610 mg/m^3) and the visual evoked potential (2400 mg/m^3) were reduced and the exposed subjects experienced light-headedness. In accidental exposure studies the most prominent effect of methylene chloride was a reversible effect on CNS (Moskowitz and Shapiro, 1952). In clinical studies irreversible damage to the CNS, with acoustic and optic illusions and hallucination, was reported in one man (Weiss, 1969) and bilateral temporal lobe degeneration in another (Barrowcliff and Knell, 1979). A case of delirium and seizures was reported following exposure to methylene chloride for four years; these effects included intermittent headaches, nausea, blurred vision, shortness of breath and transient memory disturbance. Neuropsychological and EEG examinations revealed a disfunction of the right hemisphere. All symptoms and signs cleared when workers avoided the workplace (Tariot, 1983).

2.3.6.6 Ethylene and pyropylene oxide. For ethylene oxide neurotoxic effects were observed in several animal species after acute and short-term exposure. In acute experiments ataxia, prostration, laboured breathing and occasional tonic convulsions were seen (Woodard and Woodard, 1971). In short-term inhalation experiments with rabbits and monkeys adverse effects on behaviour were seen (>200 mg ethylene oxide/m^3). In rats effects on the nervous system were also shown (730 mg ethylene oxide/m^3). In dogs the toxic effects (530 mg/m^3) were reflected in occasional slight tremors, transient weakness in the hind legs and muscular atrophy. In cynomolgus monkeys the neurotoxic effects consisted of lesions in the medulla oblongata of the brain and increased incidence of axonal bodies in the nucleus gracilis (Sprinz *et al.*, 1982). The number of animals in these experiments were rather small (Vermeire *et al.*, 1985). Propylene oxide also caused behavioural disturbances in rats, rabbits and guinea pigs (Rowe *et al.*, 1956). The neurotoxic effects of propylene oxide were studied in cynomolgus monkeys. The only treatment-related neuropathological lesions were found in the medulla oblongata of the brain, in which axonal dystrophy was observed in the nucleus gracilis (Sprinz *et al.*, 1982). In Fischer 344 rats no evidence of neurotoxicity related to the exposure of propylene oxide was observed (Young *et al.*, 1985).

2.3.6.7 Phenol. After acute exposure phenol causes demyelinisation of the peripheral nerve tissue (Liao and Oehme, 1980). The cat seems to be the most sensitive animal species. Neurohyperexcitation has also been noted (Van Koten-Vermeulen *et al.*, 1986).

2.3.7 Effects on the immune system

Specific attention to effects of VOCs on the immune system has only rarely been reported. Consequently, the available data on immunotoxic effects, immuno-suppression and immunostimulation are mainly gathered from general toxicological experiments.

2.3.7.1 Styrene. Studies on the possible effects of styrene on the immune system are very limited. Except for a slight increase in the concentration of immunoglobulin G (IgG) in the serum in few cases (Sharma *et al.*, 1981), and decreased phagocytotic activity in rodents (Izmerov, 1984) no sign of immune toxicity of styrene has been reported (Van Apeldoorn *et al.*, 1986).

2.3.7.2 Benzene. No specific immunotoxic studies have been performed with benzene but in several acute and short-term experiments leukocytopenia has been reported (Slooff, 1987).

2.3.7.3 Chlorophenols. Clear effects on the immune system have only been seen with 'technical grade' PCP; pure PCP revealed some marginal indications for immunosuppression (Kerkvliet *et al.*, 1982; NTP, 1989b). The immunosuppressive effects of 'technical grade' PCP were concluded to be related to impurities (polychlorodibenzofurans and polychlorodibenzodioxines).

2.3.7.4 Chloroform. In a short-term toxicity experiment with rats orally administered with chloroform ($\geqslant 50$ mg/kg bw) a dose-related decrease in the number of antibody-forming cells (IgM response to sheep red blood cells) was found (Munson *et al.*, 1982). In mice a mild lymphoid atrophy of the spleen was seen at higher dose levels (Jorgenson and Rushbrook, 1980).

2.3.7.5 Chlorobenzenes. Except for hexachlorobenzene (HCB), which is not a volatile chlorobenzene, no immunotoxic effects have been reported or specifically studied. HCB influenced the humoral immunity; both the primary and secondary IgM and IgG responses to tetanus toxoid were significantly increased, but no difference was found in the IgM response to *E. coli* lipopolysaccharide. Regarding cell mediated immunity no effects were found. Rats exposed pre- and postnatally to HCB had increased spleen weights and decreased thymus weights (Vos *et al.*, 1979). In contrast Loose *et al.* (1978a,b) reported that HCB suppressed the humoral and cell mediated immune response in mice.

2.3.7.6 Formaldehyde. Formaldehyde does not cause immunosuppression. In contrast it can cause contact dermatitis. Earlier observations suggesting that formaldehyde could induce dermatitis via inhalation were anecdotal and were not supported by careful immunological testing. Systemic exposure to significant levels of circulating formaldehyde can result in sensitization. It is presumed that formaldehyde combines rapidly with a variety of proteins to induce immunologic sensitivity. The most frequent sequela of this type of sensitisation has been autoimmune haemolytic anaemia. Rarely, systemic sensitivity has been associated with peripheral eosinophilic and associated sensitivity or asthmatic reactions. Elevated levels of IgE-specific antibody to a formaldehyde conjugate were noted. Formaldehyde can act as an immunogen, and IgG- and IgA-specific antibodies have been demonstrated in experimental animals. Humans orally or parenterally exposed to formaldehyde showed similar changes in immunoglobulins; for exposure to gaseous formaldehyde this was less certain (Bardonna and Montanaro, 1991).

2.3.8 Effects on the cardiovascular system

The effects of VOCs on the cardiovascular system can be divided into: (i) a direct effect on the heart as a pumping organ; and (ii) effects on the vascular system including the coronary arteries of the heart. Of particular concern in the vascular system are atherosclerotic changes. However, to study atherogenic effects of chemicals specific animal models with long-term exposure are necessary. Rats and mice are not suitable as test animals in atherosclerosis, and such experiments have not been made with VOCs. The data available on cardiotoxic and cardiovascular toxic effects of VOCs depend therefore on accidental observations, which are very limited as is illustrated in this section.

2.3.8.1 Phenol. In several animal species phenol was claimed to cause, among other acute effects, myocardium infarcts (Van Oettingen, 1949). More recent acute toxicity studies have revealed that phenol causes degeneration and necrosis of the myocardium (Deichmann and Keplinger, 1963; Liao and Oehme, 1980). Inhalatory exposure of rats, rabbits and guinea pigs to phenol for three months caused heart lesions and lesions of the blood vessels (Deichmann, 1949; Kurnatowski, 1960).

2.3.8.2 Chloroform. In chloroform-treated rats, especially male rats, effects of chloroform include polyarteritis of the mesenterial, pancreatic and other arteries, and arterioles (Reuber, 1979; Van der Heijden *et al.*, 1987). In some long-term studies mortality has been caused by pulmonary inflammation and cardiac thrombosis (NCI, 1976). Another long-term study revealed decreased cholesterol and triglyceride concentrations in the serum (Jorgenson *et al.*, 1984).

2.3.8.3 Methylene chloride. In acute toxicity studies with methylene chloride ($>60\,000$ mg/m^3) in monkeys and rabbits cardiovascular effects such as arrhythmia, tachycardia and hypotension were found (Belej *et al.*, 1974; Taylor *et al.*, 1976).

2.3.9 Effects on the endocrine system

The effects of VOCs on the endocrine system are very limited, and there have been no studies with specific emphasis on the endocrine system, such as serum hormone levels and specific histopathological examination of endocrine organs. The data available are restricted to accidental observations in regular toxicological experiments. The limited data can be explained, at least partly, by the special requirements needed for endocrine studies, such as stress-free handling, blood sampling and trigger tests.

2.3.9.1 Chloroform. In short-term (90-day) toxicity tests with chloroform (40 mg per rat per day) thyroid lesions were reported. These thyroid lesions consisted of a reduction in follicular size and colloid density, an increase in epithelial height and occasional collapse of follicles. Thyroid lesions diminished in severity during the 90-day recovery period (Chu *et al.*, 1982b).

2.3.9.2 Hexane. The effect of inhalation of an air:*n*-hexane mixture on insulin degradation in rat blood was studied *in vitro* using simultaneous monitoring of the *n*-hexane concentrations in the blood and in the inhaled air. Inhalation of *n*-hexane vapours at a concentration of 54 694 ppm (197 g/m^3) resulted in *n*-hexane saturation (0.18 mg/ml). This was associated with a statistically significant increase in insulin degradation in the blood of exposed rats compared with that of controls. *In vitro* addition of *n*-hexane to pure plasma or to a mixture of the blood elements revealed that degradation activities present in the cellular blood compartment are required for *n*-hexane-induced enhancement of insulin degradation in the blood (Klimes *et al.*, 1987). It is claimed that extensive insulin degradation in target tissues may even result in the development of serious insulin resistance in affected individuals (Misbin *et al.*, 1981; Blazar *et al.*, 1984), therefore this endocrine effect of hexane might be of importance for individuals chronically exposed to *n*-hexane (or *n*-heptane).

2.3.9.3 Phenols. Although it is known that organic phenols as present in plants have oestrogenic properties (Stob, 1983), such effects have not been reported for phenol or chlorophenols.

2.3.10 Interaction of VOCs in cases of combined exposure

Combined exposure includes two types of combination. The first type is exposure to single VOC by different routes such as oral and inhalatory, inhalatory and dermal, or oral and dermal. Special studies for this type of combined exposure have not been made. Nevertheless, based on the pharmacokinetic behaviour of VOCs by different exposure routes, the effect of such combined exposure to a VOC can be predicted and, by definition, will be not more than additive.

More complex is the situation where exposure is to two or more different VOCs, either by the same route or by different routes. In this type of exposure (co-exposure) interactions may be antagonistic, additive or synergistic. Such interactions may take place at the level of toxicokinetic behaviour or toxicity. In terms of health risks antagonistic effects of VOCs are of no direct concern, whereas additive actions and, in particular, synergistic actions are very important. Due to the complexity of studying toxicological interaction and the difficulties in quantifying the response, studies dealing with co-exposure to different VOCs are very limited.

An example of an antagonistic effect was recorded by Nylen *et al.* (1989), who found that toluene and xylene protect rats from *n*-hexane-induced testicular atrophy.

Examples of additive effects are rare. In a human volunteer study eight males were experimentally exposed to toluene, *p*-xylene and a combination of toluene and *p*-xylene in order to study the influence of co-exposure and exposure to different levels of each solvent on their uptake and elimination. The exposures were performed for hours at exposure levels lower than Swedish threshold limit values for toluene (300 mg/m^3) and *p*-xylene. During and after exposure, solvent concentrations in the blood and expired air were measured. Decreases in the blood/end-exhaled air concentration ratios were noted for both toluene and *p*-xylene when they were given in combination compared with separate exposure. The total solvent uptake relative to the exposure level decreased after exposure to the higher solvent concentrations, and the apparent clearance also decreased after exposure to the higher concentration of solvents. The changes in blood/end-exhaled air concentrations may indicate an effect of co-exposure (Wallen *et al.*, 1985). Anshelm *et al.* (1985) studied the effect of co-exposure to toluene and *p*-xylene on the human central nervous system. The results of this study indicated that after a 4 h exposure period to concentrations which did not exceed the Swedish threshold limit values, the risk of acute effects was minimal. It was also be concluded that no synergistic effects occurred at these exposure levels (Anshelm *et al.*, 1985).

In another study with ten male volunteers the neurobehavioural effects of experimental exposure to toluene (100 ppm), xylene (100 ppm) and their mixture (50 ppm of toluene and 50 ppm of xylene) were studied by means of nine psychological tests. Acute exposure to xylene produced the most adverse effect on simple reaction time and choice reaction time. Exposure to toluene affected only the memory test performance. The effect of combined exposure appeared to be weaker than the effect of exposure to xylene alone, but stronger than the effect of exposure to toluene (Dudek *et al.*, 1990). This experiment gave no proof for synergism, addition or interaction, because combined exposure to each chemical at a dose level of half that given in separate dosage experiments will not necessarily have the same or a greater effect. For example the dose response curves of xylene and toluene may differ, as the qualitative effects on the nervous system might differ. This study illustrates well the fact that is not easy to study interaction in case of combined exposure.

Kilburn *et al.* (1985) studied the neurobehavioural (disturbance of memory, mood equilibrium and sleep, combined with headaches and indigestion) and respiratory symptoms of formaldehyde and xylene exposure in histology technicians. Neurobehavioural symptoms were accompanied by irritation of the eyes, upper airways and trachea. Kilburn *et al.* (1985) were not able to conclude anything about additive or synergistic effects. The only conclusion was that formaldehyde exposure correlated better with neurobehavioural symptoms, and with respiratory and mucous membrane symptoms, than did exposure to xylene, toluene or other agents.

Abou-Donia *et al.* (1985, 1991) studied the synergism of *n*-hexane-induced neurotoxicity in hens by methylisobutyl ketone (MiBK) following subchronic (90 days) inhalation. Hens continuously exposed to 1000 ppm MiBK developed leg weakness with subsequent recovery, while inhalation of 1000 ppm *n*-hexane produced only mild ataxia. Hens exposed to mixtures of *n*-hexane and MiBK developed clinical signs of neurotoxicity, the severity of which depended on the MiBK concentration. Thus all hens exposed to 1000 ppm *n*-hexane in combination with 250, 500 or 1000 ppm MiBK progressed to paralysis. Hens continuously exposed to 1000:100 ppm *n*-hexane: MiBK showed severe ataxia which did not change during the observation period. The neurologic dysfunction in hens exposed simultaneously to *n*-hexane and MiBK was accompanied by large swollen axons and degeneration of the axons and myelin of the spinal cord and peripheral nerves. The results of this study indicated that the non-neurotoxic chemical MiBK synergised the neurotoxic action of the weak neurotoxicant *n*-hexane (Abou-Donia *et al.*, 1985). In another experiment to investigate the mechanism of MiBK synergism of *n*-hexane neurotoxicity, continuous inhalation over 50 days of 1000 ppm of both *n*-hexane and MiBK significantly induced aniline hydroxylase and cytochrome P-450 contents in hen liver microsomes. These results suggested that the synergistic action of MiBK on *n*-hexane neurotoxicity may be related to its ability to induce liver microsomal cytochrome P450, resulting in increased metabolic activation of *n*-hexane to more potent neurotoxic metabolites. Similar synergistic effects on hydrocarbon neuropathies are described for methylethyl ketone (MEK) (Altenkirch *et al.*, 1978). *In vivo* studies on the effects of MEK pretreatment on hepatic mixed-function oxidase activity and on metabolism of *n*-hexane were also performed (Robertson *et al.*, 1989). It was shown that pretreatment with MEK increased the concentrations of the metabolites 2,5-hexanedione (the proximal neurotoxin) and 2,5-dimethylfuran in the blood, sciatic nerve and testes. It was also found that the activity of 7-ethoxy-coumarin-*o*-diethylase increased (up 500%). Hence the synergistic effects of MEK on the neurotoxicity of *n*-hexane appear to arise, at least in part, from the activating effects of MEK on selected enzymes responsible for *n*-hexane activition.

2.3.11 Effects in humans: differences from observations in animals

In many cases the toxic effects of VOCs observed in humans, mostly after exposure in the occupational situation or indoor exposure, are at least qualitatively similar. Since many of these data reflect the occupational health situation, they are not discussed here but in chapter 7.

To compare quantitatively the toxic effects in humans with those in laboratory animals is often difficult, because the toxicological parameters that can be studied in humans are restricted. Histopathological alterations have not been established in humans. With the exception of a few human volunteer studies, which were mostly toxicokinetic experiments with exposure levels below the threshold limits based on animal experiments, toxicological studies of VOCs in humans seldom provide reliable and accurate exposure levels and toxicological data for use in risk assessment. Human observations and epidemiological studies provide little conclusive evidence for possible carcinogenicity. In this section only a few examples from the literature will be given. Extrapolation of animal data from carcinogenicity studies to the human is complex. Relatively recently, physiologically based pharmacokinetic models have been used to facilitate the extrapolation process of methylene chloride. Specifically, tumour incidence has been correlated with the amount of methylene chloride metabolised by the glutathione *S*-transferases (GSTs), not with the mixed-function oxidases. The mouse, the only species in which methylene chloride was carcinogenic, has significant GST-mediated methylene chloride metabolism in the liver, with a rate 12-fold greater than that of rat liver. No such metabolism was found in either hamster or human liver. A low rate of GST metabolism was also present in mouse lung tissue but not in rat or hamster lung. Along with kinetic data, these observations suggest that there is little risk of humans developing cancer following exposure to methylene chloride (American College of Toxicology, 1988).

Two epidemiologic studies have been conducted to determine the toxicity of methylene chloride for humans. No significant excess deaths were observed in either study, and there were no significant increases in malignant neoplasms or ischemic heart disease (American College of Toxicology, 1988).

For toluene it can be concluded that it acts on the human nervous system. Toluene also has an irritating effect on mucous membranes. However, no haematological disturbance has been observed during routine medical examination of humans occupationally exposed to toluene. Effects on the liver and kidneys have not been reported in persons exposed to typical occupational levels (Van der Heijden *et al.*, 1988).

2.3.12 Effects in humans: risk groups

Of interest in the human health situation is the recognition of population groups at risk. These risk groups can be more vulnerable to the exposure of chemicals including VOCs. This might be due to differences in exposure, toxicokinetic behaviour and/or toxic properties of the chemical. In some cases infants and the elderly are indicated as risk groups for the toxicity of certain chemicals. There are few examples of specific studies in laboratory animals or man in which attention has been paid to this possibly higher vulnerability of certain population groups. Nevertheless, as the toxicokinetics, and particularly the elimination of VOCs, strongly relies on the liver and kidneys, it might be expected that infants and the elderly, who commonly have reduced liver and especially kidney functions, are more at risk from some VOCs than are young adults. As several VOCs exhibit neurotoxic effects, and in newborn children or animals the nervous system is incomplete, this group of newborn subjects may be more vulnerable to neurotoxic VOCs.

The elderly or people with certain diseases have reduced organ or tissue functions, which may interact with the toxicity of VOCs. For example, elderly people with Alzheimer's disease may be more vulnerable to exposure to neurotoxic VOCs.

An example of such an interaction is shown in the study of Kanz *et al.* (1991). This study in which rats exposed to 1,1-dichloroethylene also had provoked hypothyroidism, revealed a decrease in metabolism and covalent binding, but an increase in hepatoxic lesions. Increased vulnerability of this kind might be particularly important for cases of indoor exposure to VOCs.

2.3.13 Risk assessment and the use of mathematical models

2.3.13.1 Risk assessment. The classical approach of threshold limit values and acceptable or tolerable daily intakes is based on the establishment of no-toxic-effect levels (usually obtained from toxicological studies performed in animals) and the use of safety factors. As long as the vulnerability of (at risk) groups within the total population does not exceed the margin of safety (10) used in the classical concept the risk assessment provides a safe health situation; when the exposure does exceed the established threshold limit or tolerable daily intake, or the vulnerability is higher than average, a more scientific approach for the safety evaluation is needed.

Recent developments in toxicology have moved in the direction of a more scientifically based risk assessment, making use of mathematical models to describe exposure (chapter 1) and toxicokinetic and toxicological effects at the level of organs and tissues. This approach makes use of advanced computerised systems. There are several published examples of mathematical models of VOC pharmacokinetics, a few of which will now be discussed.

2.3.13.2 Mathematical modelling. **P**hysiologically **b**ased **p**harmaco**k**inetic (PBPK) models describing the uptake, distribution, metabolism and excretion of VOCs are proposed for use in regulatory risk assessment (Andersen, 1981; Ramsey and Andersen, 1984; Clewell and Andersen, 1985; Bogen, 1988). The use of PBPK models is a promising alternative for conducting more scientifically sound extrapolations. PBPK models contain sufficient biological detail to allow pharmacokinetic behaviour to be predicted for widely differing exposure scenarios. In recent years, successful physiological models have been developed for a variety of volatile and non-volatile chemicals, and their ability to perform the extrapolations needed in risk assessment has been demonstrated. Techniques for determining the necessary biochemical parameters are readily available, and the computational requirements are now within the scope of even a personal computer. In addition to providing a sound framework for extrapolation, the predictive power of a PBPK model makes it a useful tool for more reliable dose selection before beginning large-scale studies, as well as for the retrospective analysis of experimental results.

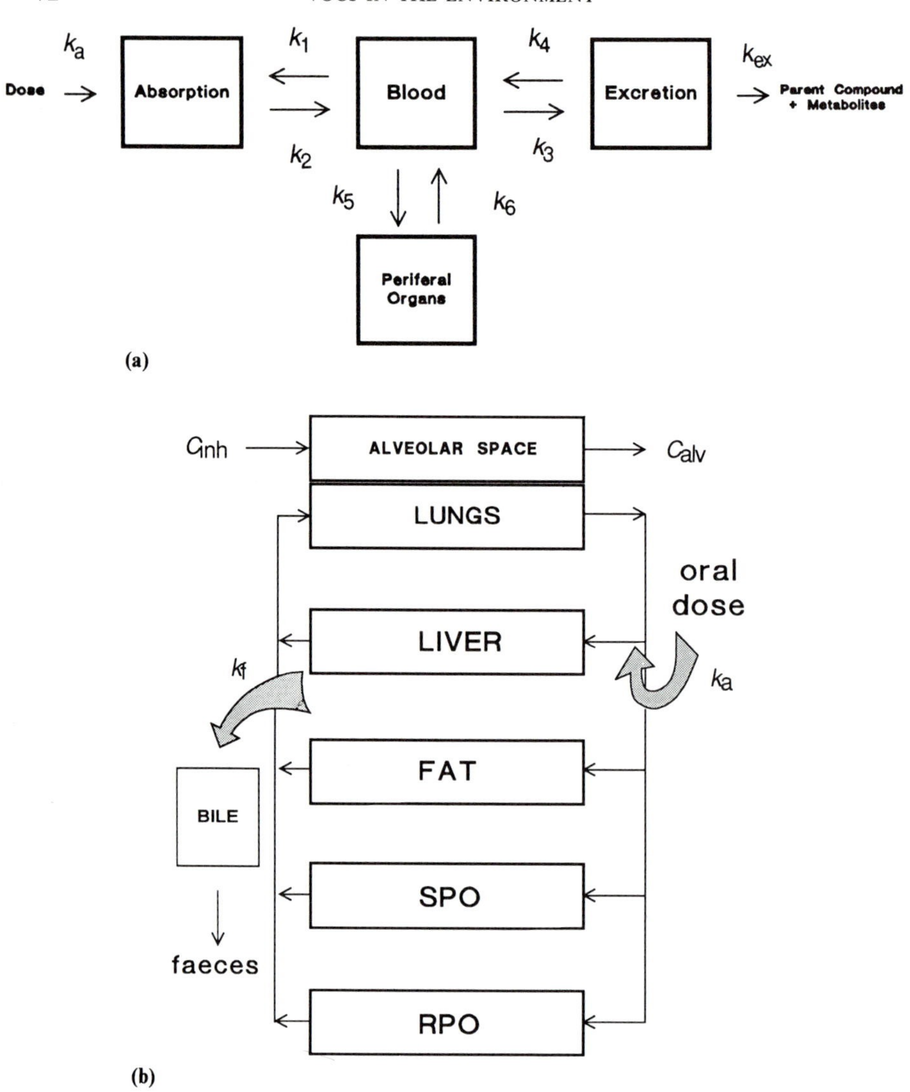

Figure 2.2 Schematic representation of (a) a four-compartment pharmacokinetic model; and (b) a **p**hysiologically **b**ased **p**harmaco**k**inetic (PBPK) model (B). In classical pharmacokinetic models the fate of a chemical in the organism is described by mass–balance equations between several organ compartments. Figure 2.2(a) shows the structure of a four-compartment pharmacokinetic model: (1) absorption compartment; (ii) blood; (iii) peripheral organs; and (iii) excretion compartment. In this model a chemical enters the organism by absorption from the gastrointestinal tract. After absorption distribution takes place over the blood and the organs connected with the blood. The chemical or its metabolites are removed from the organism by excretion from the blood into an excretion compartment (for example bile) and the faeces. The rate at which these processes occur are characterised by the rate constants k_a, k_{ex}, and k_1–k_6. Figure 2.2(b) shows the groundplan of a PBPK model. In such models mass–balance equations are described within a physiological concept of the organism (organs connected by blood flows). In this specific case the organism is represented as the sum of the following five organs or organ

(*Continued*)

PBPK models differ from conventional compartment models in that they are based to a large extent on the actual physiology of the organism, meaning that they focus on the organs connected by blood flow. Examples of such a PBPK model and a classical toxicokinetic model are shown in Figure 2.2. Actual organ and tissue groups are used, with weights and blood flows from the literature. Actual physical, chemical and biochemical constants of the compound are used (Clewell and Anderson, 1985). The PBPK model is used to estimate the extent to which humans metabolise VOCs such as trichloroethylene (TCE). A PBPK model for TCE is described by Bogen (1988). A steady-state analysis of one such model provided a simple and convenient means of predicting the relationship between the applied external dose and the corresponding toxicologically effective, metabolised dose of TCE (internal exposure level in the body). Bogen (1988) fitted a version of this PBPK model to data on human metabolism of TCE (urinary metabolites in chronically exposed workers), yielding a direct estimate of PBPK parameters governing human capacity to metabolise TCE. This estimate was shown to be consistent with others based on experimental studies of TCE metabolism in humans exposed to TCE by inhalation for short periods. These results were applied to human cancer risk assessment using rodent bioassay data on TCE-induced tumorigenesis. TCE was chosen because its metabolism, as opposed to applied dose, is directly correlated with experimentally induced chronic toxicity and carcinogenicity (Bogen, 1988).

Another example of the use of a PBPK model is the model which describes accurately the behaviour of inhaled styrene in rats and predicts the behaviour of inhaled styrene in humans (Ramsey and Andersen, 1984). This model consists of a series of mass–balance differential equations which quantify the time course of styrene concentration within four tissue groups representing: (1) highly perfused organs; (2) moderately perfused tissues such as muscle; (3) slowly perfused fat tissue; and (4) organs with high capacity to metabolise styrene (principally the liver). The pulmonary compartment of the model incorporates the uptake of styrene controlled by ventilation and perfusion rates and the blood/air partition coefficient. The metabolising tissue group incorporates saturable Michaelis–Menten metabolism controlled by the biochemical constants V_{max} and K_m. With a single set of physiological and

(Continued)

compartments: (i) lungs; (ii) liver; (iii) fat tissue; (iv) slowly perfused organs (SPO) like the muscles and the skin; and (v) richly perfused organs (RPO) like the spleen and the kidneys. These organs are connected by blood flows (solid arrows). The model allows a chemical to enter the organism by two routes, the oral route of administration (absorption from the gastrointestinal tract) or the inhalatory route (entrance of a chemical by interaction of the alveolar air into the blood). Excretion of a chemical from the organism takes place by hepatic metabolism to water-soluble metabolites, followed by excretion of the formed metabolites into the bile and, subsequently, the faeces. k_a and k_f are rate constants characterising absorption and excretion processes; C_{inh} and C_{alv} are the concentrations of the chemical in the inhaled air and in the alveolar air (exhaled air), respectively.

biochemical constants, the model adequately simulates concentrations in the blood and fat of rats exposed to different concentrations of styrene. The constants used to simulate the fate of styrene in rats were scaled up to represent constants for humans and are in good agreement with previously published experimental data. The results show that PBPK models provide a rational basis to: (1) explain the relationship between blood concentration and exposure conditions of an inhaled chemical; and (2) extrapolate this relationship from experimental animals to humans (Ramsey and Andersen, 1984).

Once the concentration at the level of the target tissue or organ compartment (internal exposure level) has been predicted by a PBPK model, the procedure could be extended by means of toxicity studies in tissue or cell systems in order to predict the toxicity. These models might then be validated by dedicated toxicological studies *in vivo*. Except for a few examples, such as TCE, models for physiologically based toxicodynamics still have to be developed.

References

Abdou-el-Makarem, M.M. *et al.* (1967) Biliary excretion of foreign compounds—benzene and its derivatives. *Biochem. J.* **105** 1269–1274.

Abdou-Donia, M.B., Lapadula, D.M., Campbell, G. and Timmons, P.R. (1985) The synergism of *n*-hexane-induced neurotoxicity by methyl isobutyl ketone following subchronic (90 days) inhalation in hens: induction of hepatic microsomal cytochrome P-450. *Toxicol. Appl. Pharmacol.* **81** 1–16.

Abdou-Donia, M.B., Hu, Z., Lapadula, D.M. and Gupta, R.P. (1991) Mechanisms of joint neurotoxicity of *n*-hexane, methyl isobutyl ketone and *O*-ethyl *O*-4-nitrophenyl phenylphosphonotioate in hens. *J. Pharmacol. Exper. Therap.* **257** 282–289.

Agrawal, A.K., Srivastava, S.P. and Seth, P.K. (1982) Effects of styrene on dopamine receptors. *Bull. Environ. Conta. Toxicol.* **29** 400–403.

Ahlborg, U.G. and Thunberg, T.M. (1978) Effects of 2,3,7,8-tetrachlorobenzo-*p*-dioxin on the *in vivo* and *in vitro* dechlorination of pentachlorophenol. *Arch. Toxicol.* **40** 55–61.

Altenkirch, H., Stoltenburg, G. and Wagner, H.M. (1978) Experimental studies on hydrocarbon neuropathies induced by methylethylketone (MEK). *J. Neurology* **219** 159–170.

Altenkirch, H., Stoltenburg-Didinger, G., Wagner, H.M., Herrmann, J. and Walter, G. (1990) Effects of lipoic acid in hexane carbon-induced neuropathy. *Neuro. Toxicol. Teratol.* **12** 619–622.

Altman, A. and Dittmer, D.S. (1974) *Biological Data Book.* Vol. 3. Federation of American Society for Experimental Biology, Bethesda, USA.

Americal College of Toxicology (1988) Final report on the safety assessment of methylene chloride. *J. Am. Coll. Toxicol.* **7** 741–835.

Amlacher, E. and Rudolph, C. (1983) Specific inhibition of nuclear DNA synthesis by carcinogens. *Acta Histochem. Suppl.* **27** 155–162.

Andersen, M.E. (1981) A physiologically based toxicokinetic description of the metabolism of inhaled gases and vapors; Analysis at steady state. *Toxicol. Appl. Pharmacol.* **60** 509–526.

Andersen, M.E., Gargas, M.L. and Ramsey, J.C. (1984) Inhalation pharmacokinetics: evaluating systemic extraction, total *in vivo* metabolism, and the time course of enzyme induction for inhaled styrene in rats based on arterial blood: inhaled air concentration ratios. *Toxicol. Appl. Pharmacol.* **73** 176–187.

Andrae, H., Hamfeldt, H., Vogl, L., Lichtmannegger, J. and Summer, K.H. (1988) 2-Nitropropane induces DNA repair synthesis in rat hepatocytes *in vitro* and *in vivo*. *Carcinogenesis* **9** 815–819.

Anon (1988) Final report on the safety assessment of methylene chloride. *J. Am. Coll. Toxicol.* **7** 748–749.

Anshelm, O., Olson, B., Gamberale, F. and Iregren, A. (1985) Co-exposure to toluene and *p*-xylene in man: central nervous functions. *Br. J. Indus. Med.* **42** 117–122.

Appelgren, L.E., Eneroth, G. and Grant, C. (1977) Studies on ethylene oxide: whole-body autoradiography and dominant lethal test in mice. In *Clinical Toxicology Proceedings*, Edinburgh, June 1976, pp. 315–317. European Society of Toxicology, Vol. 18, Excerpta Medica, Amsterdam.

Appelman, L.M. (1985) Review of literature data formaldehyde. *Report of TNO/CIVO/TNO-MT*. Halkoning, BV.

Astrand, I. (1975) Uptake of solvents in the blood and tissues of man. A Review. *Scand. J. Work Environ. Health.* **1** 199–218.

Astrand, I. and Gamberale, F. (1978) Effects on humans of solvents in the inspiratory air: A method for estimation of uptake. *Environ. Res.* **15** 1–4.

Astrand, I., Kilbom, A., Ovrum, P., Wahlberg, I. and Vesterberg, O. (1974) Exposure to styrene. 1. Concentration in alveolar air and blood at rest and during exercise and metabolism. *Scand. Work Environ. Health* **11** 69–85.

Bakke, O.M. and Scheline, R.R. (1970) Hydroxylation of aromatic hydrocarbons in the rat. *Toxicol. Appl. Pharmacol.* **16** 691–700.

Baranowska-Dutkiewicz, B. (1981) Skin absorption of phenol from aqueous solutions in man. *Int. Arch. Occup. Environ. Health* **49** 99–104.

Baranowska-Dutkiewicz, B. (1982) Skin absorption of aniline from aqueous solutions in man. *Toxicology Lett.* **10** 367–372.

Bardodej, Z. (1978) Styrene, its metabolism and the evaluation of hazards in industry. *Scand. J. Work Environ. Health* **4** 95–103.

Bardodej, Z. and Bardodejova, E. (1970) Biotransformation of ethylbenzene, styrene and alpha-methylstyrene in man. *Am. Ind. Hyg. Assoc. J.* **31** 206–209.

Bardonna, E.J. and Montanaro, A. (1991) Formaldahyde: an analysis of its respiratory, cutaneous, and immunologic effects. *Ann. All.* **66** 441–452.

Barlow, S.M. and Sullivan, F.M. (1982) *Reproductive Hazards of Industrial Chemicals—An Evaluation of Animal and Human Data*, pp. 592–599. Academic Press, New York.

Barrowcliff, D.F. and Knell, A.J. (1979) Cerebral damage due to endogenous chronic carbon monoxide poisoning caused by exposure to methylene chloride. *J. Soc. Occup. Med.* **29** 12–14.

Battig, K. and Grandjean, E. (1964) Industrial solvents and avoiding conditioning. *Arch. Environ. Health* **9** 745–749.

Baudinette, R.V., Wheldrake, J.K., Hewitt, S. and Hawke, D. (1980) The metabolism of (^{14}C) phenol by native Australian rodents and marsupials. *Aust. J. Zool.* **28** 511–520.

Behl, C.R., Linn, E.E., Flynn, G.L., Pierson, C.L., Higuchi, W.I. and Ho, N.F.H. (1983) Permeation of skin and eschar by antiseptics. I: Baseline studies with phenol. *J. Pharmaceut. Sci.* **72** 391–397.

Belej, M.A., Smith, D.G. and Aviado, D.M. (1974) Toxicity of aerosol propellants in the respiratory and circulatory systems. IV. Cardiotoxicity in monkeys. *Toxicol.* **2** 381–395.

Belvedere, G., Cantoni, L., Facchinetti, T. and Salmona, M. (1977) Kinetic behaviour of microsomal styrene monooxygenase and styrene hydratase in different animal species. *Experentia* **33** 708–709.

Benignus, V.A., Muller, K.E., Barton, C.N. and Bittikofer, J.A. (1981) Toluene levels in blood and brain of rats during and after exposure respiratory exposure. *Toxicol. Appl. Pharmacol.* **61** 326–334.

Benignus, V.A., Muller, K.E., Graham, J.A. and Barton, C.N. (1984) Toluene levels in blood and brain in rats as a function of toluene level in inspired air. *Environ. Res.* **33** 34–46.

Bergman, K. (1979) Whole-body autoradiography and allied techniques. *Scand. J. Work Environ. Health.* **5** 93–263.

Berlin, M., Holm, S., Knutson, P. and Tunek, A. (1979) Biological threshold limits for benzene based on pharmacokinetics of inhaled benzene in man. *Arch. Toxicol. Suppl.* **2** 305–310.

Berlin, M., Gage, J.C., Gullberg, B., Hom, S., Knutson, P. and Tunek, A. (1980) Breath concentration as an index of the health risk from benzene. *Scand. J. Work Environ. Health* **6** 104–111.

Bianchi, V., Nuzzo, F. and Abbondandolo, A. (1982) Scintillometric determination of DNA repair in human cell lines. A critical appraisal. *Mutat. Res.* **87** 447–463.

Blazar, B.A., Withley, Ch. B., Kitabchi, A.E., Tsai, M.Y., Santiago, J., While, N., Stentz, F.B. and Parown, D.M. (1984) *In vivo* chloroquine-induced inhibition of insulin degradation in a diabetic patient with severe insulin resistance. *Diabetes* **33** 1133–1137.

Bogen, K.T. (1988) Pharmacokinetics for regulatory risk analysis: The case of trichloroethylene. *Regul. Toxicol. Pharmacol.* **8** 447–466.

Bonnet, P., Rauolt, G. and Gadiski, D. (1979) Lethal concentration of 50 main aromatic hydrocarbons. *Arch. Mal. Prof. Med. Trav. et Soc. Med. Trav. France* **40** 805–810.

Bonnet, P., Francin, J.M., Gradiski, D., Raoult, G. and Zissu, D. (1980) Determination de la concentration lethale$_{50}$ des principaux hydrocarbures aliphatiques chlorés chez le rat. *Arch. Mal. Prof. Med. Trav. et Soc. Med. Trav. France* **41** 317–321.

Borzelleca, J.F. *et al.* (1985) Toxicological evaluation of selected chlorinated phenols. In Jolley, R.L. *et al.* (eds) *Water Chlorination: Environmental Impact and Health Effects*, Vol. 5, pp. 331–343. Lewis Publishers Inc., Chelsea.

Bos, R.P., Brauns, R.M.E., Van Doorn, R., Theeuws, J.L.G. and Henderson, P.T. (1981) Non-mutagenicity of toluene, *o*-, *m*-, and *p*-xylene, *o*-methylbenzyl alcohol and *o*-methyl benzyl sulfate in the Ames assay. *Mutat. Res.* **88** 273–279.

Bowman, F.J., Borzelleca, J.F. and Munson, A.E. (1978) The toxicity of some halomethanes in mice. *Toxicol. Appl. Pharmacol.* **44** 213–215.

Brondeau, M.T., Bonnet, P., Guenier, J.P. and De Ceaurriz, J. (1983) Short-term inhalation test for evaluating industrial hepatoxicants in rats. *Toxicol. Lett.* **19** 139–146.

Brown, V.K.H., Box, V.L. and Simpson, B.T. (1975) Decontamination procedures for skin exposed to phenolic substances. *Arch. Environ. Health* **30** 1–6.

Browning, E. (1965) *Toxicity and Metabolism of Industrial Solvents*, pp. 77–89. Elsevier, New York.

Bruckner, J.V. and Peterson, R.G. (1981a) Evaluation of toluene and acetone inhalant abuse. I. Pharmacology and pharmacodynamics. *Toxicol. Appl. Pharmacol.* **61** 27–38.

Bruckner, J.V. and Peterson, R.G. (1981b) Evaluation of toluene and acetone abuse. II. Model development and toxicology. *Toxicol. Appl. Pharmacol.* **61** 302–312.

Bushnell, P.J. (1989) Behavioural effects of acute *p*-xylene inhalation in rats: auto shaping, motor activity, and reversal learning. *Neurotoxicol. Teratol.* **10** 569–577.

Calleman, C.J., Ehrenberg, L., Jansson, B., Osterman-Golkar, S., Segerbäck, D., Svensson, K. and Wachtmeister, C.A. (1978) Monitoring and risk assessment by means of alkyl groups in hemoglobin in persons occupationally exposed to ethylene oxide. *J. Environ. Pathol. Toxicol.* **2** 242–247.

Cameron, G.R., Paterson, J.L.H., Desaram, G.S. and Thomas, J.C. (1938) The toxicity of some methyl derivatives of benzene with special reference to pseudo cumene and heavy coal tar naphta. *J. Pathol. Bacteriol.* **46** 95–107.

Cantoni, L., Salmona, M., Facchinetti, T., Pantarotto, C. and Belvedere, G. (1978) Hepatic and extrahepatic formation and hydration of styrene oxide *in vitro* in animals of different species and sex. *Toxicol. Lett.* **3** 179–186.

Capel, I.D., Millburn, P. and Williams, R.T. (1972a) Species variations in phenol metabolism. *Biochem. J.* **127** 25–26.

Capel, I.D., French, M.R., Millburn, P., Smith, R.L. and Williams, R.T. (1972b) The fate of [^{14}C]phenol in various species. *Xenobiotica* **2** 25–35.

Capel, I.D., Millburn, P. and Williams, R.T. (1974) Monophenyl phosphate a new conjugate of phenol in the cat. *Biochem. Soc. Trans.* **2** 305–306.

Caperos, J.R. *et al.* (1979) Exposition au styrene. II. Bilan de l'absorption de l'excretion et du metabolism sur des sujets humains. *Int. Arch. Occup. Environ. Health* **42** 223–230.

Carlson, G.P. (1977) Chlorinated benzene induction of hepatic porphyria. *Experienta* **33** 1627–1629.

Carlson, G.P. and Tardiff, R.G. (1976) Effect of chlorinated benzenes on the metabolism of foreign organic compounds. *Toxicol. Appl. Pharmacol.* **36** 383–394.

Carlsson, A. (1983) Exposure to toluene: Uptake, distribution and elimination in man. *Scand. J. Work Environ. Health* **8** 43–53.

Carlsson, A. and Hultengren, M. (1978) Exposure to methylene chloride. III. Metabolism of ^{14}C-labelled methylene chloride. *Scand. J. Work Environ. Health* **1** 104–108.

Carlsson, A. and Lindqvist, T. (1977) Exposure of animals and man to toluene. *Scand. J. Work Environ. Health* **3** 135–143.

Carpenter, C.P., Kinkead, E.R., Geary, D.L., Ir. Sullivan, L.J. and King, J.M. (1975) Petroleum hydrocarbon toxicity studies. V. Animal and human response to vapors of mixed xylenes. *Toxicol. Appl. Pharmacol.* **33** 343–588.

Cassidy, M.K. and Houston, J.B. (1980) *In vivo* assessment of extrahepatic conjugative metabolism in first pass effects using the model compound phenol. *J. Pharm. Pharmacol.* **32** 57–59.

Ceaurriz, J., Desiles, J.P., Bonnet, P., Marignac, B., Muller, J. and Guenier, J.P. (1983) Concentration dependent behavioural changes in mice following short-term exposure to various industrial solvents. *Toxicol. Appl. Pharmacol.* **67** 383–389.

Chang, Y.C. (1987) Neurotoxic effects of *n*-hexane on the human central nervous system; evoked potential abnormalities in *n*-hexane polyneuropathy. *J. Neurol. Neurosurg. Psych.* **50** 269–274.

Chang, Y.C. (1990) Patients with *n*-hexane induced polyneuropathy: a clinical follow up. *Br. J. Indus. Med.* **47** 485–489.

Chu, I., Secours, V., Marino, I. and Villeneuve, D.C. (1980) The accumulation of four trihalomethanes in male and female rats. *Toxicol. Appl. Pharmacol.* **52** 351–353.

Chu, I., Villeneuve, D.C., Secours, V.E. and Becking, G.C. (1982a) Toxicity of trihalomethane. I. The acute and subacute toxicity of chloroform, bromodichloromethane, chlorodibromomethane and bromoform. *J. Environ. Sci. Health* **17** 205–224.

Chu, I., Villeneuve, D.C. Secours, V.E. and Becking, G.C. (1982b) Toxicity of trihalomethane. II. Reversibility of toxicological changes produced by chloroform, bromodichloromethane, chlorodibromomethane and bromoform in rats. *J. Environ. Sci. Health* **17** 225–250.

Clewell, H.J. III. and Andersen, M.E. (1985) Risk assessment extrapolations and physiological modeling. *Toxicol. Indus. Health.* **1** 111–131.

Coan, M.L., Baggs, R.B. and Bosmann, H.B. (1982) Demonstration of direct toxicity of phenol on kidney. *Chem. Pathol. Pharmacol. Res. Comm.* **9** 137–143.

Coate, W.B., Lewis, T.R. and Busey, W.M. (1977) Chronic inhalation exposure of rats, rabbits and monkeys to 1,2,4-trichlorobenzene. *Arch. Environ. Health* **32** 249–255.

Cohr, K.H. and Stokholm, J. (1979) Toluene: a toxicologic review. *Scand. J. Work Environ. Health* **5** 71–99.

Conaway, C.C., Hussain, N.S., Way, B.M. and Fiola, E.S. (1991) Evaluation of secondary nitro alkanes, their nitronates, primary nitro alkanes, nitro carbinols, and other aliphatic nitro compounds in the Ames *Salmonella* assay. *Mutat. Res.* **261** 197–207.

Conca, G.L. and Maltagliati, A. (1955) Study of the percutaneous absorption of benzene. *Med. Lavoro* **46** 194–198.

Condie, L.W. *et al.* (1983) Comparative renal and hepatotoxicity of halomethanes: bromodichloroform, bromoform, chloroform, dibromochloromethane and methylene chloride. *Drug Chem. Toxicol.* **6** 563–578.

Conning, D.M. and Hayes, M.J. (1970) The dermal toxicity of phenol: an investigation of the most effective first-aid measures. *Br. J. Ind. Med.* **27** 155–159.

Cornish, H.H. and Ryan, R.R. (1965) Metabolism of benzene in non-fasted, fasted and arylhydroxylase inhibited rats. *Toxicol. Appl. Pharmacol.* **75** 767–771.

Courtney, K.D., Andrews, J.E., Springer, T. *et al.* (1986) A perinatal study of toluene in CD-1 mice. *Fundam. Appl. Toxicol.* **6** 145–154.

Cumming, R.B. and Michaud, T.A. (1979) Mutagenic effects of inhaled ethylene oxide in male mice. *Environ. Mutagen.* **1** 166–167.

Cunningham, M.L. and Matthews, H.B. (1991) Relationship of hepatocarcinogenicity and hepatocellular proliferation induced by mutagenic noncarcinogens vs. carcinogens. *Toxicol. Appl. Pharmacol.* **110** 505–513.

Dansette, P.M., Makedonska, V.B. and Jerina, D.M. (1978) Mechanism of catalysis for the hydration of substitution styrene oxides by hepatic epoxide hydrase. *Arch. Biochem. Biophys.* **187** 219–228.

De Flora, S., Zannachi, P., Camoira, A., Bennicelli, C. and Badolati, G.S. (1984) Genotoxic activity and potency of 135 compounds in the Ames reversion test and in a bacterial DNA-repair test. *Mutat. Res.* **133** 161–198.

Deichmann, W.B. (1944) Phenol studies. V. The distribution, detoxification, and excretion of phenol in the mammalian body. *Arch. Biochem.* **2** 345–355.

Deichmann, W.B. (1949) Local and systemic effects following skin contact with phenol. A review of the literature. *J. Ind. Hyg. Toxicol.* **31** 146–154.

Deichmann, W.B. and Keplinger, M.L. (1963) Phenols and phenolic compounds. In Patty, F.A. (ed.) *Industrial Hygiene and Toxicolgy*, pp. 2567–2627. Interscience Publishers, New York.

Deichmann, W.B., Witerup, S. and Dierker, M. (1952) Phenol studies. XII. The percutaneous and alimentary absorption of phenol by rabbits with recommendations for the removal of phenol from the alimentary tract or skin of persons suffering exposure. *J. Pharmacol. Exp. Ther.* **105** 265–272.

Delbressine, L.P.C. (1981) Stereoselective oxidation of styrene to styrene oxide in rats as measured by mercaptiric acid excretion. *Xenobiotica* **11** 589–594.

Denk, B., Filer, J.G., Kessler, W., Shen, J. and Oesterle, D. (1990) Dose-dependent emergence of peneoplastic foci in rat livers after exposure to 2-nitropropane. *Arch. Toxicol.* **64** 329–331.

Desi, I. Kovacs, F., Zahumenszky, Z. and Ralogh, A. (1967) Maze learning in rats exposed to xylene intoxication. *Psychopharmacolgia* **11** 224–230.

De Vries, T., Tonkelaar, E.M. den., Danse, L.H.J.C. and Leeuwen, F.X.R. van (1983) Onderzoek naar de subacute toxiciteit van xyleen bij ratten. Report nr. 61810001. National Institute of Public Health and Environmental Protection, Bilthoven, The Netherlands.

DGMK (1985) *Effects of Toluene on Man and Animals.* DGMK project 174-7, Hamburg (Deutsche Gesellschaft für Minerolwissenschaft und Kohlechemie EV).

Divincenzo, G.D., Yanna, F.J. and Astill, B.D. (1972) Human and canine exposure to methylene chloride vapor. *Amer. Indus. Hyg. Assoc. J.* **33** 125–35.

Divicenzo, G.D. and Kaplan, C.J. (1981) Effect of exercise or smoking on the uptake, metabolism, and excretion of methylene chloride vapor. *Toxicol. Appl. Pharmacol.* **59**, 141–148.

Doctor, H.J. and Zielhuis, R.L. (1967) Phenol excretion as a measure of benzene exposure. *Ann. Occup. Hyg.* **10** 317–326.

Donner, M., Maki-Paolkakanen, J., Norpa, N., Sorsa, M. and Vainio, H. (1980) Genetic toxicology of xylenes. *Mutat. Res.* **74** 170–171.

Dudek, B., Gralewicz, K., Jakubowski, M., Kostrzewski, P. and Sokal, J. (1990) Neuro behavioural effects of experimental exposure to toluene, xylene and their mixture. *Pol. J. Occupat. Med.* **3**(1) 109–116.

Dutkiewicz, T. and Tyras, H. (1967) A study of the skin absorption of ethylbenzene in man. *Br. J. Ind. Med.* **24** 330–332.

Dutkiewicz, T. and Tyras, H. (1968) Skin absorption of toluene, styrene and xylene by man. *Br. J. Ind. Med.***25** 243–246.

Duverger-Van Bogaert, M. *et al.* (1978) Determination of oxide synthetase and hydratase activities by a new highly sensitive gas chromatographic method using styrene and styrene-oxide as substrates. *Biochim. Biophys.* **526** 77–84.

Egle, J.L. and Gochberg, B.J. (1976) Respiratory retention of inhaled toluene and benzene in the dog. *J. Toxicol. Environ. Health* **1** 531–538.

Ehrenberg, L., Hiesche, K.D., Osterman-Golkar, S. and Wennberg, I. (1974). Evaluation of genetic risks of alkylating agents: tissue doses in the mouse from air contaminated with ethylene oxide. *Mutat. Res.* **24** 83–103.

El Masri, A.M., Smith, J.N. and Williams, R.T. (1956) Studies in detoxication. The metabolism of alkylbenzenes, *n*-propylbenzene, and *n*-butylbenzene with further observations on ethylbenzene. *Biochem. J.* **64** 50–56.

Elovaara, E., Engström, K. and Vainio, H. (1984). Metabolism and disposition of simultaneously inhaled *m*-xylene and ethylbenzene in the rat. *Toxicol Appl. Pharmacol.* **75** 446–478.

Engström, K. (1978) Styrene in subcutaneous adipose tissue after experimental and industrial exposure. *Scand. J. Work Environ. Health* **4** 119–120.

Engström, K. and Bjurström, R. (1977) Exposure two methylene chloride. Content in subcutaneous adipose tissue. *Scand. J. Work Environ. Health* **3**, 215–224.

Engström, K., Haerkoenen, H., Kalliokoski, P. and Rantanen, J. (1976) Urinary mandelic acid concentration after occupational exposure to styrene and its use as a biological exposure test. *Scand. J. Work. Environ. Health* **2** 21–26.

Engström, K., Astrand, I. and Wigaeus, E. (1978a) Exposure to styrene in a polymerization plant. Uptake in the organism and concentration in subcutaneous adipose tissue. *Scand. J. Work Environ. Health* **4** 324–329.

Engström, K., Bjurström, R., Astrand, I. and Oevrum, P. (1978b) Uptake distribution and elimination of styrene in man. *Scand. J. Work Environ. Health* **4**, 315–323.

EPA (1982) Health assessment document for dichloromethane (methylene dichloride). *US EPA-6000/8-82-004.* PB82 185257, cited by Van Apeldoorn *et al.* (1987)

EPA (1983) Health assessment document for toluene. *US EPA-600/8-82-008F (Final report).* NTIS, Springfield, USA. PB 84-100056.

EPA (1984) Health assessment document for chlorinated benzenes. *US EPA-600/8-84-015A. (Final report).*

Ernst, M.R., Klesmer, R., Huebner, R.A. and Martin, J.E. (1961) Susceptibility of cats to phenol. *J. Am. Vet. Med. Assoc.* **138** 197–199.

Falk, A., Löf, A., Hagberg, M., Wigaeus, E., Hjelm, W.E. and Wang, Z. (1991) Human exposure to 3-carene by inhalation: Toxicokinetics effects, effects on pulmonary function and occurrence of irritation and CNS symptoms. *Toxicol. Appl. Pharmacol.* **110** 198–205.

Fernandez, J.G. and Caperos, J.R. (1977) Exposition au styrene. I. Etude experimentale de l'absorption et de l'excretion pulmonaire sur des sujets humains. *Int. Arch. Occup. Environ. Health* **40** 1–12.

Fielder, R.J. (1982) *Toxicity Review No. 4: Benzene.* Health and Safety Services, Her Majesty's Stationery Office, London.

Fiola, E.S., Czermiak, R., Castonquay, A., Conaway, C.C. and Rivenson, A. (1987a) Assay of 1-nitropropane, 2-nitropane, 1-azoxypropane and 2-azoxypropane for carcinogenicity by gavage in Sprague-Dawley rats. *Carcinogenesis* **8**, 1947–1949.

Fiola, E.S., Conaway, C.C., Biles, W.T. and Johnson, B. (1987b) Enhanced mutagenicity of 2-nitropropane nitronate with respect to 2-nitropropane—possible involvement of free radical species. *Mutat. Res.* **179** 15–22.

Fishbein, I. (1985a) An overview of environmental and toxicological aspects of aromatic hydrocarbons. II. Toluene. *Sci. Total Environ.* **42** 267–288.

Fishbein, L. (1985b) An overview of environmental and toxicological aspects of aromatic hydrocarbons. III Xylene. *Sci. Total Environ* **43** 165–183.

Flickinger, C.W. (1976) The benzenediols: catechol, resorcinol and hydroxyquinone; a review of the industrial toxicology and current industrial exposure limits. *Am. Ind. Hyg. Assoc. J.* **37** 596–606.

Gad-El-Karim, N.N., Harper, B.L. and Legator, M.S. (1984) Modifications in the myeloclastogenic effect of benzene in mice with toluene, phenobarbital, 3-methylcholanthrene, Arochlor 1254 and SKF-525A. *Mutat. Res.* **135** 225–243.

Garton, G.A. and Williams, R.T. (1949) Studies in detoxication. The fates of phenol, phenylsulphuric acid and phenylglucuronides in the rabbit in relation to the metabolism of benzene. *Biochem. J.* **45** 158–163.

Gbodi, T.A. and Oehme, F.W. (1978) The fate of phenol, *o*-phenolphenol and diisophenol in rats. *Toxicol. Appl. Pharmacol.* **45** 219–362.

George, E., Burlinson, B. and Gatehouse, D. (1989). Genotoxicity of 1- and 2-nitropropane in the rat. *Carcinogenesis* **10**, 2329–2334.

Gerarde, H.W. and Ahlstrom, D.B. (1966) Toxicologic studies on hydrocarbons. XI. Influence of dose on the metabolism of mono-*n*-alkyl derivatives of benzene. *Toxicol. Appl. Pharmacol.* **9** 185–190.

Gessner, P.K., Parke, D.V. and Williams, R.T. (1961) The metabolism of ^{14}C-labelled ethylene glycol. *Biochem. J.* **79** 482–489.

Gill, D.P. and Ahmed, A.E. (1981) Covalent binding of [^{14}C] benzene to cellular organelles and bone marrow nucleic acids. *Biochem. Pharmacol* **30** 1127–1131.

Gill, D.P., Kempen, R.R., Nash, J.B. and Ellis, S. (1979) Modifications of benzene myelotoxicity and metabolism of phenobarbital, SKF-525A and 3-methylcholanthrene. *Life Sci.* **25** 1633–1640.

Gomez-Arroyo, S. and Villalobos-Pietrini, R. (1981) Induction of chromosome abnormalities in *Vicia faba* by a thinner containing toluene, benzene and *n*-hexane. *Mutat. Res.* **85** 244–245.

Greenlee, W.F., Gross, E.A. and Irons, R.D. (1981) Relationship between benzene toxicity and the disposition of ^{14}C-labelled benzene metabolites in the rat. *Chem. Biol. Interac.* **33** 285–299.

Griffin, T.B., Coulston, T. and Stein, A.A. (1980) Chronic inhalation exposure of rats to vapors of 2-nitropropane at 25 ppm. *Ecotoxicol. Environ. Safety* **4** 267–281.

Griffin, T.B., Stein, A.A. and Coulston, F. (1982) Inhalation exposure of rats to vapors of 1-nitropropane at 100 ppm. *Ecotoxicol Environ. Safety* **6** 268–282.

Groger, W.K.L. and Grey, T.F. (1979) Effect of chloroform on the activities of liver enzymes in rats. *Toxicol.* **14** 23–28.

Guillemin, M.P. and Bauer, D. (1979) Human exposure to styrene. III. Elimination kinetics of urinary mandelic and phenylglyoxylic acids after single experimental exposure. *Int. Arch. Occup. Environ. Health.* **44** 249–263.

Guo, N., Conaway, C.C., Hussain, N.S. and Fiola, E.S. (1990) Sex and organ differences in oxidative DNA and RNA damage due to treatment of Sprague-Dawley rats with acetoxime or 2-nitropropage. *Carcinogenesis* **11** 1659–1662.

Haas-Jobelius, M., Ziegler-Skylalakis, K. and Andrae, H. (1991) Nitroreduction in not involved in the genotoxicity of 2-nitropropane in cultured mammalian cells. *Mutagenesis* **6** 87–91.

Hadidian, Z., Fredrickson, T.N., Weisburger, E.K., Weisburger, J.H., Glass, R.M. and Manlet, N. (1968) Tests for chemical carcinogens. Report on the activity of aromatic amines, nitrosamines, quinolines, amides, epoxides, aziridines and purine antimetabolites. *J. Nat. Cancer. Inst.* **41** 985–1025.

Harkonen, H. (1978) Styrene, its experimental and clinical toxicology. *Scand. J. Work Environ. Health.* **4** 104–113.

Hattula, M.L., Wasenius, M., Krees, R., Arstila, A.U. and Kihlström, M. (1981) Acute and short-term toxicity of 2,3,4,6-tetrachlorophenol in rats. *Bull. Environ. Contam. Toxicol.* **26** 795–800.

Haun, C.G. *et al.* (1972) Continuous animal exposure to low levels of dichloromethane. In *Proceedings of the 3rd Annual Conference on Environmental Toxicology*, pp. 199–208. Wright-Patterson Air Force Base, Ohio, Aerospace Medical Research Laboratory, AMRL-TR-130, paper no. 12, USA.

Hayes, W.C., Kirk, H.D., Gushow, T.S. and Young, J.T. (1988) Effect of inhaled propylene oxide on reproductive parameters in Fischer 344 rats. *Fund. Appl. Toxicol* **10** 82–88.

Heller, V.G. and Pursell, L. (1938) Phenol-contaminated waters and their physiological action. *J. Pharmacol. Exp. Ther.* **63** 99–107.

Hesse, J. M., Speijers, G.J.A. and Taalman, R.D.F.M. (1991) Integrated criteria document: chlorobenzenes effects. *Appendix to Report nr. 710401015.* National Institute of Public Health and Environmental Protection, Bilthoven, The Netherlands.

Heywood, R., Sortwell, R.J., Noel, P.R.B., Street, A.E. Prentice, D.E., Roe, F.J.C., Wadsworth, P.F., Worden, A.N. and Van Abbe, N.J. (1979) Safety evaluation of toothpaste containing chloroform. III. Long-term study in beagle dogs. *J. Environ. Pathol. Toxicol.* **2** 835–851.

Himnan, D.J. (1984) Tolerance and reserve tolerance to toluene inhalation: effects of openfield behaviour. *Pharmacol. Biochem. Behav.* **21** 625–631.

Hine, C. and Zuidema, H. (1970) The toxicological properties of hydrocarbon solvents. *Ind. Med. Surg* **39** 215–220.

Hinkel, G.K. and Kintzel, H.W. (1968). Phenol poisoning of a newborn through skin resorption. *Dtsch. Gesundheitsw.* **23** 2420–2422.

Hinz, G., Gohlke, P. and Burck, D. (1980). The effect of simultaneous oral application of ethanol and styrene. *J. Hyg. Epidemiol. Microbiol. Immunol.* **24** 262–263.

Hobara, T., Kobayashi, H., Higashihara, E. Kawamoto, T. and Sakai, T. (1984). Acute effects of 1,1,1-trichloroethane, trichloroethylene, and toluene and the hematologic parameters in dog. *Arch. Environ. Contam. Toxicol.* **13** 589–593.

Hollingworth, R.L., Rowe, V.K., Oyen, F., McCollister, D.D. and Spencer, H.C. (1956a) Toxicity of ethylene oxide determined on experimental animals. *Arch. Ind. Health* **13** 217–227.

Hollingworth, R.L., Rowe, V.K., Oyen, F., Hoyle, H.R. and Spencer, H.C. (1956b) Toxicity of paradichlorobenzene. *Am. Med. Assoc. Arch. Ind. Health* **14** 138–147.

Hollingworth, R.L., Rowe, V.K., Oyen, F., Torkelson, T.R. and Adams, E.M. (1958) Toxicity of *O*-dichlorobenzene. *Am. Med. Assoc. Arch. Ind. Health* **17** 180–187.

Horiguichi, S. and Inoue, K. (1977) Effects of toluene on the wheel-turning activity and peripheral blood findings in mice—an approach to the maximum allowable concentration of toluene. *J. Toxicol. Sci.* **2** 363–372.

Houston, J.B. and Cassidy, M.K. (1982) Sites of phenol conjugation in the rat in sulfate metabolism and sulfate conjugation. In Taylor, G.I. and Francis, T. (eds) *Proceedings of an International Workshop, Noordwijkerhout, The Netherland* 20–23 September 1981, pp. 270–278. Elsevier, Amsterdam.

Huang, C.C., Shih, F.S., Cheng, S.Y., Chen, S.S. and Tchen, P.H. (1991) *n*-Hexane polyneuropathy in a ball-manufacturing factory. *J. Occupat, Med.* **33** 139–142.
Hunter, C.G. and Blair, D. (1972). Benzene: Pharmacokinetic studies in man. *Ann. Occup. Hyg.* **15** 193–199.
Husain, R., Srivastava, S.T., Mushtaq, M. and Seth, P.K. (1980) Effect of styrene on levels of serotonin, noradrenalin and activity of acetylcholinesterase and monoamineoxidase in rat brain. *Toxicol. Lett.* **7** 47–50.
IARC (1982a) *IARC Monographs on the Evaluation of the Carcinogenic Risk of Chemical to Humans. Some Industrial Chemicals and Dyes.* Vol. 25, pp. 345–389. IARC, Lyon.
IARC (1982b) *IARC Monographs on the Evaluation of the Carcinogenic Risk of Chemicals to Humans. Benzene* Vol. 29. IARC Lyon.
IARC (1986) *IARC Monographs on the Evaluation of the Carcinogenic Risk of Chemicals to Humans. Final Draft on Methylene Chloride.* Vol. 41 IARC Lyon.
IARC (1987) *IARC Monographs on the Evaluation of the Carcinogenic Risks to Humans. Overall Evaluations of Carcinogenecity: an Updating of IARC Monographs.* Vols 1–42, suppl. 7 IARC, Lyon, France.
Ikeda, M. and Imamura, T. (1974) Evaluation of hippuric, phenylglyoxylic and mandelic acids in urine as indices of styrene exposure. *Int. Arch. Arbeitsmed.* **32** 93–101.
Ikeda, M. and Ohtsuji, H. (1971) Phenobarbital-induced protection against toxicity of toluene and benzene in the rat. *Toxicol. Appl. Pharmacol.* **20** 30–43.
IPCS/WHO (1986) Principles of toxicokinetics studies. *Environmental Health Criteria*, Vol. 57, pp. 11–166. WHO, Geneva, Switzerland.
Irons, R.D., Heck, H.A. and Moore, B.J. (1979) Effects of the principal of hydroxy metabolites of benzene on micro-tubule polymerization. *Toxicol. Appl. Pharmacol.* **51** 300–409.
Izmerov, N.F. (1984) (ed.) *Styrene. Scientific Reviews of Soviet Literature on Toxicity and Hazards of Chemicals*, Vol. 49. Pub Moscow.
Jacobson, K.H., Hackley, E.B. and Feinsilver, L. (1956) The toxicity of inhaled ethylene oxide and propylene oxide vapors. *Arch. Ind. Health* **13** 237–244.
Jaeger, R.J. *et al.* (1979) Toxicity and biochemical changes in rats after inhalation exposure to 1,1-dichloroethylene, bromobenzene, styrene, acrylonitrile or 2-chlorobutadiene. *Toxicol. Appl. Pharmacol.* **29** 81–82.
James, M.O., Fouts, J.R. and Bend, J.R. (1976). Hepatic and extrahepatic metabolism, *in vitro*, of an epoxide (8-^{14}C-styrene oxide) in the rabbit. *Biochem. Pharmacol.* **25** 187–193.
James, S.D. and White, D.A. (1967) The metabolism of phenetylbromide, styrene and styrene oxide in the rabbit and rat. *Biochem. J.* **104** 914–921.
Janus, J.A., Taalman, R.D.F.M. and Theelen, R.M.C. (1990) Integrated criteria document: chlorophenols effects. *Appendix To Report nr. 710401003.* National Institute of Public Health and Environmental Protection, Bilthoven, The Netherlands.
Johnson, R.L., Gehring, P.J., Kociba, R.J. and Schwertz, B.A. (1973). Chlorinated dibenzodioxins and pentachlorophenol. *Environ. Health Perspect.* **5** 171–175.
Johnstone, R.T. and Miller, S.E. (1960) *Occupational Diseases and Industrial Medicine.* W.B. Saunders and Co., Philadelphia.
Jones, A.R. and Wells, G. (1981). The comparative metabolism of 2-bromoethanol and ethylene oxide in the rat. *Xenobiotica* **11** 763–770.
Jörgensen, N.K. and Cohr, K.H. (1981) *n*-Hexane and its toxicologic effects. *Scand. J. Work. Environ. Health.* **7** 157–168.
Jörgenson, T.A. and Rushbrook, C.J. (1980) Effects of chloroform in the drinking water of rats and mice. Ninety-day subacute toxicity study. US. EPA-600/1-80-030, NTIS, PB80-219108, cited by Van der Heijden *et al.* (1987).
Jörgenson, T.A., Rushbrook, C.J. and Jones, D.C.L. (1982) Dose–response study of chloroform carcinogenesis in the mouse and rat: status report. *Environ. Health Perspect.* **46** 141–149.
Jörgenson, T.A. *et al.* (1984) *Carcinogenicity of Chloroform in Drinking Water to Male Osborne-Mendel Rats and Female B6C3FI Mice.* SRI International, 333 Ravenswood Avenue, Menlo Park, CA 95052, cited by Van der Heijden *et al.* (1987).
Kale, P.G. and Baum, J.W. (1983) Genetic effects of benzene in *Drosophila melanogaster* males. *Environ. Mutag.* **5** 223–226.
Kankaanpaa, J.T.J., Elovaara, E., Hemminki and Vaino, H. (1980) The effect of maternally

inhaled styrene on embryonal and foetal development in mice and chinese hamsters. *Acta Pharmacol. Toxicol.* **47** 127–129.

Kanz, M.F., Taj, Z. and Treinen Moslen, M. (1991) 1,1-Dichloroethylene hepatotoxicity: Hypothyroidism decreases metabolism and covalent binding but not injury in the rat. *Toxicol.* **70** 213–229.

Kao, J., Bridges, J.W. and Faulkner, J.K. (1979) Metabolism of [^{14}C]phenol by sheep, pig and rat. *Xenobiotica* **9** 141–147.

Kashin, L.M., Matotchenko, V.M., Malinina-Putsenko, V.P., Mikhailovskaia, L.F. and Shmuter, L.M. (1980). Experimental and clinico-hygienic investigations of methylene chloride toxicity. *Vrach. Delo.* **35** 100–103.

Kerkvliet, N.I., Baecher-Steppan, L. and Schmitz, J.A. (1982). Immunotoxicity of pentachlorophenol (PCP): increased susceptibility to tumor growth in adult mice fed technical PCP-contaminated diets. *Toxicol. Appl. Pharmacol.* **62** 55–64.

Khinkova, L. (1974). Experimental data on the toxicity of some organic solvents used in the furniture industry. *Tr. Inst. Khig. Okhr. Tr. Prof. Zabol.* **22** 133–140.

Kilburn, K.H., Seidman, B.C. and Warshaw, B. (1985) Neurobehavioural and spiratory symptoms of formaldehyde and xylene exposure in histology technicians. *Arch. Environ. Health* **40** 229–233.

Kimura, E.T., Ebert, D.M. and Dodge, P.W. (1971) Acute toxicity and limits of solvents residue for sixteen organic solvents. *Toxicol. Appl. Pharmacol.* **19** 699–704.

Kissling, M. and Speck, B. (1972) Further studies on experimental benzene induced aplastic aneamia. *Blut.* **25** 97–103.

Kliesch, L.L. and Adler, I.D. (1987) Micro nuclens test in bone marrow of mice treated with 1-nitropropane, 2-nitropropane and cisplatin. *Mutat. Res.* **192** 181–184.

Klimes, I., Zorad, S., Svabova, E. and Macho, L. (1987) Effect of *n*-hexane inhalation on insulin degradation in rat blood *in vitro*. *Endocrinologia Experimentalis* **21** 209–217.

Knudsen, I., Verschuuren, H.G., Tonkelaar, E.M. den., Kroes, R. and Helleman, P.F. (1974). Short-term toxicity of pentachlorophenol in rats. *Toxicol.* **2** 141–152.

Kociba, R.J., Leong, B.K.J. and Hefner, R.E. Jr. (1981) Subchronic toxicity study of 1,2,4-trichlorobenzene in the rat, rabbit and beagle dog. *Drug Chem. Toxicol.* **4** 229–249.

Koga, K. (1978) Distribution, metabolism and excretion of toluene in mice. *Folia Pharmacol. Jpn.* **74** 687–698.

Korzak, Z., Sokal, J.A., Wasiela, T. and Swiercz, R. (1990) Toxic effects of acute exposure to particular xylene isomers in animals. *Pol. J. Occupat. Med.* **3** 221–226.

Koster, H.J., Halsema, I., Scholtens, E., Knippers, M. and Mulder, G.J. Dose-dependent shifts in the sulfation and glucuronidation of phenolic compounds in the rat *in vivo* and in isolated hepatocytes. The role of saturation of phenol-sulfotransferase. *Biochem. Pharmacol.* **30** 2569–2575.

Kurnatowski, A. (1960) Pathomorphological changes of some inner organs of guinea pigs induced by the prolonged inhalation of phenol vapor. *Lodz. Towzrz. Naul. Wydzial.* IV **18** 43–57.

Kurppa, K. and Vainio, H. (1981) Effects of methylene dichloridemethane inhalation on blood carboxyhaemoglobin concentration and drug metabolizing enzymes in rat. *Res. Commun. Chem. Pathol. Pharmacol.* **32** 535–544.

Laborde, J.B. and Kimmel, C.A. (1980) The tetratogenecity of ethylene oxide administered intravenously to mice. *Toxicol. Appl. Pharmacol.* **56** 16–22.

Lam, C.W., Galen, J.F. and Pierson, D.L. (1990) Mechanism of transport and distribution of organic solvents in blood. *Toxicol. Appl. Pharmacol.* **104** 117–129.

Leibman, K.C. (1975) Metabolism and toxicity of styrene. *Environ. Health Perspect.* **11** 115–119.

Leibman, K.C. and Ortiz, E. (1969) Oxidation of styrene in liver microsomes. *Biochem. Pharmacol.* **18** 552–554.

Lewin, J.F. and Cleary, W.T. (1982) An accidental death caused by the absorption of phenol through skin. A case report. *Forensic. Sci. Int.* **19** 177–179.

Lewis, T.R. and Tatken, R.L. (1982) *Registry of Toxic Effects of Chemicals Substances.* NOSH, WL 3675000.

Lewis, T.R., Ulrich, C.E. and Busey, W.M. (1979) Subchronic inhalation toxicity of nitromethane and 2-nitropropane. *J. Environ. Pathol. Toxicol.* **2** 233–249.

Liang, J.C., Hsu, H.C. and Henry, J.E. (1983) Cytogenetic assays for mitotic poisons: the grasshopper embryo system for volatile liquids. *Mutat. Res.* **113** 467–479.

Liao, J.T.F. and Oehme, F.W. (1980) Literature reviews of phenolic compounds. I. Phenol. *Vet. Hum. Toxicol.* **22** 160–164.

Liao, J.T.F. and Oehme, F.W. (1981) Tissue distribution and plasma protein binding of [^{14}C]phenol in rats. *Toxicol. Appl. Pharmacol.* **57** 220–225.

Löf, A., Gullstrand, E. and Nordqvist, M.B. (1983) Tissue distribution of styrene, styrene glycol and more polar styrene metabolites in the mouse. *Scand. J. Work Environ. Health* **9** 419–430.

Löf, A., Gullstrand, E., Lundgren, E. and Nordqvist, M.B. (1984). Occcurrence of styrene-7,8-oxide and styrene glycol in the mouse after the administration of styrene. *Scand. J. Work Environ. Health* **10** 179–187.

Loose, L.D., Silkworth, J.B., Pittman, K.A., Benitz, K.F. and Mueller, W. (1978a) Impaired host resistance to endotoxin and malaria in polychlorinated biphenyl- and hexachlorobenzene-treated mice. *Infect. Immun.* **20** 30–35.

Loose, L.D., Pittman, K.A., Benitz, K.F., Silkworth, J.B., Mueller, W. and Coulston, F. (1978b) Environmental chemical-induced immune dysfunction. *Ecotoxicol. Environ. Safety* **3** 173–198.

Loprieno, N., Presciuttini, S., Sbrana, I., Stretti, G., Zaccaro, L., Abbondandolo, A., Bonatti, S. and Mazzaccaro, A. (1978) Styrene and styrene oxide. II. Point mutations, chromosome aberrations and DNA repair induction analyses *Scand. J. Work Environ. Health* **4** 169–178.

Lorenzana-Jimenez, M. and Sallas, M. (1983) Neotanal effects of toluene on the locomotor behavioural development of the rat. *Neurobehav. Toxicol. Teratol.* **5** 295–299.

Low, C.K., Meeks, J.R. and Mackerer, C.R. (1989) Health effects of the alkylbenzenes. II xylenes. *Toxicol. Indust.* **5** 85–105.

Mahrlein, W., Bergert, K.D., Frenzel, E., Frenzel, M. and Niederdorf, A.H.Z. (1983) Zur inhalativen Belastung durch Styren. *Arbeitsmed. Inf.* **10** 17–18.

McChesney, E.W., Goldberg, L., Parekh, C.K., Russell, J.C. and Min, B.N. (1971) Reappraisal of the toxicity of ethylene glycol. II. Metabolism studies in laboratory animals. *Food Cosmet. Toxicol.* **9** 21–30.

McCollister, D.D., Lockwood, D.T. and Rowe, V.K. (1961) Toxicologic information on 2,4,5-trichlorophenol. *Toxicol. Appl. Pharmacol.* **3** 63–70.

McKenna, M.J., Zempel, J.A. and Braun, W.H. (1982) The pharmacokinetics of inhaled methylene chloride in rats. *Toxicol. Appl. Pharmacol.* **65** 1–10.

Mehta, R., Hirom, P.C. and Millburn, P. (1978) The influence of dose on the pattern of conjugation of phenol and 1-naphthol in non-human primates. *Xenobiotica* **8** 445–452.

Mikiska, A. (1960) Determination of the electrical excitability of the cerebral motor cortex and its use in pharmacology and toxicology. I. Methods, control experiments and physiological foundation. *Arch. Gewerbe Pathol. Gwerbe Hyg.* **19** 286–299.

Mikiskova, H. (1960) Determination of the electrical excitability of the cerebral motor cortex and its use in pharmacology and toxicology. II. The effects of benzene, toluene and xylol in guinea pigs. *Arch. Gewerbe Pathol. Gewerbe Hyg.* **19** 300–309.

Miller, J.J., Powell, G.M., Olavesen, A.H. and Curtis, C.G. (1976) The toxicity of dimethoxyphenol and related compounds in the cat. *Toxicol. Appl. Pharmacol* **38** 47–57.

Minor, J.K. and Becker, B.A. (1971) A comparison of the teratogenic properties of sodium salycate, sodium benzoate and phenol. Abstracts 10th Annual Meeting Society Toxicity No. 40. *Toxicol. Appl. Pharmacol.* **19** 373–374.

Misbin, R.I., Almira, E.C. and Cleman, M.W. (1981) Insulin degradation in serum of a patient with apparent insulin resistance. *J. Clin. Endocr. Metab.* **52** 177–180.

Morimoto, K., Koizumi, A., Tachibana, Y. and Dobashi, Y. (1976) Inhibition of repair of radiation-induced chromosome breaks: effect of phenol in cultured human leukocytes. *Jap. J. Ind. Health.* **18** 478–479.

Morimoto, K., Wolff, S. and Koizumi, A. (1983) Induction of sister-chromatid exchanges in human lymphocytes by microsomal activation of benzene metabolites. *Mutat. Res.* **119** 355–360.

Morisset, Y., P'An, A. and Jegier, Z. (1979) Effect of styrene and fibre glass on small airways of mice. *J. Toxicol. Environ. Health* **5** 943–956.

Morley, R., Eclestron, D.W., Douglas, C.P., Greville, W.E., Scott, D.J. and Anderson, H. (1970) Xylene poisoning: A report on one fatal case and two cases of recovery after prolonged unconsciousness. *Brit. Med. J.* **3** 442–443.

Moser, V.C. and Balster, R.L. (1985) Acute motor and lethal effects of inhaled toluene, 1,1,1-trichloroethane, halothane, and ethanol in mice: effects of exposure duration. *Toxicol. Appl. Pharmacol.* **77** 285–291.

Moskowitz, S. and Shapiro, H. (1952) Fatal exposure to methylene chloride vapor. *Am. Ind. Hyg. Occup. Med.* **6** 116–123.

Munson, A.E., Sain, L.E., Sanders, V.M. Fauffmann, B.M., White, K.L., Page, D.G., Barnes, D.W. and Borzelecca, F. (1982) Toxicology of organic drinking water contaminants: trichloromethane, bromodichloromethane, chlorodibromomethane and tribromomethane. *Environ. Health Perspect.* **46** 117–126.

Murray, F.J., John, J.A., Balmer, M.F. and Schwetz, B.A. (1978) Teratologic evaluation of styrene given to rats and rabbits by inhalation or by gavage. *Toxicol.* **11** 335–343.

Mutti, A., Falzoi, M., Romanelli, A. and Franchini, I. (1984) Regional alterations of brain catecholamines by styrene exposure in rabbits. *Arch. Toxicol.* **55** 173–177.

Nawrot, P.S. and Staples, R.E. (1979) Embryo-fetal toxicity and teratogenicity of benzene and toluene in the mouse (abst.). *Teratol.* **19** 41A.

NCI (1976) Report on carcinogenesis bioassays of chloroform. *NTIS PB-264-018*, as cited by Van der Heijden *et al.* (1987).

NCI (1979) Bioassay of 2,4,6-T3CP for possible carcinogenicity. *NCI-CG-TR-155*, National Cancer Institute, Technical report series No. 155. US Department of Health, Education and Welfare, Public Health Service, National Institute of Health, Bethesda DHew publications No. (NIH) 79-1711.

Nitsche, K.D. *et al.* (1982) *Methylene Chloride: A Two Year Inhalation Toxicity and Oncogenicity Study.* Toxicology Research Laboratory, Health and Environmental Sciences, Dow Chemicals, Midland, Michigan, USA, cited by Van Apeldoorn *et al.* (1987).

Nomiyama, K. and Nomiyama, H. (1974a) Respiratory retention, uptake and excretion of organic solvents in man. *Int. Arch. Arbeitsmed.* **32** 75–83.

Nomiyama, K. and Nomiyama, H. (1974b) Respiratory elimination of organic solvents in man. *Int. Arch. Arbeitsmed.* **32** 85–91.

Norpoth, K., Witting, U. and Springorum, M. (1974) Induction of microsomal enzymes in the rat liver by inhalation of hydrocarbon solvents. *Int. Arch. Arbeitsmed.* **33** 315–321.

NTP (1984) Toxicology and carcinogenesis studies of propylene oxide in F344/N rats and B63FI mice (inhalation studies). *National Toxicity Program Technical Reports Series, No. 267*, NIH Publication No. 35-2527, Research Triangle Park; North Carolina, USA.

NTP (1985a) Toxicology and carcinogenesis studies of chlorobenzene (CAS Bo. 108-90-7) in F344/N rats and B6C3FI mice (gavage studies). *National Toxicology Program, Technical Reports Series, No. 261.* NIH Publication No. 86-2517, US Department of Health and Human Services.

NTP (1985b) Toxicology and carcinogenesis studies of 1,2-dichlorobenzene (*o*-dichlorobenzene, CAS No. 95-50-1) in F344/N rats and B6C3FI mice (gavage studies). *National Toxicology Program, Technical Reports Series, No 255.* NIH Publication No. 86-2511, US Department of Health and Human Services.

NTP (1987) Toxicology and carcinogenesis studies of 1,4-dichlorobenzene (*p*-dichlorobenzene, CAS No. 106-46-7) in F344/N rats and B6C3FI mice (gavage studies). *National Toxicology Program, Technical Reports Series, No. 319.* NIH Publication No. 87-2575, US Department of Health and Human Services.

NTP (1989a) Toxicology and carcinogenesis studies of 2,4-dichlorophenol (CAS No. 120-83-2) in F344/N rats and B6C3FI mice (feed studies). *National Toxicology Program Technical Report Series No. 353 June 1989.* US Department of Health and Human Services, Public Health Service, National Institute of Health Service, National Institutes of Health. NIH Publication No. 89-2808.

NTP (1989b) Toxicology and carcinogenesis studies of two pentachlorophenol technical-grade mixtures (CAS No. 87-86-5) in B6C3FI mice (feed studies). *National Toxicology Program, Technical Report Series No. 349, March 1989.* US Department of Health and Human Services, Public Health Service, National Institute of Health. NIH Publication No. 89-2804.

Nylen, P., Ebendal, T., Eriksdotk., Nilsson, M., Hanson, T.P., Henschen, A., Johnson, A.C., Kronevi, T., Kvist, U., Sjorstrand, N.O., Hoglund, G. and Olson, L. (1989) Testicular atrophy and loss of nerve growth factor-immunoreactive germ cell line in rats exposed to *n*-hexane and a protective effect of simultaneous exposure to toluene or xylene. *Arch. Toxicol.* **63** 296–307.

Oedkvist, L.M., Larsby, B., Tham, R. and Hyden, D. (1982) Vestibulo-oculomotor disurbances caused by industrial solvents. *Acta Otolaryngol.* **94** 487–493.

Oehme, F.W. and Davis, L.E. (1970) The comparative toxicity and biotransformation of phenol. *Toxicol. Appl. Pharmacol.* **17** 283–285.

Ohashi, Y., Nakai, Y., Harada, H., Horiguchi, S. and Teramoto, K. (1982) Studies on industrial styrene poisoning (part XIII), Experimental studies about the damage on the mucosa membrane of respiratory tracts of rats exposed to styrene; contemporary observations of the changes in ciliary functions and fine structure. *Jpn. J. Ind. Health* **24** 360–372.

Ohtsuji, H. and Ikeda, M. (1971) Metabolism of styrene in the rat and the stimulation effect of phenobarbital. *Toxicol. Appl. Pharmacol.* **18** 321–328.

Ohtsuji, H. and Ikeda, M. (1972) Quantitative relationship between atmospheric phenol vapour and phenol in the urine of employees in Bakelite Factories. *Br. J. Ind. Med.* **29**, 70–73.

Osterman-Golkar, S. (1983) Tissue doses in man: implications in risk assessment. In Hayes, A.W. (ed.) *Developments in the Science and Practice of Toxicology; Proceedings of the Third International Congress on Toxicology*, San Diego, California.

Pacifici, G.M. and Rane, A. (1982) Metabolism of styrene-oxide in different human fetal tissues. *Drug Metab. Dispos.* **10** 302–305.

Pacifici, G.M., Collizi, C., Giuliani, L. and Rane, A. (1983a) Cytosolic epoxide hydrolase in fetal and adult human liver. *Arch. Toxicol.* **54** 331–341.

Pacifici, G.M., Lindberg, B., Glaumann, H. and Rane, A. (1983b) Styrene oxide metabolism in Rhesus monkey liver: enzyme activities in subcellular fractions and in isolated hepatocytes. *J. Pharmacol. Exp. Ther.* **226** 869–875.

Painter, R.B. and Howard, R. (1982) The hela DNA-synthesis inhibition as a rapid screen for mutagenic carcinogens. *Mutat. Res.* **92** 427–437.

Pantarotto, C. *et al.* (1982) Arene oxides in styrene metabolism, a new perspective in styrene toxicity. *Scand. J. Work Environ. Health* **4** 67–77.

Parke, D.V. and Williams, R.T. (1953) Studies in detoxication—The metabolism of benzene containing ^{14}C benzene. *Biochem. J.* **54** 231–238.

Parkki, M.G., (1978) The role of glutathione in the toxicity of styrene. *Scand. J. Work Environ. Health* **4** 53–59.

Paterson, S. and Mackay, D. (1989) Correlation of tissue, blood and air partition coefficients of volatile organic chemicals. *Br. J. Indus. Med.* **46** 321–328.

Peterson, R.G. and Bruckner, J.V. (1978) Measurement of toluene levels in animals tissues. In Sharp, C.W. and Carrol, L.T., (eds) *Voluntary Inhalations of Industrial Solvents*, Vol. 24, pp. 33–42. National Institute for Drug Abuse, Rockville, MD, USA.

Pericin, C. and Thomann, P. (1979) Comparison of the acute toxicity of clioquinol, histamine and chloroform in different strains of mice. *Arch. Toxicol. Suppl.* **2** 371–373.

Pero, R.W., Bryngelsson, T., Hoegstedt, and Akeson, B. (1982) Occupational and *in vitro* exposure to styrene assessed by unscheduled DNA synthesis in resting human lymphocytes. *Carcinogenesis* **3** 681–685.

Pezzoli, G., Barbieri, S., Ferrante, C., Zecthinella, A. and Foa, K. (1989) Parkinsonism due to *n*-hexane exposure. *Lancet* **2** 874–880.

Pezzoli, G., Ricardi, S., Massotto, C., Mariani, C.B. and Carenzi, A. (1990). n-Hexane induces Parkinsonism in rodents. *Brain Res.* **531** 355–357.

Piotrowski, J.K. (1971) Evaluation of exposure to phenol, absorption of phenol vapour in the lungs and through the skin and excretion of phenol in urine. *Br. J. Ind. Med.* **28**, 172–178.

Plotnick, H.B. and Weigel, W.W. (1979) Tissue distribution and excretion of ^{14}C-styrene in male and female rats. *Res. Common. Chem. Pathol. Pharmacol.* **24** 515–523.

Poirier, M.C., De Cicco, B.T. and Lieberman, M.W. (1975) Nonspecific inhibition of DNA repair synthesis by tumor promoters in human diploid fibroblasts damaged with *N*-acetoxy-2-acetylaminofluorene. *Cancer Res.* **35** 1392–1397.

Pryor, G.T., Dickson, J., Feeney, E. and Rebert, C.S. (1984) Hearing loss in rats first exposed to toluene as weanlings or as young adults. *Neurobehav. Toxicol. Teratol.* **6** 111–119.

Pullin, T.G., Pinkerton, M.N., Johnston, R.V. and Killan, D.J. (1978) Decontamination of the skin of swine following phenol exposure: a comparison of the relative efficacy of water versus polyethylene glycol/industrial methylated spirits. *Toxicol. Appl. Pharmacol.* **43** 63–76.

Putz, V.R., Johnston, B.L. and Setzer, J.V. (1976) A comparative study of the effects of carbon monoxide and methylene chloride on human performance. *J. Environ. Pathol. Toxicol.* **2** 97–112.

Pyykko, K., Tahti, H. and Vapaatalo, H. (1977) Toluene concentrations in various tissues of rats after inhalation and oral administration. *Arch. Toxicol.* **38**, 169–176.

Ramli, J.B. and Wheldrake, J.F. (1981) Phenol conjugation in the desert hopping mouse, *Notomys alexis. Comp. Biochem. Physiol. C.* **69**, 379–381.

Ramsey, J.C. and Andersen, M.E. (1984) A physiological based description of the inhalation pharmacokinetics of styrene in rats and humans. *Toxicol. Appl. Pharmacol.* **73** 159–175.

Ramsey, J.C. and Young, J.D. (1978) Pharmacokinetics of inhaled styrene in rats and humans. *Scand. J. Work. Environ. Health* **4** 84–91.

Ramsey, J.C., Young, J.D., Karbowski, R.J., Chenoweth, M.B., McCarty, L.P. and Braun, W.H. (1980) Pharmacokinetics of inhaled styrene in human volunteers. *Toxicol. Appl. Pharmacol.* **53** 54–63.

Rao, K.S., Johnston, K.A. and Henck, J.W. (1982) Subchronic dermal toxicity study of trichlorobenzene in the rabbit. *Drug Chem. Toxicol.* **5** 438–444.

Restani, P. and Galli, G.L. (1991) Oral toxicity of formaldehyde and its derivatives. *Crit. Rev. Toxicol.* **21** 315–328.

Reuber, M.D. (1979) Carcinogenicity of chloroform. *Environ. Health Perspect.* **31** 171–182.

Rickert, D.E., Baker, T.S., Bus, J.S., Barrow, C.S. and Irons, R.D. (1979). Benzene disposition in the rat after exposure by inhalation. *Toxicol. Appl. Pharmacol.* **49**, 417–423.

Riihimaki, V. and Pfaffli, P. (1978) Percutaneous absorption of solvent vapors in man. *Scand. J. Work Environ. Health* **4** 73–85.

Robbiano, L., Mattioli, F. and Brambilla, G. (1991) DNA fragmentation by 2-nitropropane in rat tissues and effects of the modulation of biotransformation processes. *Cancer Lett.* **57**, 61–66.

Robert, C.S., Sorenson, S.S., Howd, R.A. and Pryor, B.T. (1983) Toluene-induced hearing loss in rats evidenced by the brainstem auditory-evoked response. *Neurbehav. Toxicol. Teratol.* **5** 59–62.

Robertson, P. Ir., White, E.L. and Bus, J.S. (1989) Effects of methyl ethyl ketone pretreatment on hepatic mixed function oxidase activity and on *in vivo* metabolism of *n*-hexane. *Xenobiotica* **19**(7) 721–729.

Robinson, K.S., Kavlock, R.J., Chernoff, N. and Gray, L.E. (1981) Multigeneration study of 1,2,4-trichlorobenzene in rats. *J. Toxicol. Environ. Health* **8** 489–500.

Roscher, E., Ziegler-Skylakakis, E.L. and Aotrae, U. (1990) Involvement of different pathways in the genotoxicity of nitropropanes in cultured mammalian cells. *Mutagenesis* **5** 375–380

Rowe, V.K., Hollingworth, R.L., Oyen, J., McCollister, D.D. and Spencer, H.C. (1956) Toxicity of propylene oxide determined on experimental animals. *Arch. Ind. Health* **13** 228–236.

RTECS (1989) *Registry of Toxic Effects of Chemical Substances.* US Department of Health and Human Services, Public Health Service, Center for Disease Control, National Institute for Occupational Safety and Health.

Ruedemann, R. and Deichmann, W.B. (1953) Blood phenol levels after topical application of phenol containing preparations. *JAMA* **152** 506–509.

Ryan, A.J., James, M.O., Ben-Zvi, Z., Law, F.C.P. and Bend, J.R. (1976) Hepatic and extrahepatic metabolism of ^{14}C-styrene oxide. *Environ. Health Perspect.* **17** 135–144.

Sagai, M. and Tappel, A.L. (1979) Lipid peroxidation induced by some halomethanes as measured by *in vivo* pentane production in the rat. *Toxicol. Appl. Pharmacol.* **49** 283–291.

Sahu, S.C. and Lowther, D.K. (1981) Pulmonary reactions to inhalation of methylene chloride: Effects on lipid peroxidation. *Toxicol. Lett.* **8** 253–256.

Sahu, S., Lowther, D. and Ulsamer, A. (1980) Biochemical studies on pulmonary response to inhalation of methylene chloride. *Toxicol. Lett.* **7** 41–45.

Saita, G. (1976) Benzene induced hypoplastic anemias and leukemias. *Excerpta Medica Amsterdam* **32** 127–146.

Salmona, M., Pachecka, J., Cantoni, L., Belvedere, G., Mussini, E. and Garattini, S. (1976) Microsomal styrene monooxygenase and styrene epoxide hydrase activities in rats. *Xenobiotica* **6** 585–591.

Sandage, C. (1961) Tolerance criteria for continuous inhalation exposure to toxic material. *ASD Technical Report 61-519 (I)* pp. 1–29. Mo Midwest Research Institute, Kansas City. Ref.: NIOSH (1976) nr. 180.

Sandell, J., Parkki, M.G., Marniemi, J. and Aitio, A. (1978) Effects of inhalation and cutaneous exposure to styrene on drug metabolizing enzymes in the rat. *Res. Commun. Chem. Pathol. Pharmacol.* **19** 109–118.

Sandmeyer, E.E. (1982) Aromatic hydrocarbons. In Clayton, G.D. Clayton, F.E. (eds) *Patty's Industrial and Hygiene Toxicology*, pp. 3260–3283, John Wiley and Sons, New York.

Sasmore, D.P., Mitoma, C., Tyson, C.A. and Johnston, J.S. (1983) Subchronic inhalation toxicity of 1,3,5-trichlorobenzene. *Drug. Chem. Toxicol.* **6** 241–258.
Sato, A. and Nakajima, T. (1997) A structure–activity relationship of some chlorinated hydrocarbons. *Arch. Environ. Health.* **34** 69–75.
Savolainen, H. and Vainio, H. (1977) Organ distribution and nervous system binding of styrene and styrene oxide. *Toxicol* **8** 135–141.
Savolainen, H. Helojoki, M. and Tengen-Junnila, M. (1980) Behavioural and glial cell effects of inhalation exposure to styrene vapour with special reference to interactions of simultaneous peroral ethanol intake. *Acta Pharmcol. Toxicol.* **46** 51–56.
Sbrana, I., Lascialfari, D., Rossi, A.M., Loprieno, N., Bianchi, M. Tortoreto, M. and Pantarotto, C. (1983) Bone marrow cell chromosomal aberrations and styrene biotransformation in mice given styrene on a repeated oral schedule. *Chem. Biol. Interact.* **45** 349–357.
Schaumburg, H.H. and Spencer, P.S. (1978) Environmental carbons produce degeneration in cat hypothalamus and optic tract. *Science* **199** 199–200.
Schaumburg, H.H. and Spencer, P.S. (1979) Clinical and experimental studies of distal axanopathy—a frequent form of brain and nerve damage produced by environmental chemical hazards. *Ann. R. Y. Acad. Sci.* **329** 14–29.
Segerbäck, D. (1983). Alkylation of DNA and hemoglobin in the mouse following exposure to ethene and ethylene oxide. *Chem. Biol. Interact.* **45** 139–151.
Seppalainen, A.M., Laine, A., Salmi, T., Riihimaki, V. and Verlekala, E. (1989) Changes induced by short-term xylene exposure in human evoked potentials. *Int. Arch. Occup. Environ. Health* **61** 443–449.
Sexton, R.J. and Henson, E.V. (1949) Dermatological injuries by ethylene oxide. *J. Ind. Hyg. Toxicol.* **31** 297–300.
Sharma, R.P., Smith, F.A. and Gehring, P.J. (1981) Styrene epoxide intermediate as a possible stimulant of lymphocytic function. *J. Immunopharmacol.* **3** 67–78.
Sherwood, R.J. (1972a) Benzene: The interpretation of monitoring results. *Ann. Occup. Hyg.* **15** 409–421.
Sherwood, R.J. (1972b) Evaluation of exposure to benzene vapour during the loading of petrol. *Brit. J. Indus. Med.* **29** 65–69.
Sherwood, R.J. (1976) *Comparative methods of biologic monitoring of benzene exposure.* Lecture presented at the International Workship on Toxicology of Benzene, Paris, 9–11 November 1976.
Sherwood, R.J. and Carter, F.W.G. (1970) The measurement of occupational exposure to benzene vapour. *Ann. Occup. Hyg.* **13** 125–146.
Shigeta, S., Misawa, T. and Aikawa, H. (1980) Effects of concentration and duration of toluene exposure on sidman avoidance in rats. *Neurobehav. Toxicol.* **2** 85–88.
Shugaev, B.B. and Yaroslavl, B. Sc. (1969). Concentrations of hydrocarbons in tissues as a measure of toxicity. *Arch. Environ. Health.* **18** 872–882.
Simmons, J.E., Allis, J.W., Grosse, E.C., Seely, J.C., Robertson, B.L. and Berman, E. (1991) Assessment of the hepatotoxicity of acute and short-term exposure to inhaled *p*-xylene in F-344 rats. *J. Toxicol. Environ. Health* **32** 295–306.
Slooff, W. (1987) Integrated critera document: benzene. *Report nr. 75846001.* National Institute of Public Health and Environmental Protection, Bilthoven, The Netherlands.
Smith, B.R., Anda, J. van, Fouts, J.R. and Bend, J.R. (1983a) Estimation of the styrene-7,8-oxide detoxifying potential of epoxide hydrolase in glutathione-depleted perfused rat livers. *J. Pharmacol. Exp. Ther.* **227** 491–498.
Smith, E.N. and Carlson, G.P. (1980) Various pharmacokinetics parameters in relation to enzyme-inducing abilities of 1,2,4-trichlorobenzene and 1,2,4-tribromobenzene. *J. Toxicol. Environ. Health.* **6** 737–749.
Smith, J.H., Maita, K., Sleight, S.D. and Hook, J.B. (1983). Mechanism of nephrotoxicity. I. Time course of chloroform in mice. *Toxicol. Appl. Pharmacol.* **70** 457–479.
Smyth, H.F. Jr., Weil, C.R., West, J.S. and Carpenter, C.P. (1969) Exploration of joint toxic action: twenty-seven industrial chemicals intubated in rats in all possible pairs. *Toxicol. Appl. Pharmacol.* **14** 340–347.
Snellings, W.M., Zelenak, J.P. and Weil, C.R. (1982a) Effects of reproduction of Fischer 344 rats exposed to ethylene oxide by inhalation for one generation. *Toxicol. Appl. Pharmacol.* **63** 382–388.

Snellings, W.M., Maronpot, R.R., Zelenak, J.P. and Laffoon, C.P. (1982b) Teratology study in Fischer 344 rats exposed to ethylene oxide by inhalation. *Toxicol. Appl. Pharmacol.* **64** 476–481.

Snyder, R. (1974) Relation of benzene metabolism to benzene toxicity. In Braun, D. (ed.) *Symposium on the Toxicity of Benzene and Aryl Benzenes.* pp. 44–53. Pittsburg Industrial Health Foundation, Pittsburg, KS, USA.

Snyder, R., Lee, E.W. and Kocsis, J.J. (1978) Binding of labelled benzene metabolites to mouse liver and bone marrow. *Res. Commun. Chem. Pathol. Pharmacol.* **30** 191–194.

Snyder, R., Longacre, C.L. and Witmer, C.M. (1981) Biochemical toxicology of benzene. In Hodgson, E., Bend, J.R. and Philpot, R.M. (eds) *Reviews in Biochemical Toxicology*, Vol. 3, pp 123–153. Elsevier/North-Holland Publishing Co., New York.

Spencer, H.C., Irish, D.D., Adams, E.M. and Rowe, V.K. (1942) The response of laboratory animals to monomeric styrene. *J. Ind. Hyg. Toxicol.* **24** 295–301.

Spencer, P.S. and Schaumburg, H.H. (1977) Ultra structural studies of the dying-back process. IV. Differential vulnerability of PNS and CNS fibers in experimental central-peripheral distal axonal pathies. *J. Neuro-Pathol. Exp. Neurol.* **36** 300–320.

Spencer, P.S., Couri, D. and Schaumburg, H.H. (1980) n-Hexane and methyl *n*-butyl ketone. In Spencer, P.S. and Schaumburg, H.H. (eds) *Experimental and Clinical Neurotoxicology. Williams and Wilkins, Baltimore.*

Sprinz, H. *et al.* (1982) Neuropathological evaluation of monkeys exposed to ethylene and propylene oxide. Prepared for NIOSH by Midwest Research Institute, Kansas City, Missouri, USA, PB 83-134817, cited in Vermeire *et al.* (1985).

Srbova, J., Teisinger, J. and Skramovsky, B. (1950) Absorption and elimination of inhaled benzene in man. *Arch. Ind. Hyg. Occup. Med.* **2** 1–8.

Steele, J.W., Yagen, B., Hernandez, O., Cox, R.H., Smith, B.R. and Bend, J.R. (1981) The metabolism and excretion of styrene oxide glutathione conjugates in the rat by isolated perfused liver lung and kidney preparations. *J. Pharmacol. Exp. Ther.* **219** 35–41.

Stewart, R.D., Dodd, H.C., Baretta, E.D. and Schaffer, A.W. (1968) Human exposure to styrene vapor. *Arch. Environ. Health* **16** 656–662.

Stob, M. (1983) Naturally occurring food toxicants: estrogens. In Rechcigal, M. (ed.) *CRC Handbook of Naturally Occurring Food Toxicants.* CRC Press, Inc, Boca Raton, Florica, USA.

Stoltenburg-Didinger, G., Altenkirsch, H. and Wagner M. (1990) Neurotoxicity of organic solvents mixtures: Embryotoxicity and fetotoxicity. *Neurotoxicol. Teratol.* **12** 585–589.

Stott, W.T. and McKenna, M.J. (1984) The comparative absorption and excretion of chemical vapors by the upper, lower, and intact respiratory tract of rats. *Fund. Appl. Toxicol.* **4** 594–604.

Susten, A.S., Dames, B.L. and Niemeier, R.W. (1986) *In vivo* percutaneous absorption studies of volatile solvents in hairless mice. I. Description of a skin-depot. *J. Appl. Toxicol.* **6** 43–46.

Susten, A.S., Niemeier, R.W. and Simon, S.D. (1990) *In vivo* percutaneous absorption studies of volatile organic solvents in hairless mice. II. Toluene, ethylbenzene and aniline. *J. Appl. Toxicol.* **10**(3) 217–225.

Szejnwald-Brown, H., Bishop, D.R. and Rowan, C.A. (1984) The role of skin absorption as a route of exposure for volatile organic compounds (VOCs) in drinking water. *Am. J. Pub. Health* **74** 479–484.

Tariot, P.H. (1983) Delirium resulting from methylene chloride exposure: case report. *J. Clin. Psychiatry* **44** 340–342.

Tatrai, E., Hudal, A. and Ungvary, G. (1979) Simultaneous effect on the rat liver of benzene, toluene, xylene and CCL_4. *Acta Physiol. Acad. Sci. Hung.* **53**, 261–262.

Täuber, U. (1983) In *UBA-Bericht 6/82: Luftqualitatskriterien fuer Benzol* pp. 15–22. Erich Schmidt Verlag, Berlin.

Taylor, G.J., Drew, R.T., Lores, E.M.Jr. and Clemmer, T.A. (1976) Cardiac depression by haloalkane propellants, solvents and inhalation anesthetics in rabbits. *Toxicol. Appl. Pharmacol.* **38** 379–387.

Teisinger, J., Bergerova-Fiserova, V. and Kurdna, F. (1952) The metabolism of benzene in man. *Pracovni Lekarstvi* **4** 175–176.

Teramoto, K. (1978) Studies on styrene poisoning from the standpoint of occupational health. (I) Surveys on the actual conditions of plastic factories using styrene. *J. Osaka City Med. Cent.* **27** 249–262. (Abstract).

Teramoto, K. and Hariguchi, S. (1981) Distribution, elimination and retention of styrene in rats. *J. Toxicol. Sci.* **6** 13–18. (Abstract).

Til, H.P., Woutersen, R.A., Feron, V.J., Hollanders, V.H.-M., Falke, H.E. and Clary, J.J. (1989) Two-year drinking-water study of formaldehyde in rats. *Food Chem. Toxicol.* **27** 77–87.

Timchalk, C., Dryzga, M.D., Smith, F.A. and Bartels, M.J. (1991) Disposition and metabolism of (14C) 1,2-dichloropropane following oral and inhalation exposure in Fischer 344 rats. *Toxicol.* **68** 291–306.

Toftgard, R., Nilsen, O.G. and Gustafson, J.A. (1982) Dose-dependent induction of rat liver microsomal enxymatic activities after inhalation of toluene and dichloromethane. *Acta Pharmacol. Toxicol.* **51** 108–114.

Toftgard, R., Haaparanta, T. and Halpert, J. (1986a) Rat lung and liver cytochrome P-450 isoenzymes involved in the hydroxylation of *m*-xylene. *Toxicol.* **30** 225–321.

Toftgard, R., Haaparanta, T., Eng, L. and Halpert, J. (1986b) Rat lung and liver microsomal cytochrome P-450 isoenzymes involved in the hydroxylation of *n*-hexane. *Biochem. Pharmacol.* **35** 3733–3738.

Torkelson, T.R., Oyen, F. and Rowe, V.K. (1976) The toxicity of chloroform as determined by single and repeated exposure of laboratory animals. *Am. Ind. Hyg. Assoc. J.* **37** 697–705.

Tsuruta, H. (1975) Percutaneous absorption of organic solvents. (I) Comparative study of the *in vivo* percutaneous absorption of chlorinated solvents in mice. *Indust. Health* **13** 227–236.

Tsuruta, H. (1977) Percutaneous absorption of organic solvents. (II) A method for measuring the penetration rate of chlorinated solvents in rough excised rat skin. *Indust. Health* **15** 131–139.

Tsuruta, H. (1982) Percutaneous absorption of organic solvents. (III) On the penetration rates of hydrophobic solvents through the excised rat skin. *Indust. Health* **20** 335–345.

Ungvary, G.Y. (1985) The possible contribution of industrial chemicals (organic solvents) to the incidence of congenital defects caused by teratogenic drugs and consumer goods; an experimental study. In *Prevention of Physical and Mental Congenital Defects, Part B: Epidemiology, Early Detection and Therapy, and Environmental Factors* pp. 295–300. Alan R. Liss Inc., New York.

Vainio, H. and Elovaara, E. (1979) The interaction of styrene oxide with hepatic cytochrome P450 *in vitro* and effects of styrene oxide inhalation on xenobiotic biotransformation in mouse liver and kidney. *Biochem. Pharmacol.* **28** 2001–2004.

Van Apeldoorn, M.E., Van der Heijden, C.A., Van Leeuwen, F.X.R., Huldy, H.J., Besemer, A.C., Lanting, R.W., Jol, A., Heyna-Merkus, E., Bergshoeff, G., Venselaar, J., Van der Most, P.F.J., De Vrijer, F., Van der Plaat, J., Knaap, A.G.A.C., Huijgen, C., Duiser, J.A. and De Jong, P. (1986) Integrated criteria document: Styrene. *Report nr. 738513003.* National Institute of Public Health and Environmental Protection, Bilthoven, The Netherlands.

Van Apeldoorn, M.E., Vermeire, T., Van der Heijden, C.A., Van Gestel, C.A.M., Heyna-Merkus, E. Knaap, A.G.A.C. and Krajnc, E.I. (1987) Integrated criteria document: Methylene chloride. *Appendix to Report nr. 758473002.* National Institute of Public Health and Environmental Protection, Bilthoven, The Netherlands.

Van Bogaert, M., Rollmann, B., Noel, G., Roberfroid, M. and Mercier, M. (1978) A very sensitive gas chromatographic method for the evaluation of styrene oxidase and styrene oxide hydratase activities. Toxicological Aspects of Food Safety. *Arch. Toxicol. Suppl.* **1** 295–298.

Van der Heijden, C.A., Speijers, G.J.A., Ros, J.P.M., Huldy, H.J., Besemer, A.C., Lanting, R.W., Maas, R.J.M., Heyna-Merkus, E., Bergshoeff, G. Gerlofsma, A., Mennes, W.C., Van der Most, P.F.J., De Vrijer, F.L., Janssen, P.C.J.M., Knaap, A.G.A.C., Huijgen, C., Duiser, J.A. and De Jong, P. (1987) Integrated criteria document: Chloroform . *Report nr. 738513009.* National Institute of Public Health and Environmental Protection, Bilthoven, The Netherlands.

Van der Heijden, C.A., Mulder, H.C.M., De Vrijer, F.L., Woutersen, R.A., Davis, J.A., Vink, G.J., Heyna-Merkus, E., Janssen, P.J.C.M., Canton, J.H. and Van Gestel, C.A.M. (1988) Integrated criteria codument: Toluene. *Appendix to Report nr. 758473010.* National Institute of Public Health and Environmental Protection, Bilthoven, The Netherlands.

Van Koten-Vermeulen, J.E.M., Van der Heijden, C.A., Van Leeuwen, F.X.R., Huldy, H.J., Besemer, A.C., Lanting, R.W., Ros, J.P.M., Canton, J.H., Heyna-Merkus, E., Bergshoeff, G., Maas, R.J.M., Van der Most P.F.J., De Vijer, F., Van der Plaat, J., Knaap, A.G.A.C., Huijgen, C., Duiser, J.A. and De Jong, P. (1986) Integrated criteria document: Phenol. *Report nr. 738513002.* National Institute of Public Health and Environmental Protection, Bilthoven, The Netherlands.

Van Oettingen, W.F. (1949) Phenol and its derivatives: The relation between their chemical constitution and their effect on the organism. *Nat. Inst. Health. Bull.* No. 190.

Van Raalte, H.G.C. and Grasso, P. (1982) Hematological, myelotoxic, clastogenic, carcinogenic and leukemogenic effects of benzene. *Regul. Tox. Pharmacol.* **2**, 153–176.

Vergieva, T. *et al.* (1979) A study on the embryonic action of styrol. *Hyg. Zdrav.* **XIII** 39–43 (in Russian).

Vermeire, T., Esch, G.J. van, Heijden, C.A. van der., Heyna-Merkus, E., Knaap, A.G.A.C. and Leeuwen, F.X.R. van (1985) Ethylene oxide criteria documents: Air effects. *Report nr. 668310.* National Institute of Public Health and Environmental Protection, Bilthoven, The Netherlands.

Vermeire, T., Heijden, C.A., van der Gestel, C.A.M., van., Heyna-Merkus, E., Knaap, A.G.A.C. and Krajnc, E.I. (1988) Integrated criteria document: Propylene oxide. *Appendix to Report nr. 758473008.* National Institute of Public Health and Environmental Protection, Bilthoven, The Netherlands.

Veulemans, H. and Masschelein, R. (1978) Experimental human exposure to toluene. II. toluene in venous blood during and after exposure. *Int. Arch. Occup. Environ. Health* **42** 105–117.

Veulemans, H. and Masschelein, R. (1979) Experimental human exposure to toluene. III. Urinary hippuric acid excretion as a measure of individual solvent uptake. *Int. Arch. Occup. Environ. Health* **43** 53–62.

Vos, J.G., Van Logten, M.J., Kreeftenberg, J.G., Steerenberg, P.A. and Kruizinga, W. (1979) Effect of hexachlorobenzene on the immune system of rats following combined pre- and postnatal exposure. *Drug Chem. Toxicol.* **2** 61–72.

Wallen, M., Holm, S. and Byfalt-Nordquist, M. (1985) Co-exposure to toluene and *p*-xylene in man: Uptake and elimination. *Br. J. Industry. Med.* **42** 111–116.

Warholm, M., Guthenberg, C. and Mannervik, B. (1983) Molecular and catalytic properties of glutathione transferase from human liver: an enzyme efficiency conjugating epoxides. *Biochem.* **22** 3610–3617.

Watabe, T., Ozama, N. and Yoshikawa, K. (1981a) Stereochemistry in the oxidative metabolism of styrene by hepatic microsomes. *Biochem. Pharmacol.* **30** 1695–1698.

Watabe, T., Hiratsuka, A., Ozama, N. and Isobe, M. (1981b) Glutathione-S-conjugates of phenyloxirane. *Biochem. Pharmacol.* **30** 390–392.

Watabe, T., Hiratsuka, A. and Aizawa, T. (1981c) Metabolic activation and inactivation of carcinogens and mutagens involving epoxides and their ultimate form. *J. Pharmacobio. Dyn.* **4** 50–55.

Watabe, T., Ozama, N. and Hiratsuka, A. (1983) Studies on metabolism and toxicity of styrene. VI. Regio selectivity in glutathione-S-conjugation and hydrolysis of racemic, *R*- and *S*-phenyloxiranes in rat liver. *Biochem. Pharmacol.* **32** 777–785.

Weinstein, R.S. and Diamond, S.S. (1972) Hepatoxicity of dichloromethane with continuous inhalation exposure at a low level. In *Proceedings of the 3rd Annual Conference on Environmental Toxicity*, pp. 209–222. Wright-Patterson Air Force Base, Ohio Aerospace Medical Research Laboratory, AMRL-72-130, Paper No. 13, USA.

Weinstein, R.S., Boyd, D.D. and Back, K.C. (1972) Effects of continuous inhalation of dichloromethane in the mouse: Morphologic and functional observations. *Toxicol. Appl. Pharmacol.* **23** 660–679.

Weiss, von G. (1969) Toxische Enzephalose beim beruflichen Umgang mit Methylen Chlorid. *Zentralbl. Arbeitsm. Arbeitsschutz* **17** 282–285.

Weitering, J.G., Krijgsheld, K.R. and Mulder, G.J. (1979) The availability of inorganic sulphate as a rate limiting factor in the sulphate conjugation or xenobiotics in the rat. *Biochem. Pharmacol.* **28** 757–762.

Wheldrake, J.F., Baudinette, R.V. and Hewitt, S. (1978) The metabolism of phenol in a desert rodent *Notomys alexis. Comp. Biochem. Physiol.* **61C** 103–107.

WHO (1987a) *Environmental Health Criteria 71: Pentchlorophenol.* IPCS International Program on Chemical Safety, World Health Organization, Geneva.

WHO (1987b) *Air Quality Guidelines of Europe: Formaldehyde.* WHO Regional Office for Europe. WHO Regional Publication, European Series No. 23, 11/49, 426.

WHO (1989a) *Environmental Health Criteria 93: Chlorophenols other than PCP.* IPCS International Program on Chemical Safety, World Health Organization, Geneva.

WHO (1989b) *Environmental Health Criteria 87: Formaldehyde.* IPCS International Program on Chemical Safety, World Health Organization, Geneva.

Wigaeus, E., Lof, A., Bjurström, R. and Nordqvist, B. (1983) Exposure to styrene. Uptake, distribution, metabolism and elimination in man. *Scand. J. Work. Environ. Health* **9** 419–488.

Williams, R.T. (1963) *Detoxications Mechanisms.* Chapman and Hall, London.
Withey, J.R. (1976). Quantitative analysis of styrene monomer in polystyrene and foods including some preliminary studies of the uptake and pharmacodynamics of the monomer in rats. *Environ. Health Perspect.* **17** 125–133.
Withey, J.R. (1978) The toxicology of styrene monomer and its pharmacokinetics and distribution in the rat. *Scand. J. Work Environ. Health* **4** 31–40.
Withey, J.R. and Collins, P.G. (1977) Pharmacokinetics and distribution of styrene monomer in rats after intravenous administration. *J. Toxicol. Environ. Health* **3** 1011–1020.
Withey, J.R. and Collins, P.G. (1979) The distribution and pharmacokinetics of styrene monomer in rats by the pulmonary route. *J. Environ. Pathol. Toxicol.* **2** 1329–1342.
Wolff, M.A., Rowe, M.A., McCollister, D.D., Hollingworth, R.L. and Oyen, F. (1956) Toxicological studies of certain alkylated benzenes and benzene. *Arch. Ind. Health.* **14** 387–389.
Wolff, D.L., Zeller, H.J., Zschenderlein, R. and Hinz, G. (1979) Behavioural, electroneurographical and histological investigations in rats concerning the combined action of styrene and ethanol. *Activ. Nerv. Suppl.* **21** 260–261.
Woodard, G. and Woodard, M. (1971) Toxicity of residuals from ethylene oxide sterilization. *Proceedings of the 1971 HIA Technical Symposium.* Health Industries Association, Washington DC, USA.
Young, J.T. *et al.* (1985) Propylene oxide: assessment of neurotoxic potential in male rats. Mammalian and Environmental Toxicology Research Laboratory, Health and Environmental Science USA, Midland, Michigan 48640 cited by Vermeire *et al.* (1988).
Zitting, A., Heinonen, T. and Vainio, H. (1980) Glutathione depletion in isolated rat hepatocytes caused by styrene and the thermal degradation products of polystyrene. *Chem. Biol. Interact.* **31** 313–318.

3 VOCs and air pollution

P. CICCIOLI

3.1 Introduction

The presence of volatile organic components (VOCs) on the earth can be dated back to the distant beginnings of life when the appearance of plants and microorganisms led to the conversion of gases (produced by geochemical processes) into organic molecules. Since that time, VOCs have taken part in the carbon cycle by regulating the content of organic compounds in air, soil and water reservoirs. Before humans entered the picture, natural cleansing action was able to control the quality and quantity of VOCs in air by dispersion, chemical conversion and deposition processes. Through the complex physical and biogeochemical equilibria established in the biosphere, VOCs became part of the transmission belt converting carbon dioxide into organic material and back to inorganic carbon. The appearance of humans has gradually changed the natural balance. Since the explosion of the industrial revolution in the 19th century in Europe and North America, the amount of waste material released by man-made activities has increased to such a point that inevitable side-effects are now felt. The tremendous hazard represented by semivolatile organic chemicals in air became dramatically evident to the public at the beginning of this century (1915–1918) when highly toxic chlorinated molecules produced by synthesis were first used by the German army as chemical weapons. The numerous accidents that occurred revealed the lack of knowledge on the fundamental processes responsible for the dispersion and deposition of gases and aerosols, and the factors influencing meteorology (Sutton, 1971). This led, in 1921, to the setting up at Porton Down in the UK of a centre for chemical weapons. A group of scientists was charged with studying the meteorology of the low troposphere and dispersion of gas and aerosols in air. Ironically, it was through field experiments intended for chemical war that it was possible to later derive the diffusion equation still used today for predicting plume dispersion of non-reactive gaseous components released from large point sources (Sutton, 1971). The wide adoption of masks using carbon materials to retain toxic VOCs was another positive input from these experiences; the same basic principle is exploited today for the collection of VOCs in air. The discovery by Haagen-Smit and Fox (1956) that ozone production was responsible for the so-called 'Los Angeles smog', and that the smog was caused by the photochemical oxidation of mixtures containing VOCs and nitrogen oxides (NO_x), provided the first compelling evidence that volatile organic molecules were also able to act as precursors of secondary pollution. Subsequently it became clear that the widespread use of fossil fuels for energy production and the increased demand for new chemical products to make life more comfortable would unavoidably be associated with a drastic change in the quality of the atmosphere; thus efforts were made to keep VOC emission under control. To achieve this goal efficiently, research on VOCs was undertaken. This research was designed to evaluate emission, study the physical and chemical transformations occurring in air, and assess long- and short-term effects on human health and the environment (US Department of Health, Education and Welfare (USDHEW), 1970). The development of analytical tools for the accurate determination of VOCs, together with mathematical models for predicting aerometric concentrations of both primary and secondary pollutants, became an important part of these efforts. Based on the results of these investigations, VOCs were included in the list of 'criteria' pollutants used for defining the quality of the atmosphere (USDHEW, 1970; Finalyson-Pitts and Pitts, 1986). Although the complete picture of sources, sinks and reaction mechanisms involving VOCs has not yet been completed, the knowledge accumulated in the last forty years has enabled better assessment of the role that VOCs play in the atmospheric environment, and identification of new effects associated with their emission. The use of mathematical models capable of describing the complex physicochemical processes occurring in air allows the prediction of short- and long-term effects associated with VOC emission and the adoption of adequate control strategies to protect human health and the environment. In this chapter the current knowledge about the emission, transformation and effects of VOCs in the atmospheric environment will be reviewed. Strategies and methods currently used for the identification and accurate quantification of VOCs in air and emission sample will be analysed and critically discussed.

3.2 VOC emission

Global estimates indicate that approximately 235 million tons (metric) per year of VOCs are released into the atmosphere by man-made sources (Cullis and Hirschler, 1989). An additional input of approximately 153 million tons per year of methane comes from man-controlled emissions. The magnitude of global emission of natural hydrocarbons is still controversial: figures varying from 830 (Zimmermann, 1977) to 2100 million tons per year have been estimated (Volz *et al.*, 1981). The major part of this natural source, mainly comprising terpenes and isoprene, arises in tropical regions, while 212 to 330 million tons per year are thought to be emitted from temperate regions. The uncertainty concerning global estimates illustrates well the intrinsic difficulties associated with the evaluation of VOC emission. Consequently, identification of sources, together with detailed information on their spatial distribution, density and emission rates, remains a fundamental task to establish the extent to which VOCs are affecting the quality of the atmosphere. The availability of figures on the amount of VOCs released in a given region during a certain period of time is the basic tool for decision makers to identify highly polluted areas and take action to reduce emission. Modelling studies performed in different scenarios provide emission data and facilitate selection of the best control or abatement strategies for local, regional or global scales. The ideal situation would be real-time knowledge of the amount of each organic component released in a given parcel of air by any existing source. In practice this is virtually impossible because of: (1) the number and type of sources emitting VOCs; (2) the large differences in the chemical composition of VOCs; and (3) the possibility that changes in emission occur in space and time. In some instances, lack of knowledge of all factors affecting the emission process makes it difficult to decide whether the actual measurements can be taken as truly representative of the source investigated. To give an idea of the difficulties encountered in evaluating VOC emission, it should be recalled that more than 200 different organic compounds can be present in some sources, and their identification and accurate quantification is not easy even for dedicated laboratories with sophisticated analytical techniques and skilled personnel. Consequently, continuous and accurate knowledge of the amount of each component present in VOC emission sources applies only to a very limited number of existing sources. For this reason, continuous and semi-continuous instrumentation for evaluating the total or non-methane VOC content in air and emission sources has been developed and used in the last two decades (see section 3.5). As instruments of this type can easily be connected to the stacks of industrial plants or tail-pipes of cars, changes in emission with time can be monitored. Information on gas and vapour composition can be obtained by submitting a limited but selected number of samples to more complex gas chromatographic (GC) or gas chromatographic–mass spectrometric (GC–MS) determination (see sections 3.4 and 3.5). In addition to the difficulties encountered in analytical procedures, other complicating factors can affect the evaluation of VOC emission. A typical example is represented by autovehicular exhaust emission, the composition of which depends on the type of engine used, the operating conditions, the fuel adopted and the car mileage. Similar problems are encountered in the evaluation of biogenic sources, as the amount and type of organic components released by plants change with the season and are influenced by some parameters (water supply, solar radiation intensity, temperature and soil composition) which are not easy to control in the field. The photochemical reactivity and ubiquitous occurrence of some components often make it difficult, if not impossible, to identify VOC sources through aerometric measurements and to derive their emission rates from simple source–receptor models; the only way to obtain information on VOC emission from certain types of sources is to set up laboratory experiments. When doing this, several assumptions, which approach reality only to a certain extent, must be made and specific protocols prepared and followed in order to provide reliable figures. Where possible, protocols should be harmonized on a national and international basis and quality control adopted to avoid discrepancies in the results obtained. In spite of the increasing efforts made in all these fields, the evaluation of VOC emission is still affected by uncertainty depending on the source considered, the procedures followed for the measurement of VOCs and the assumptions made in estimating emissions if empirical factors are used. Uncertainties are thus considered an inevitable and unavoidable part of the inventory process (Seinfeld, 1988). Moreover, most of the techniques used to quantify uncertainties still rely on the judgment of the inventory specialists. The way in which such uncertainty will influence predictions is also a function of the resolution in space and time required by the model adopted for evaluating the effects. Short-term predictions of photochemical oxidants require high resolution in space and time, therefore emission data must be provided on a hourly basis within grids varying from 2 to 50 km depending on the type of model considered (local, regional) (Hov *et al.*, 1986). For long-term predictions, annual averaged emissions provided on a national basis can be sufficient. Since computational models describe the chemistry of VOCs in simplified form by using some specific organic compounds as surrogates for an entire class of components (Hov *et al.*, 1986, 1987; Seinfeld, 1988) the final uncertainty will also depend on the accuracy of evaluating the composition of the VOCs and on the procedures followed for their allocation into surrogate classes. For these reasons different techniques have been developed to derive VOC emission as a function of: (1) the information available; (2) the effects requiring investigation; and (3) the model used. In all cases, a basic distinction is made between point and area sources. Point sources are all stationary emission sources (facilities, activities or plants) for which individual records of geographical position, type of process, stacks parameters and operating conditions (i.e. exit gas temperature, velocity and volumetric flow rate) are maintained by local agencies and up-dated on a regular basis (USEPA, 1979a). The term area sources is used to identify an aggregation of sources which because of their size, number or mobility cannot easily be classified as point sources. Emissions from area sources are assumed to be spread over a certain geographical area such that averaged values in time and space are used for each square of the grid system into which the area has been divided (USEPA, 1979a). Examples of area sources include all self-propelled sources for VOCs (motor vehicles, vessels, locomotives and aircraft), small heating appliances, indoor activities and vegetation. For point sources for which a large part of the emission is measured, the use of empirical factors derived from activity levels is restricted to uncontrolled emission (e.g. fugitive or evaporation losses). Emission factors are, however, necessary for evaluating net VOC emissions from area sources (USEPA, 1979a). These two classes include all source categories normally used in the emission inventories.

Allocation of sources within proper categories is useful for several reasons. By knowing which activities are contributing most to VOC emission, source category reduction plans can be developed and applied to specific areas. Moreover, emission data can be better verified through the analysis of economical and statistical indicators such as production, consumption, transport and storage of goods emitting hydrocarbons; alternatively, when experimental data are not available, they may be calculated by using emission factors. Emission sources can be grouped into categories according to economic activities (e.g. power plants, dry-cleaning, agriculture, etc.) and/or type of emission-generating process (e.g. combustion, storage and transfer of fuel in power plants) (Lubkert and De Tilly, 1989). When doing this, it is desirable that sources are allocated in such a way that emission scenarios can be evaluated and emission rates measured or calculated homogeneously for each source category (Lubkert and de Tilly, 1989). It is also important to design the VOC inventory in such a way that different models can be used and economic assessment of the cost benefits associated with a certain abatement strategy derived.

3.2.1 VOC sources and source categories: estimates and spatial distribution

For several years the United States Environmental Protection Agency (USEPA) has prepared specific documents to help local agencies prepare VOC emission inventories (USEPA, 1977, 1979a, 1980). Guidelines are given to allocate point and area sources into categories, estimate emissions for each of them, and collect and store data in a format that can be readily interfaced with the available types of model. Four basic elements have been identified in the preparation of VOC emission inventories: (1) planning; (2) data collection; (3) analysis; and (4) report. Once the type of emission inventory that best satisfies a certain control strategy has been identified, guidelines are given for the selection of source categories and methods for determining the emission distribution in space and time are described. VOC profiles for 175 types of processes are reported in order to allow the best evaluation of VOC emission in cases where emission data are not directly available or are difficult to obtain. Recommendations as to how experience and common sense should be used for successful completion of a given inventory are given. Examples of how land-use maps can be used to develop apportioning factors for those area sources that are distributed over various land-use classifications are provided. Various methodologies for incorporating spatial and temporal resolution for area source categories in a detailed emission inventory and for making projections in future years are suggested.

Procedures for interfacing available transportation models with emission factor models are described. Based on this consolidated experience, a comprehensive emission inventory for VOCs has been developed (Shareef *et al.*, 1988; Zimmermann *et al.*, 1988). Currently used by the National Acid Precipitation Assessment Program (NAPAP) for evaluating acid deposition in contiguous portions of the USA, this inventory provides data for approximately 600 individual man-made organic components. Each component is identified by a standard code (since the 1970s the USEPA has used a storage and retrieval of aerometric data (SAORAD) system) and annual emissions in kilotons per year are given, together with the percentage of the total emission associated with different area and point sources. Identification codes are used to allocate area and point sources into categories. Naturally emitted components will shortly be added to this inventory so that detailed emission for more than 1000 organic compounds will be available. However, these numbers cannot be treated by Eulerian models such as the regional acid deposition model (RADM), developed at NCAR (Chang *et al.*, 1987) and used in the USA, or the regional transport model for photochemical oxidants (RTM III) (Liu *et al.*, 1984), the European version of which (PHOXA) (Stern and Builtjes, 1986) has been applied to Northern Europe. A large part of computer capability is absorbed by extensive description of the physical and chemical processes occurring in the three layers of each grid in which the entire domain (usually more than 1000 × 1000 km) is split, therefore no more than 20 organic components can be used as input data. To restrict the inventory to a limited number of surrogate species representative of all emitted compounds, specific aggregation procedures taking into account the reactivity of each compound as a function of the chemical description adopted by the model (parametrization, lumped mechanism, etc.) have been developed (Middleton *et al.*, 1990). By including the rate of reaction with OH radicals in the inventory, weighting factors can be estimated so that all compounds can be allocated into the 32 reactivity classes that can be managed by implemented versions of the NAPAP model. Specific codes have been assigned to each compound to prevent ambiguity in the allocation procedure. Today, and for some years to come, 32 surrogate VOC components represent the upper limit that can be used by Eulerian models, therefore the only way to fully exploit all the information contained in extensive emission inventories is to restrict the physical description to a single-layer parcel of air moving along the trajectory followed by air pollutants and their precursors. In this way, a large part of the computer memory can be devoted to chemistry and more than 70 organic components can be treated in a single run (Derwent, 1990). This does not ensure, however, that the results obtained are more reliable than with Eulerian models as they may be affected by other sources of uncertainties arising from more extensive parametrization of heterogeneous reactions, and by transport and deposition processes. In any case, redundant information on VOC emission is always useful for validating models.

The level of information on VOC emission in Europe is not as good as that existing in the USA. Only few countries are able to produce detailed emission inventories, whereas others have only a limited data base. To fill this gap, different European organizations (Economic Commission for Europe of the United Nations (UN-ECE), Commission of the European Communities (CEC), Organization for Economic Co-operation and Development (OECD)), each one pursuing different goals, have started the preparation of European emission inventories (Lubkert and de Tilly, 1989). However, differences in emission factors, source categories and spatial resolution make it extremely difficult to compare data (Lubkert and Zierock, 1989). To improve the quality of each inventory, avoid duplication of work, promote exchange of information, and make better use of available research, a proposal has been made to start a concerted action through worksharing. Today, close co-operation exists between these organizations in order to harmonize data and to make them compatible. Regardless of the approach taken, all European inventories are based on total VOC emission (Veldt and Bakkum, 1987; Bouscarin, 1989; Lubkert and de Tilly, 1989). A substantial part of the work is devoted to the calculation of emission factors for point and area sources for which emission showed marked deviation from the US data base and procedures. Specifically this was needed for vehicular emission because of differences in size and type of engines used, fuel composition adopted and number of control devices installed in European and US vehicles. Figure 3.1 shows a schematic view of the sources and source categories selected by the OECD for preparing the VOC emission inventory for Western Europe (OECD, 1990). A total of approximately 80 source categories, including natural emission, were identified. In order to meet the requirements of other ongoing UN-ECE programs such as the European Monitoring Environmental Program (EMEP) for long-range transport of photochemical pollutants and acidic species (OECD, 1977), the inventory was prepared for a grid resolution of 50 × 50 km. Figure 3.2 shows the spatial distribution of VOC emission in Western Europe, while the amounts emitted by the main sources present in each country are shown in Figure 3.3. Data refer to the year 1980. From these figures it is possible to identify the sites where the highest emission density occurred and to relate VOC levels to man-made activities. A total of 17.9 million tons per year of man-made VOCs (excluding methane, which is relatively unreactive) were released by the OECD European countries in 1980. This value was relatively close to that estimated for the USA in the same year (approximately 20 million tons) (Middleton *et al.*, 1990). Also similar were the percentage contributions of the various sources to man-made emission, the dominant sources being vehicles, power plants and solvent emission (see Figure 3.4(a)). Differences were, however, observed in the contributions of biogenic sources, with US estimates being twice as high (approximately 70%) (Lamb *et al.*, 1987) as those evaluated for Western Europe (30.6% according to the data shown in Figure 3.4(b)). Direct comparison of the relative importance of biogenic versus man-made emission is difficult because emission factors adopted in the European inventory (Zimmermann, 1979; Veldt and Bakkum, 1987) were higher than those reported in the study of Lamb *et al.* (1987). Data from the European emission inventory have already been used in regional models to assess the effects of different emission reduction scenarios on the existing levels of ozone (OECD, 1990). The conclusion was that a cut of at least 50% in man-made VOC emission was needed to approach the new air quality guidelines issued by the World Health Organization (WHO), according to which hourly maximum levels of ozone ranging from 75 to 100 ppbv should not be exceeded for more than 1 h, while 8 h averaged values should fall in the 50–60 ppbv range (WHO, 1986). Based on these indications, some European governments have taken action to reduce VOC emission. Among them, the Italian government has recently issued new guidelines (DPR, 1990) in which limits not to be exceeded have been established for all existing industrial and heating plants. Such limits apply to approximately 200 individual organic components, with restrictions on the total VOC content in cases where the emission does not contain any of the components listed. Regular checks must be performed to verify compliance with national regulations, therefore the law also provides a useful means of gaining information on those VOC sources whose 1980 emission was estimated through empirical factors. Action was taken because Italy was the fourth highest contributing country to man-made VOC emission in Europe (approximately 1.6 million tons per year) and little change had occurred from 1980 to 1985. An additional reduction is now expected because of the EC directive requiring that all new cars released on the market are equipped with exhaust control devices for VOCs. In the meantime, tax-reduction policies and circulation restriction have been combined with the introduction of less polluting fuels in order to reduce the high levels of photochemical pollution in Europe where mixing ratios as high as 175 ppbv of ozone are reached (OECD, 1990). More important benefits are expected following the application of the UN–ECE protocol for reducing VOC emission (UN–ECE, 1991) signed by many national governments. As an immediate target, all members have committed themselves to cut VOC emission for the year 1999 by 30% (using the 1988 estimates as reference levels). If this occurs, a substantial reduction in short-term ozone peaks in many areas of Europe should be observed. Although mainly designed to reduce the effects associated with transboundary pollution of photochemical oxidants, the protocol extends to some VOCs (methane, CFC) for which control may affect stratospheric ozone depletion and global warming (see sections 3.3.2 and 3.3.3). The document suggests that the best available technologies which are economically feasible should be applied to reduce VOC emission; exchange of information and technologies between parties is also necessary. The provision of emission data for all reactive species with a high potential to produce photochemical oxidants (ozone, peroxyacetyl

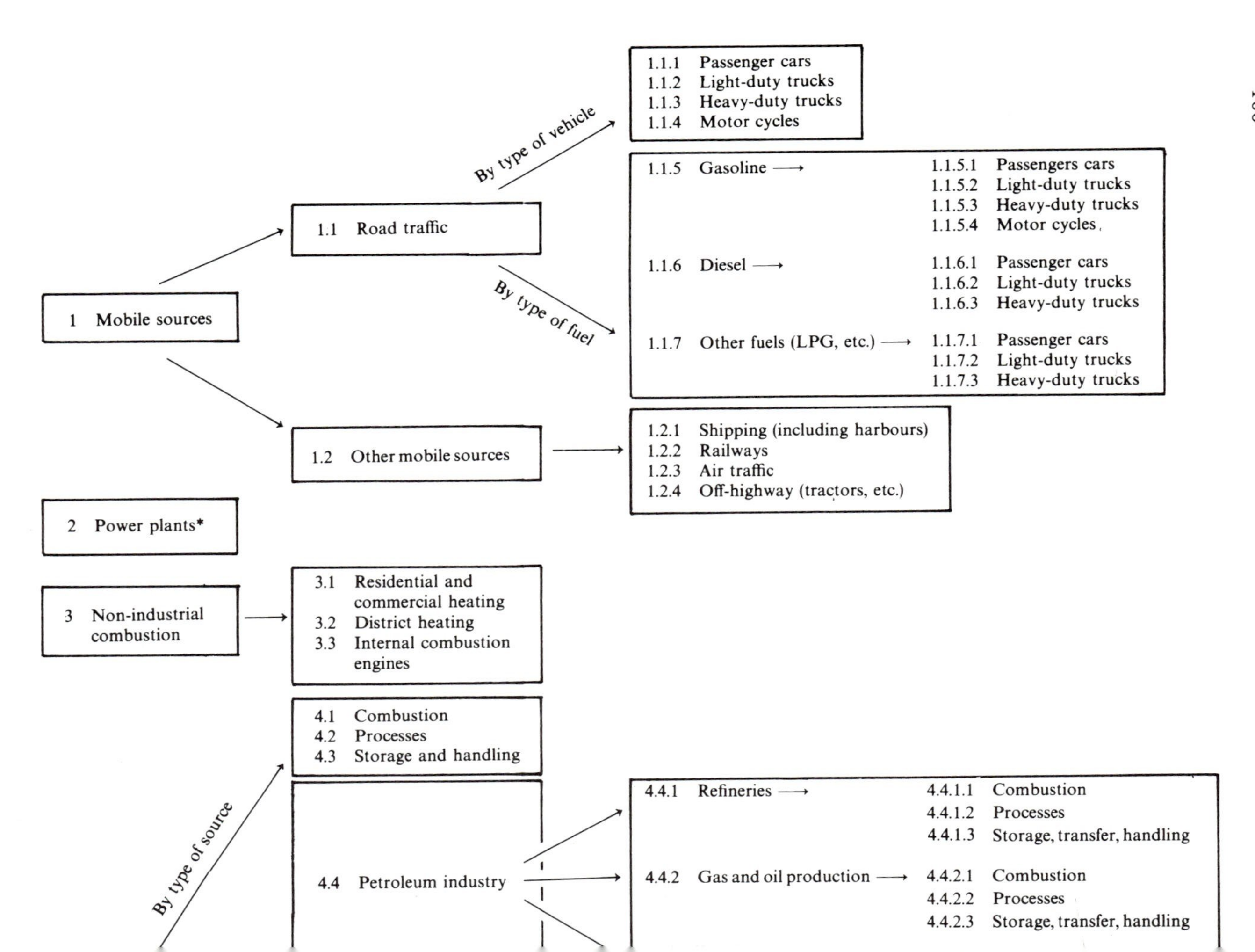

1 Mobile sources
1.1 Road traffic
By type of vehicle
1.1.1 Passenger cars
1.1.2 Light-duty trucks
1.1.3 Heavy-duty trucks
1.1.4 Motor cycles
By type of fuel
1.1.5 Gasoline →
1.1.5.1 Passengers cars
1.1.5.2 Light-duty trucks
1.1.5.3 Heavy-duty trucks
1.1.5.4 Motor cycles
1.1.6 Diesel →
1.1.6.1 Passenger cars
1.1.6.2 Light-duty trucks
1.1.6.3 Heavy-duty trucks
1.1.7 Other fuels (LPG, etc.) →
1.1.7.1 Passenger cars
1.1.7.2 Light-duty trucks
1.1.7.3 Heavy-duty trucks
1.2 Other mobile sources
1.2.1 Shipping (including harbours)
1.2.2 Railways
1.2.3 Air traffic
1.2.4 Off-highway (tractors, etc.)
2 Power plants*
3 Non-industrial combustion
3.1 Residential and commercial heating
3.2 District heating
3.3 Internal combustion engines
By type of source
4.1 Combustion
4.2 Processes
4.3 Storage and handling
4.4 Petroleum industry
4.4.1 Refineries →
4.4.1.1 Combustion
4.4.1.2 Processes
4.4.1.3 Storage, transfer, handling
4.4.2 Gas and oil production →
4.4.2.1 Combustion
4.4.2.2 Processes
4.4.2.3 Storage, transfer, handling

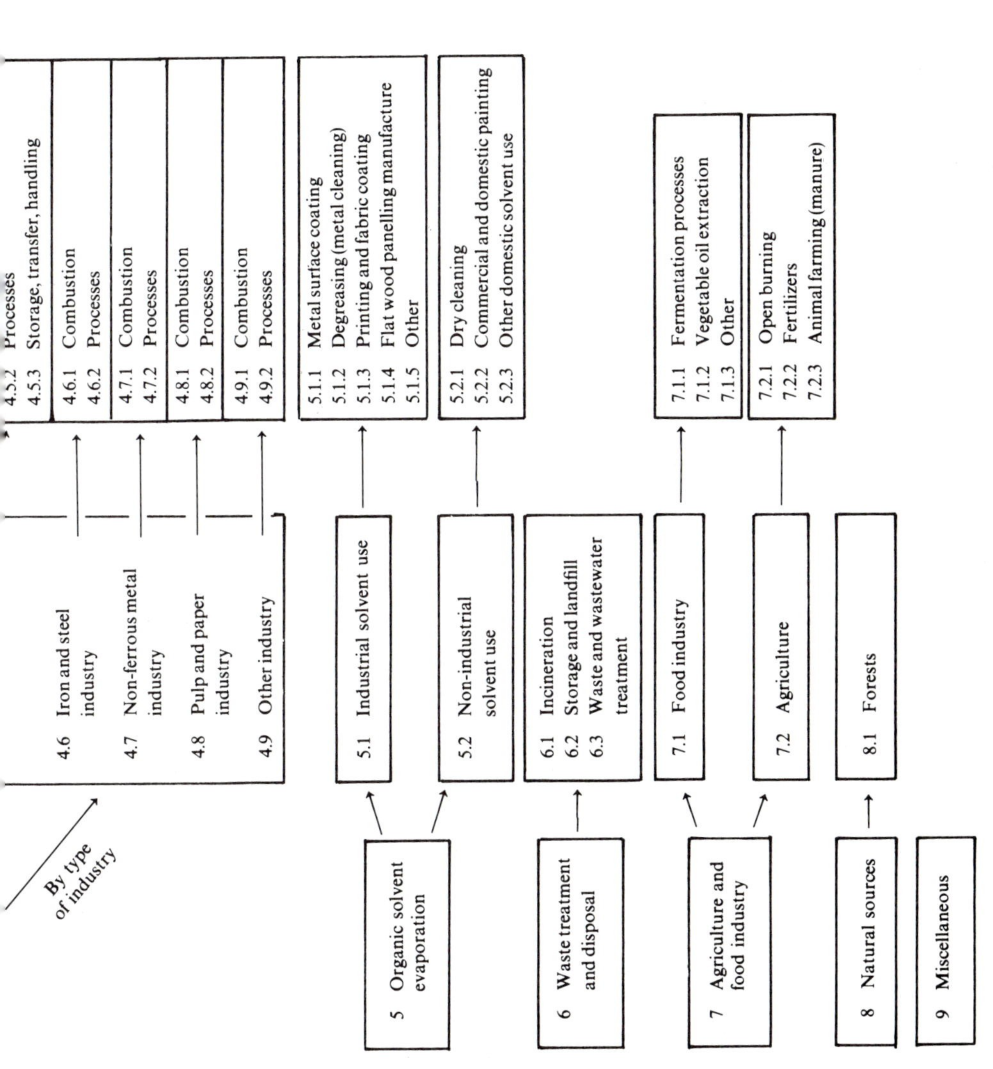

Figure 3.1 Source and sources categories used in the VOC emission inventory prepared by the OECD for Western Europe. After OECD (1990).

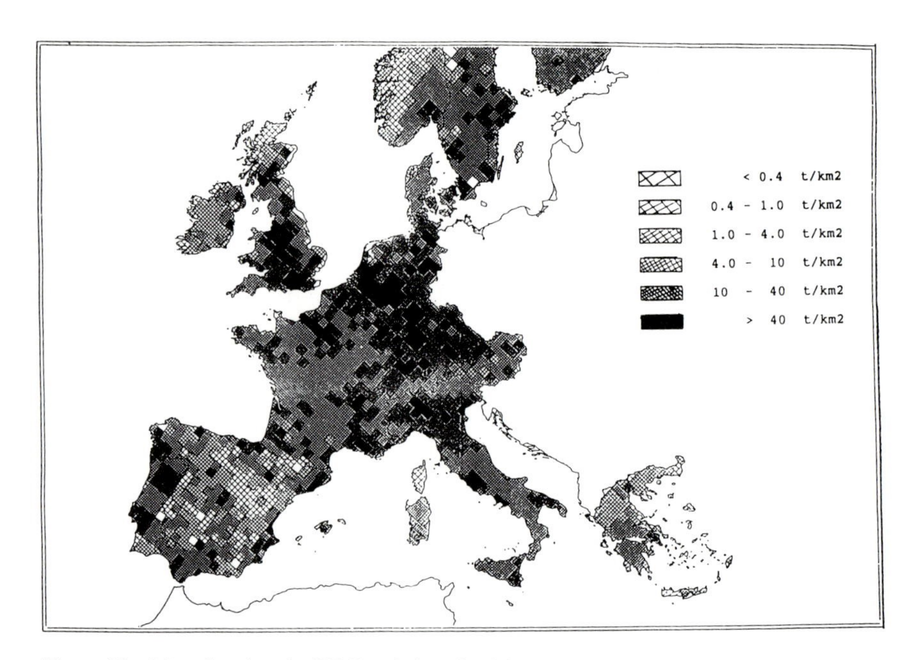

Figure 3.2 Map showing the VOC emission densities occurring in Europe in 1980. Grids of 50 × 50 km have been used for preparing the inventory. From OECD (1990).

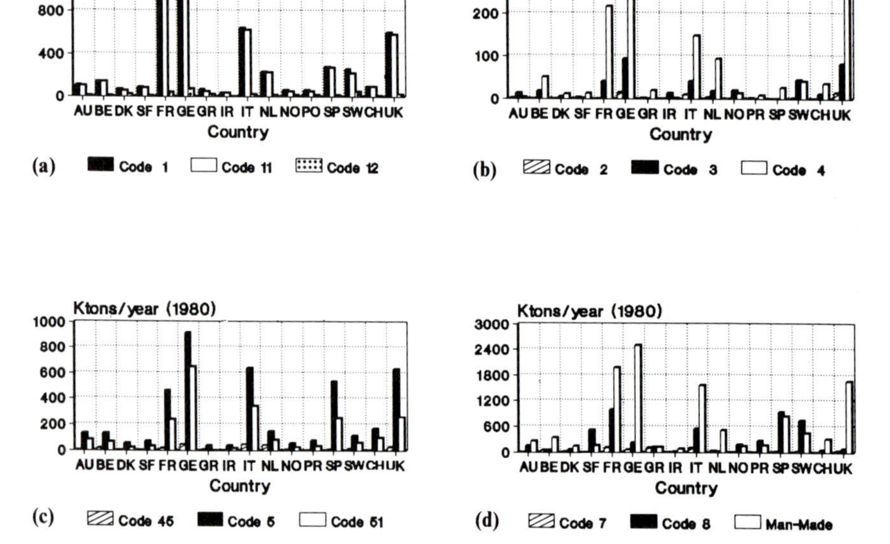

Figure 3.3 National man-made and natural VOC emissions in Europe according to the main sources and source categories reported in Figure 3.1. (a) Road traffic and other mobile sources; (b) power plants, non-industrial combustion and industry; (c) chemical and petroleum industry, solvent evaporation and industrial solvent use; (d) agriculture and food industry, natural sources and total man-made emissions. Data refer to the year 1980. Redrawn and adapted from OECD (1990).

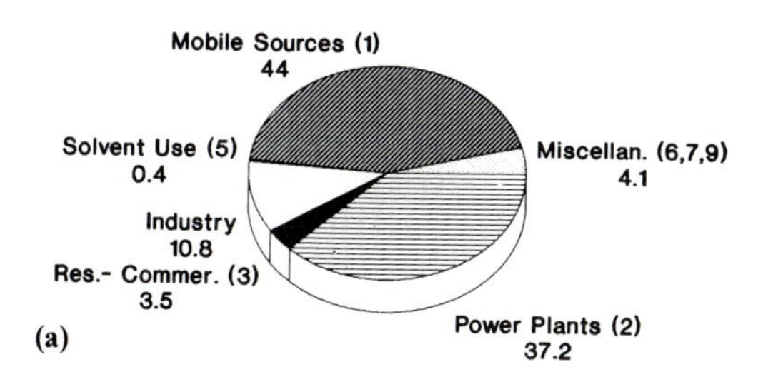

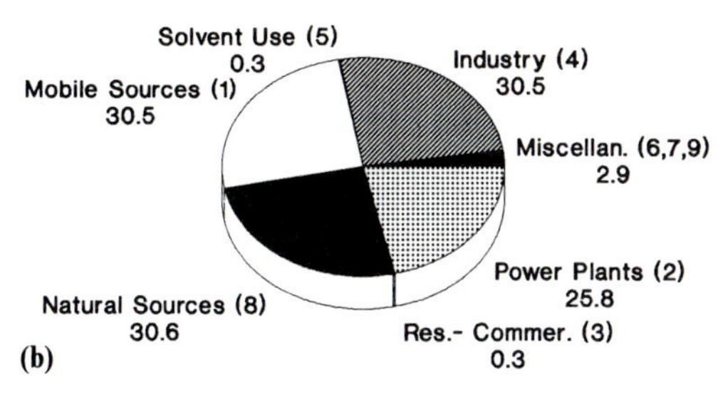

Figure 3.4 VOC emissions by major source sectors in 1980. (a) Man-made; (b) total. (Average shares for OECD Europe.) After OECD (1990).

nitrate (PAN)) (see section 3.3.2) is recommended. This is a step forward to improve existing emission inventories in Europe, and make them closer and possibly compatible with the US data base.

3.2.2 Future trends in the evaluation of VOC emission

The recognition that each organic component affects the quality of the atmosphere in a different way implies that emission data must be provided on a individual basis. In this sense, the US approach can be taken as the reference point for future VOC emission inventories. The large number of compounds to be quantified and the uncertainties still associated with their measure requires, however, that close co-operation is established between local authorities, industry and the scientific community. Since evaluation of VOC emission and testing of the best available techniques necessary for VOC reduction apply to numerous types of existing source, extensive laboratory and field experiments, covering a wide range of individual compounds, must be performed. This requires sufficient funding for developing instrumentation, facilities and protocols. To save resources and avoid duplications, joint projects supported by national and international organizations should be encouraged. Again, the approach followed in the USA (Schuetzle *et al.*, 1991) for evaluating the impact on air quality of reformulated gasolines and methanol-based fuels can be taken as an example. Three car producers and fourteen oil companies have joined together in an Auto/Oil Air Quality

Research Program (AQIRP) aimed at meeting the requirement of a recent law for improving air quality. To be compatible with modelling studies, detailed chemical speciation data on VOC emission were required for more than 8000 samples over a period of nine months. A complete test facility allowing the simultaneous determination of running loss and evaporative, exhaust and engine emissions was used. In addition to total VOCs, various classes of hydrocarbons were analysed. Individual emission factors for paraffins, olefins, aromatics and oxygenated compounds containing one to twelve carbon atoms were obtained. A total of 151 components, accounting for more than 95% of the total emission, were detected; 70 were identified and quantified with 98% accuracy. While details on the analytical system are not pertinent to this section, the basic concept (speciation of individual components and definition of emission factors through well-designed testing facilities) seems perfectly to match the requirements for making an accurate VOC emission inventory. As the source categories defined by the OECD are similar to those identified by the USEPA, it is possible to envisage joint projects for those area sources where little differences in composition exists. A possible area of international co-operation can be identified in the evaluation of individual VOCs emitted by natural sources, particularly vegetation. It should be recalled that most existing inventories use the same data. Such inventories are based on pioneering work carried out in the US (Zimmermann, 1977, 1979; Tingey *et al.*, 1978) and mainly restrict their attention to isoprene and terpene components. However, 367 different compounds have been reported to be emitted by vegetation (Graedel *et al.*, 1976). Among them, semivolatile aldehydes were found to be particularly abundant in forest areas (Yokouchi *et al.*, 1990; Ciccioli *et al.*, 1992a). By participating in the photochemical cycle, such compounds can also contribute to increased levels of ozone and acidic species in the atmosphere. The availability of test facilities enabling botanists, plant physiologists, environmental chemists and meteorologists to join together and, through robust protocols, to evaluate emission of all components emitted by vegetation should drastically reduce present uncertainties. Sharing of costs and experience through international programmes would extend the evaluation of naturally emitted VOCs to a large number of samples so that homogeneous data from tropical to temperate regions can be obtained. Although natural (or man-made naturally associated) emission does not directly affect human health, its evaluation is of vital importance for establishing the extent to which man-made VOC emission must be reduced. Naturally emitted VOCs are, in some cases, as photochemically reactive as man-made ones, therefore they can be a source of pollution in remote areas whenever sufficient levels of NO_x or PAN are associated with polluted air masses transported into the site. Without detailed information on natural emission, it is not possible to determine whether present reduction plans will enable ozone levels to be maintained below the limits fixed by WHO. Although a certain decline in photochemical oxidants has been observed in the USA (USEPA, 1984), the results obtained from 30 years of control programmes were lower than expected and ozone levels in many areas of the USA still exceed national ambient air quality standards (Chock and Heuss, 1987). Another possible item for international co-operation is represented by the evaluation of dimethylsulphide (DMS) emitted by the marine environment. Present estimates indicate that approximately 30 to 40 million tons per year of sulphur are released into the atmosphere as DMS, an amount approximately equal to 50% of all sulphur emitted by natural sources (Bates *et al.*, 1987). As oxidation of DMS ultimately leads to acidic sulphate aerosols (Barnes *et al.*, 1991) it is of great importance to obtain reliable figures.

The few examples reported clearly indicate that the measurement of VOC emissions will present a great challenge for chemists in years to come, with a large amount of work still required to prepare emission inventories based on individual components. It should not be forgotten, however, that a 'perfect' inventory will probably never exist, and that some degree of uncertainty must always be accepted.

3.3 VOC effects

Once emitted, VOCs undergo four main processes: (1) dispersion; (2) physical transformations; (3) chemical reactions; and (4) deposition (Schroeder and Lane, 1988). Dispersion accounts for all diffusion and transport processes caused by turbulent motions and air mass circulation arising from local and global forces. It provides a substantial dilution of emission by distributing VOCs over different air parcels of the atmospheric boundary layer (ABL). Physical transformations, such as condensation into particles, adsorption on their surface or solution in water droplets, remove VOCs from the gaseous phase through equilibrium processes. Chemical reactions convert VOCs into organic or inorganic molecules which, in turn, undergo dispersion and physical transformations. Deposition transfers VOCs from the atmosphere into soil and water reservoirs by slow sedimentation or fast scavenging processes. The balance between the rates of emission, dispersion, physical transformations, chemical reactions and deposition determines the time spent by a component in the atmosphere, the actual distance travelled in the ABL and the aerometric concentrations reached (Schroeder and Lane, 1988). As such rates differ dramatically from compound to compound, residence times of VOCs ranging from minutes to hundred of years can be observed, and concentrations from ppmv to fractions of pptv measured (Finalyson-Pitts and Pitts, 1986; Hov *et al.*, 1986, 1987). The longer the time spent in the atmosphere, the higher the distance travelled by a compound before deposition occurs; there is also the possibility of VOCs affecting the biosphere by changing the secular equilibria established on the earth. Organic compounds characterized by few days' residence time can be dispersed and accumulated over an entire continent (synoptic-scale long-range transport) whereas those having residence times longer than months can be distributed throughout the whole troposphere and, by diffusion, can reach the stratospheric layers (global-scale long-range transport) (Beilke and Gravenhorst, 1980; Hov *et al.*, 1986, 1987). The fact that a rapidly reacting organic component is completely consumed within few kilometres of the emission leaving its source does not necessarily imply that the effects are restricted to that area. If the product formed is characterized by a residence time longer than that of its precursor, it can act as propagating agent of secondary effects in space and time. Particularly important is the case where the products formed have a definite impact on living organisms and the environment. By considering that some VOCs are toxic to humans, animals and plants, there are numerous ways in which accumulation in the air can affect human life and the environment. This explains why VOCs have been included by some environmental agencies in the list of 'criteria' pollutants. According to legislation, this means that VOCs must be considered as pollutants having adverse effects on public health and welfare, and that their presence in ambient air results from numerous and diverse stationary sources. As 'criteria' pollutants are the fundamental indices used for defining the quality of the atmosphere, primary and secondary air quality standards have been established (Finalyson-Pitts and Pitts, 1986). While primary standards are mainly designed to protect public health with an adequate margin of safety, secondary standards are designed to protect public welfare. According to US and Canadian legislation, deleterious effects can be felt whenever concentrations of non-methane VOCs exceed 0.24 ppmv of carbon from 6 to 9 am (Finalyson-Pitts and Pitts, 1986). In Italy, the 3h standard for non-methane VOCs has been fixed at a value of 200 $\mu g/m^3$ (DPR, 1983). However, in contrast to the air quality standards defined for SO_2, NO_2, ozone, CO, fluoride, lead and suspended particulate matter, the VOC standard applies only when hourly averaged concentrations of ozone exceed 200 $\mu g/m^3$ more often than once a month. Being related to ozone, the VOC standard is clearly intended for protecting health and welfare from photochemical pollution and does not take into account other short- and long-term effects associated with VOC emission. Due to the specific mechanism of the law, values exceeding 1.5–2 ppmv of VOC methane equivalent are currently measured in the major Italian cities; VOC standards are often exceeded from late April to September. Regardless of the various actions taken by the Italian government to maintain VOCs within the air quality standards, the question remains as to whether the non-methane VOC content can be still considered as an adequate index to express the quality of the atmosphere. Although more than ten years of experience worldwide has clearly shown the severe limitations associated with the adoption of standards based on the non-methane VOC content, there are considerable difficulties in the selection and identification of organic components for defining new air quality standards for VOC. Possible sources of controversy can be better understood through a detailed analysis of the effects associated with VOC emission.

3.3.1 VOCs as air toxics

The first aspect to be considered is whether exposure to VOCs at levels existing in the atmosphere does represent a serious threat to human health by inducing mutagenic, carcinogenic or teratogenic effects, or by causing acute and chronic diseases. If we restrict the discussion to cases where no accidental disaster causing fugitive emission of high amounts of highly toxic organic chemicals has taken place, a well-defined correlation between human diseases and exposure to the levels of VOCs present in the air has yet to found. Undoubtedly, whenever air quality standards are exceeded some risk exists but evaluation is made difficult by the fact that inhabitants of industrialized countries may spend as much as 80 to 90% of their time participating in indoor activities which also involve exposure to VOCs. To obtain information on the array of chemicals and levels to which individuals can be exposed, Shah and Singh (1988), starting in 1980, expanded the national ambient data base prepared for the USEPA (Brodzinsky and Singh, 1980) to include all organic compounds detected at outdoor and indoor sites through the various monitoring activities carried out in the USA. The final data base consisted of 320 organic components, with 261 measured in the outdoor air and 66 measured indoors. By analysing the ambient data collected by various agencies located in 300 cities from 42 states, it was found that nearly 50% of the organic compounds were falling within the concentration range 0.001–1 ppbv. A sizeable fraction showed concentrations lower than 0.001 ppbv, whereas the median concentration exceeded 1 ppbv in only 10% of cases. Among the 66 potentially more toxic components only 10 compounds showed median daily concentrations higher than 1 ppbv. Particularly abundant were formaldehyde, formic acid, 2-propenal, propanal and phenol, the concentrations of which ranged from 4 to 8 ppbv. Benzene, acetaldehyde, 1,3-dimethylbenzene and butanal occurred at concentrations of approximately 1.5 ppbv, whereas 1,3-butadiene slightly exceeded 0.3 ppbv. It must be said, however, that some of the compounds reported in the list (namely formaldehyde, acetaldehyde and formic acid) are also formed by photochemical degradation of reactive VOCs and may be included in the effects discussed in section 3.3.2. Comparison with indoor data supplied by 30 cities from 16 different states showed that daily median concentrations of many compounds were higher in indoor sites. For some VOCs the indoor median concentrations were lower than outdoor values, and scattering between lower and upper quartiles was suggestive of episodic bursts where outgassing of solvents from new building materials had occurred. In spite of the uncertainty associated with the statistical treatment of data provided by various sources using different sampling and analytical techniques, the observation that indoor VOC pollution exceeds outdoor pollution has been confirmed by further studies (Yocom, 1982; De Bortoli *et al.*, 1985; Wallace *et al.*, 1987; Pitts *et al.*, 1989). Consequently it may be more effective in causing human diseases (Sexton and Hayward, 1987). We should also not forget that indoor and outdoor exposure to VOCs are just two of the possible ways in which toxic organic compounds, such as chlorinated solvents and arenes, can penetrate the body. Ingestion of food or drinking water and dermal adsorption through handling of materials where toxic organics are dissolved are also important in determining the net amount absorbed, therefore any attempt to establish correlations between atmospheric exposure to toxic VOCs and human diseases requires extreme caution. If the quantitative risk assessment based on the carcinogenic potency of compounds, expressed as unit risk value (USEPA, 1987), is applied to the data reported by Shah and Singh (1988) formaldehyde and 1,3-butadiene can be seen to account for approximately 70% of carcinogenic risk, whereas much of the rest comes from benzene, trichloro- and tetrachloroethylene and dichloromethane. The chances of a chronically exposed individual developing cancer can be estimated to be 1 to 2 per 10 000 at most—a relatively low number when compared with occupational exposure. The approach taken is of course very crude as people can actually be exposed to much higher VOC levels if they live close to strongly emitting sources such as chemical plants, refineries, refuelling stations or busy roads where traffic jams often occur. A more detailed analysis of the same VOC data, carried out by reporting the daily median aerometric concentrations of 20 organic compounds as a function of the type of site where they were collected (urban, suburban, rural and remote areas), showed that people living in large cities are exposed to levels of potentially toxic VOCs ten times higher than those occurring in rural areas (Shah and Sing, 1988), thus the risk may become serious. In the case of benzene and toluene daily median concentrations of 0.46 and 0.36 ppbv, respectively were measured in rural sites compared with 1.8 and 2.88 ppbv in urban areas. The same single order of magnitude difference between urban and semi-rural sites was measured during monitoring campaigns in Rome and the suburban area of Montelibretti, located 30 km from the city centre. Examples of the diurnal profiles recorded for benzene and toluene when national air quality standards were exceeded in the two sites are shown in Figure 3.5. Attention was focused on benzene and aromatics as these are the most widespread toxic components in air, resulting from vehicular emission (see Figure 3.3(a)). In many countries vehicular emission accounts for approximately 50% of benzene, toluene and other alkylbenzenes present in the air. Consequently, specific attention has been paid worldwide to reduce their emission. In Italy, the Ministry of the Environment together with petroleum industries has decided to reduce the benzene content in gasoline from the 5% figure suggested by the EC directives to less than 3.8% (CONCAWE,

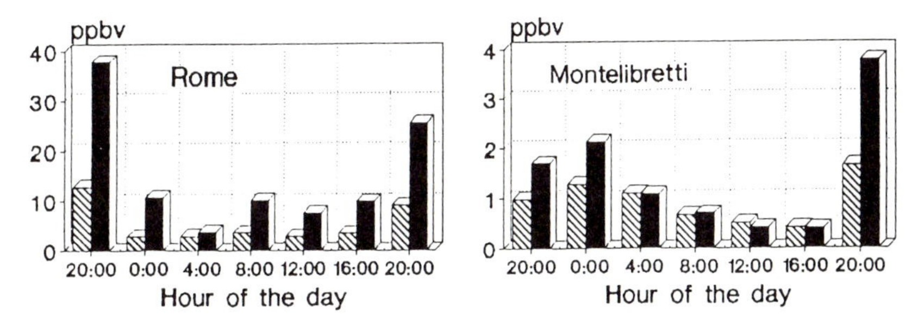

Figure 3.5 Differences in the levels of benzene and toluene to which individuals living in urban (Rome) and suburban/semi-rural areas (Montelibretti) may be exposed. Approximately 340 kg/h of benzene are estimated to be emitted in the area where monitoring was performed.

1991). Moreover, the maximum concentration of benzene allowed to be present in the stacks of industrial plants is 1.5 ppmv when the total mass flow is equal to or higher than 25 g/h (DPR, 1990). If other toxic substances known to be carcinogens, mutagens and teratogens (e.g. acrylonitrile, 1,3-butadiene, 1,2-dibromoethane, 1,2-dichloroethane and vinylchloride) are also present in the emission, the limit of 5 mg/m^3 applies to the total amount of toxic organic substances emitted. The need for such action becomes clear if we consider that present day levels of benzene and aromatics measured in Rome are only slightly different from those measured in the 1970s and 1980s (Ciccioli *et al.*, 1976; Possanzini, 1981) and that daily averages are still higher than those measured in most polluted areas of the USA and Canada (Dann *et al.*, 1989). Although tailpipe emission and fugitive evaporation in Italy has decreased in the last ten years because of the better quality of cars, the number of circulating vehicles in large cities has increased to such an extent (Gaudioso *et al.*, 1991) that the situation can only be improved by diminishing the benzene content in gasoline and adopting three-way catalysts. It is possible, however, that further benzene reductions might be necessary in future. The decision made in California to limit the benzene content of gasoline to 0.8% from January 1993 (CONCAWE, 1991) clearly indicates that atmospheric levels of this compound are still too high and, in spite of the large action taken in the last 30 years, further efforts are needed to reduce the potential risk it might represent to human health. The effects of VOC accumulation in the atmosphere are not restricted to potential carcinogenicity, mutagenicity and teratogenicity, as irritation of the eyes and nose have been also associated with the presence of certain components in air. Formaldehyde, acrolein and, to a less extent, acetaldehyde are known to be eye irritants and can induce lacrimation above certain levels (CCMS–NATO, 1974). Formaldehyde appears to be detectable by optical chronaxy at concentrations of the order of 60 ppbv, although threshold eye irritation can occur at lower levels (> 10 ppbv). Acrolein can be detected by odour and eye irritation at levels close to 250 ppbv, whereas threshold optical chronaxy occurs at approximately 750 ppbv. Acetaldehyde is much less irritating than formaldehyde and almost no physiological effects are thought to occur in ambient air. Although large amounts of oxygenated components are formed by photochemical oxidation of rapidly reacting VOCs, they are also emitted by anthropogenic sources. In addition to their emission from chemical plants and oil refineries, aldehydes with one to six carbon atoms are present in vehicular exhaust emission at a total level close to 1% (Schuetzle *et al.*, 1991), while high amounts of formaldehyde are emitted by building materials made from urea-formaldehyde bonding resins, consumer products and indoor combustion. Levels as high as 100 ppbv has been measured in 'energy efficient' homes with relatively low ventilation rates, solely from gas-fired stoves (Pitts *et al.*, 1989). Nose irritation, annoyance and headaches are other symptoms frequently associated with the occurrence of high VOC levels in the atmosphere. The high nervous sensitivity to certain substances (e.g. mercaptans and ketones) makes it possible to detect them by odour at ppbv levels. Oil refineries, paper industries (Kraft mill pulp) and waste disposals emanate such strong odours that the location of the source can be identified by the human nose. This is why olfactometric methods are still endorsed by some environmental agencies (ASTM 1981; Perrin *et al.*, 1986) and used as a simple means of identifying sources and estimating air concentrations. Continuous exposure to high levels of intense odorous organic compounds may result in reduced olfactometric sensitivity and headaches (Goldsmith and Friberg, 1987), thus reducing the ability of individuals to perform their usual activities.

When considering the effects of VOCs we should not forget possible damages to plants. In this respect, ethylene, which is the most abundant component in air based on mole volume, has long been known to behave as a strong phytotoxic component (Stahal, 1969; USEPA, 1986). It can cause growth retardation, epinasty, abscission of leaves, senescence of leaves and flowers, irregular opening of flowers and fading. Its effects are most evident in flower crops and have been known since the beginning of this century. Although average levels in air do not seem to represent a potential danger to plant growth, local effects can reduce crop production during the flowering season.

3.3.2 VOCs as precursors of photochemical smog and acid deposition

The key role played by VOCs in determining photochemical smog pollution was identified as early as the 1950s (Haagen-Smit and Fox, 1956). At that time photochemical pollution was believed to be mainly restricted to the Los Angeles basin and for this reason it was usually termed 'Los Angeles smog'. The term smog was coined because one of the most obvious effects of this type of pollution was reduced visibility associated with the production of fine particles. Twenty years later, smog was already recognized to be a serious problem plaguing many of the largest cities in the USA (US Department of Health, Education and Welfare ((USDHEW), 1970), Europe

(Guicherit and Van Dop, 1977), Japan and Australia (OECD, 1975). Today, it is considered to be one of the most important transboundary pollution problems in the troposphere. When VOCs are mixed with nitrogen oxides (NO_x) and irradiated by ultraviolet (UV) light, a complex chain of reactions converts them into products generally indicated as photochemical pollutants (Demerjan *et al.*, 1974; Cox, 1979; Finlayson-Pitts and Pitts, 1986; Carlier and Mouvier, 1988). The atmosphere then behaves as a photochemical reactor whereby the input of reagents is provided by natural and anthropogenic emission and energy is supplied by the UV component of sunlight. Mixing of reagents and products is provided by diffusion and eddies; the height of the mixing layer and geographical contours define the volume of the reactor. Advection provides exchange of matter with contiguous parcels of air. The type of reactions occurring in the reactor are mainly radical and the chain ultimately leads to the production of ozone, aldehydes, hydrogen peroxide, peroxyacetyl nitrate (PAN), organic and inorganic acids, and fine particles. Among these, ozone is considered to be the most important because of the large concentration reached (in some instances $\gg$200 ppbv) and the wide range of effects it may have on human health, plant growth, materials and climatic change. Its reactivity toward important chemical moieties present in the body can induce changes of cells and fluids of the respiratory tract, causing alterations in the mechanical functions of the lungs and structural injuries of specific types of cells in the bronchial region (Tilton, 1989). As a result, coughing, shortness of breath, and nose and throat irritation coupled with discomfort and pain when breathing deeply have been observed in humans. Exposure to increasing concentrations of ozone, from 80 to $>$240 ppbv, results in a decrease in the forced expiratory volume at 1 s (FEV 1 s) and reduced human performance under stressed conditions (USEPA, 1986; Tilton, 1989). This happens because ozone can affect cilia, alter mucus secretion and reduce viability, bactericidal activity and interferon production of alveolar macrophages—all located in the region linking the respiratory tract with the tracheobronchial region, where host defense against foreign particles mainly occurs. Ozone can also penetrate the body deeply to affect the centriacinar region by injuring epithelial cells, increasing collagen in the interalveolar septa and changing the structure of distal airways in the terminal tract of the bronchus (bronchioles). Young and healthy adults represent the portion of population most at risk from transient lung function impairment (Tilton, 1989). Insurgence of such effects is a cause of discomfort and reduces the capability to perform activities. Ozone can also enter plant leaves through stomata, penetrating the intercellular space and passing into the liquid phase of cells where it can react with the active sites of the cell membrane, cytoplasm and cellular organelles (Tingey and Taylor, 1982). By injuring parts of the cell it can inhibit photosynthetic activity and allocation and translocation of sucrose from the shoots to the roots (USEPA, 1986). Ozone is also able to enhance evolution of phytotoxic ethylene in plants. Because of these biochemical and physiological effects, it can reduce biomass, growth and yield of vegetation and cause visible injuries (USEPA, 1986). Occurrence of visible injuries is actually exploited to monitor ozone in the air by using particularly sensitive species (i.e. Bel W-3 tobacco cultivar, dicotyledons and monocotyledons) as bioindicators. The appearance of darker coloured areas, indicative of the presence of necrotic tissue in the interveinal space, is taken as a measure of photochemical pollution by ozone. Estimates indicate that a 10% increase in the mortality of sensitive trees and a 25% reduction in their height can be observed following continuous exposure to high ozone levels (USEPA, 1986). It has been calculated that ozone damage to crops (soybean, kidney bean, corn and peanuts) in the USA resulted in a loss in productivity of $3 billion per year (Hileman, 1982). Attack of the double bonds of elastomers, fibres and dyes by ozone may result in the disruption of molecular structure, leading to cracking of elastomers and peeling, erosion and discoloration of painted surfaces (USEPA, 1986). In the USA damage to elastomers accounts for $1.5 billion losses each year (Mueller and Stickney, 1970). Effects associated with the presence of the other photochemical pollutants in air (PAN, aldehydes and hydrogen peroxide) have not been estimated with accuracy. Although PAN is certainly a phytotoxic component (Taylor, 1969) and a strong mutagenic agent, levels in very polluted atmospheres seldom exceed 20–30 ppbv (Altshuller, 1983a; Temple and Taylor, 1983) and it is believed that the effects on human health and vegetation are small when compared to those caused by ozone. The same is true for hydrogen peroxide. The potential risk represented by formaldehyde of photochemical origin is hard to evaluate as the source of this component is often man-made. Although PAN, aldehydes and hydrogen peroxide are not regarded as 'criteria' pollutants their monitoring is essential to enable better understanding of the degree of reactivity in the atmosphere, and the transport phenomena. The great importance attributed to photochemical pollution is linked to the fact that the residence time of ozone in the atmosphere (Hov *et al.*, 1978) is long enough (a few days) that, once formed in a certain area, it can propagate over regions covering several European countries and American states. This usually occurs during the so-called photochemical smog episodes observed when anticyclonic weather is established over air parcels characterized by intense VOC and NO_x emission (usually urban and highly industrialized areas). Since such favourable conditions can last for 2 to 5 days before a breakdown situation occurs in the ABL (Smith and Hunt, 1978), polluted air masses can be transported far away from the original place where production started. If deposition losses of photochemical oxidants are totally or partly compensated by their production occurring along the trajectory followed by the air masses, high levels of ozone ($\gg$100 ppbv hourly averages) and PAN ($>$10 ppbv hourly averages) can be recorded in sites thousands of kilometres distant from one another (Wolff *et al.*, 1982; Levy *et al.*, 1985). Only rapid changes in weather conditions, combined with turbulent mixing and rapid scavenging of pollutants, can reduce atmospheric levels down to hourly concentrations of 40–60 ppbv. It is important to note that the highest values of ozone and photochemical oxidants are usually reached 30 to 100 km away from the core of high emission. This is because ozone and PAN near the source are scavenged by reaction with the high amounts of NO released by combustion and vehicular sources. To understand the role played by VOCs in photochemical smog formation it is useful to refer to Figure 3.6 where the main daytime reactions are reported in the form of a cycle (Ciccioli and Cecinato, 1992). To simplify

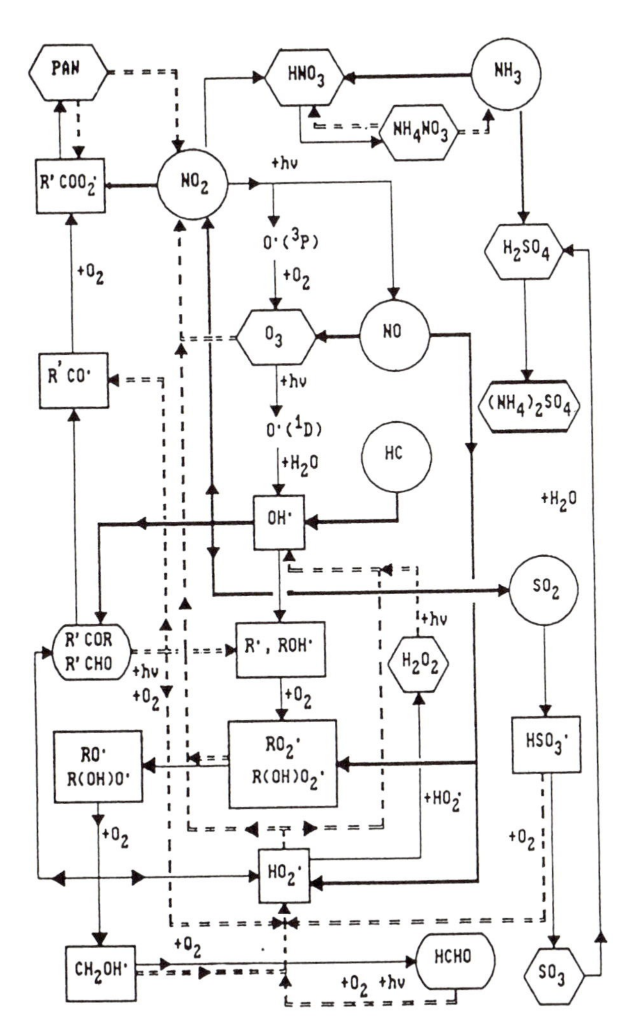

Figure 3.6 Schematic diagram of the main daytime reactions contributing to photochemical smog formation and dry acid deposition. For explanation see text. From Ciccioli and Cecinato (1992).

this figure only the main reactions are shown, no distinction is made between heterogeneous and homogeneous reactions, and lumped mechanisms are often used to describe processes. Circles identify reagents, while rectangles and hexagons are used for products. Rectangles with curved edges indicate compounds that can either be formed or directly emitted. Lines marked with arrows are used to indicate reactions between compounds enclosed in boxes. Lines emerging from boxes lead to reaction products. Dashed lines are used for those products that are pumped back into the cycle. For a detailed description of the reactions rates and mechanisms the reader should refer to specialized literature (Demerjan *et al.*, 1974; Finlayson-Pitts and Pitts, 1977, 1986; Cox, 1979; Atkinson and Lloyd, 1982; Calvert and Stockwell, 1983; Carlier and Mouvier, 1988). If, for the moment, we disregard some of the reactions involving NO_2 and SO_2, we can summarize the various processes starting from the photostationary state linking NO, NO_2 and ozone. In unpolluted atmospheres such as those existing before humans appeared on the earth, both VOCs and NO_x were originated by natural processes (e.g. biogenic emission, accidental forest fires, volcanic emission), and maximum hourly concentrations of ozone present in the atmosphere were probably in the order of 10–20 ppbv (Hov *et al.*, 1986). By injecting into the cycle increasing amounts of NO_x and VOCs, man has slowly but constantly increased the ozone levels. Some historical records indicate that background levels in Central Europe are now 100% higher than those occurring a century ago (Hov *et al.*, 1986). As more VOCs are pumped into the cycle more oxidized organic species, capable of converting NO to NO_2, are produced by reaction with OH radicals. Since these radicals, which are the real motor of the whole cycle, are continuously restored by photolysis of oxidized products or by reaction with NO, the final result is accumulation of ozone, PAN, aldehydes (especially formaldehyde) and hydrogen peroxide in the air during daytime hours. In strongly photochemically polluted atmospheres, VOCs can also be consumed by reaction with ozone, leading to organic acids through Grieeges intermediates (Atkinson and Carter, 1984). The diurnal profiles shown in Figure 3.7 illustrate well the working mechanism of the photochemical cycle. Again, observations refer to the suburban area of Montelibretti located downwind of the city of Rome. After sunrise, VOCs of natural and anthropogenic origin are consumed by reaction with OH radicals and a substantial decrease in their concentration is observed. Oxidation processes first produce aldehydes and ketones, which, by photolysis or reaction with OH radicals inject peroxy radicals (RO_2) into the cycle thus promoting formation of ozone and PAN. Rapid production of photochemical pollutants gives rise to their accumulation in air. Maximum values are usually reached during the afternoon. After sunset, when most (not all) of the atmospheric reactivity ceases because of disappearance of sunlight, VOCs are no longer consumed and increases in their aerometric concentrations take place in spite of decreased emission. Besides the low night-time reactivity, VOC accumulation

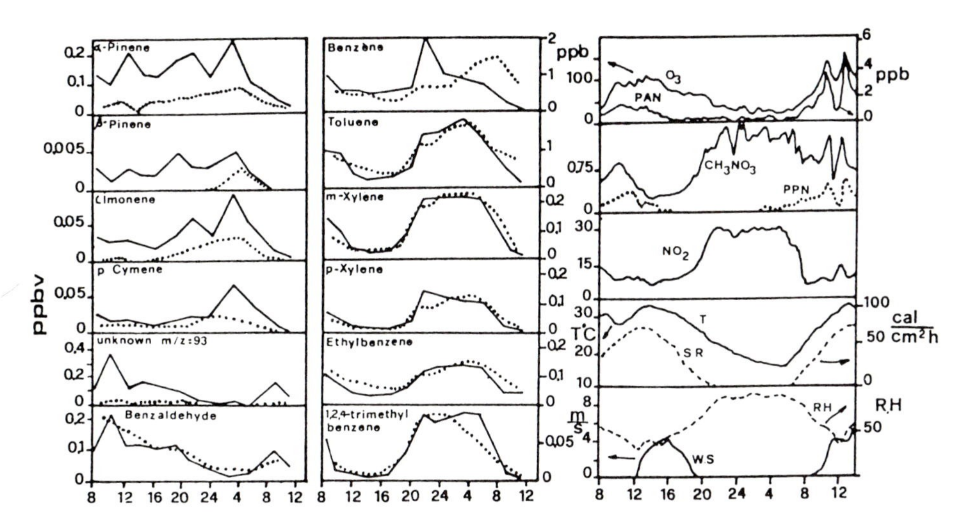

Figure 3.7 Diurnal profiles followed by man-made and naturally emitted VOCs, aldehydes, PAN, PPN, methyl nitrate and NO_2 during a photochemical smog observed in the Tiber valley in July 1984. T, SR, RH and WS are used to indicate temperature, solar radiation intensity, relative humidity and wind speed, respectively. Solid lines refer to ground-based measurements, dotted lines refer to determinations carried out above the tree canopy (12m). From Ciccioli *et al.* (1986).

in air is also promoted by meteorological factors, the most important of which is the progressive lowering of the inversion layer at night. This greatly reduces the area over which VOCs are dispersed. As secondary pollutants and radical species are rapidly scavenged by night-time homogeneous and heterogeneous reactions and deposition processes, the lowest ozone and PAN concentrations are usually reached late at night. In Figure 3.7 the influence of transport is also evident. It is revealed by the sharp ozone and PAN peaks recorded during the mid-afternoon hours of the second day of monitoring. This type of event is always associated with the circulation of sea breezes, which push polluted air masses deep inside the Tiber valley where Montelibretti is located (Ciccioli *et al.*, 1989). Although the ozone levels recorded during this episode (150 ppbv) were far above the air quality standards defined in Italy, they do not represent the highest levels that can be observed in the ABL. Daily maximum 1 h concentrations approaching 350 ppbv were reported in some areas of the Los Angeles basin subjected to severe photochemical smog episodes (Chock and Heuss, 1987). Although the continuous observations made in the USA from the mid-1970s to the mid-1980s are indicative of a slow decrease in the annual second highest daily maximum 1 h concentrations of ozone, levels exceeding 125 ppbv were still recorded in 183 urban sites of the USA a few years ago (Chock and Heuss, 1987). According to the data collected by the OXIDATE network (Grennfelt *et al.*, 1988), present levels of ozone in

Northern Europe are similar to those measured in the USA. In all cases, they are still far higher than the limits established by the WHO to protect health and vegetation. While it is clear that meeting national or international air quality standards for ozone necessarily implies that VOC and NO_x emissions must be controlled, the way to do this is not as simple as it might appear. From our present knowledge of the photochemical cycle, we can say that, in general, effective control of VOCs will decrease the ozone maxima, whereas NO_x emission reductions may result in either an increase or a decrease (Seinfeld, 1988). The counteracting effects of VOCs and NO_x in timing the chemistry of the photochemical cycle (the slowing down of VOC reduction can be balanced by the speeding-up effect of NO_x reduction) make it difficult to predict the effect of combined control of VOCs and NO_x emissions; in general, however, lowering of the ozone peak is expected (Seinfeld, 1988). Whether the reduction of ozone can be better accomplished by VOC control alone or by combined control of VOCs and NO_x is a complex question, with the solution requiring the use of computational models. For areas characterized by a clearly defined urban core and simple paths followed by the air masses, empirical trajectory models (such as the EKMA model used in the USA) are capable of generating curves (isopleths) of daily maximum ozone levels as a function of the concentrations of non-methane VOCs and NO_x present in air (Finlayson-Pitts and Pitts, 1986). Based on the specific form of the isopleths, it can be decided which is the most effective approach for reducing maximum ozone levels generated downwind of the emitting sources. Simple models are, however, limited in domain (urban airsheds) and do not allow the development of control strategies for single or small groups of emission sources (Seinfeld, 1988). Due to the transboundary features of photochemical pollution, the use of Lagrangian or Eulerian models incorporating detailed descriptions of physical and chemical processes is preferred, despite the large amount of input data and detailed inventories required. Such models make it possible to simulate scenarios where VOC reduction is applied to different sources and source categories, and to investigate control effects on those areas (rural, forest and remote) subjected to long-range transport of photochemical oxidants (Seinfeld, 1988). Models are not only useful for deciding if and to what extent the emission of non-methane VOCs should be reduced, but also allow the measurement of the reactivity of individual VOCs under the prevailing conditions leading to photochemical smog pollution. This information is essential because the photochemical reactivity of VOCs is a complex function which is impractical (and sometimes impossible) to derive from smog chamber studies due to the extensive experiments necessary to simulate all situations occurring in the atmosphere (Seinfeld, 1988). To understand the meaning of the term VOC 'reactivity' relative to photochemical smog pollution, it is useful to analyse the main factors affecting the cycle shown in Figure 3.6. Since some VOCs are more reactive towards OH radicals than others (Atkinson, 1986), and each one has to compete for OH radicals to form the species necessary for the production of ozone and PAN in the presence of nitric oxide, the composition of the organic emission is crucial in determining the final amount of photochemical pollutants that can be formed. If we regard the whole atmosphere as a three-dimensional assembly of interconnected photochemical flow reactors, each one characterized by its own chemical and physical equilibrium, the reactivity of a given VOC mixture will also depend on the boundary conditions established in each reactor. This means that the same organic species can produce different amounts of photochemical pollutants depending upon the temperature, pressure, solar radiation intensity and relative humidity existing in the air. The situation is further complicated by the fact that changes in reactivity occur because of the continuous input and removal of reagents and products due to advection, diffusion and deposition processes. Since advection depends on the speed of moving air masses, whereas deposition rates change as a function of the recipient surface (e.g. type of soil and its coverage, presence of watersheds and their nature), a large number of variables must be considered when evaluating the VOC reactivity. Available models can take into account (with some unavoidable approximations) many of the chemical and physical processes mentioned above, therefore they can be used to derive the reactivity of individual VOCs under the atmospheric conditions prevailing in a given area (Carter and Atkinson, 1988; Lowi and Carter, 1990; Derwent and Jenkin, 1991). First of all, a set of scenarios that are fully representative of the situations leading to photochemical pollution and its transport must be identified for the area under investigation. The composition and rates of VOC emission must be accurately evaluated for each grid into which the area has been divided and the model validated through repeated field observations. Simulations in which one of the VOCs present in the original emission is set to zero can then be performed. A decrease in ozone levels, proportional to the role played by that specific component in the whole cycle, will be observed and data exploited for deriving the hydrocarbon reactivity. By using a Lagrangian model, Derwent and Jenkin (1991) have been able to evaluate the reactivity of 69 different hydrocarbons for various trajectories leaving or reaching the British Isles. Ozone gradients were measured at regular time points (e.g. every 24h starting from mid-afternoon on the first day) for each trajectory followed by the air masses during typical photochemical smog episodes lasting five days. To normalize ozone gradients with respect to VOC emission, data were divided for the hydrocarbon emission integrated up to the same time points at which ozone gradients were determined (i.e. mid-afternoon). The hydrocarbon reactivity was expressed in terms photochemical ozone creation potential (POCP) units calculated with respect to ethylene as a reference compound (POCP = 100). POCP values for some selected VOCs are reported in Table 3.1. Numbers were obtained by averaging POCP values measured by simulating different trajectories and

Table 3.1 Photochemical ozone creating potentials (POCPs) of some VOCs calculated for the British Isles using a Lagrangian model. Selected values from Derwent and Jenkin (1991); compounds marked with an asterisk from Derwent (1991).

VOC	POCP	VOC	POCP
Alkanes		**Aromatic hydrocarbons**	
Methane	1	Benzene	20
Ethane	10	Toluene	55
Propane	40	Ethylbenzene	60
n-Pentane	40	*o*-Xylene	65
Isopentane	30	*m*-Xylene	105
n-Hexane	50	*p*-Xylene	90
2,3-Dimethylbutane	40	1,2,3-Trimethylbenzene	115
Branched C12 alkanes	40	1,2,4-Trimethylbenzene	120
		C10-Trisubstituted benzene	115
Cycloalkanes			
Cyclopentane*	50	**Oxygenated hydrocarbons**	
Methylcyclopentane*	50	Formaldehyde	40
Cyclohexane*	25	Acetaldehyde	55
		Propionaldehyde*	60
Olefins		Acrolein	120
Ethylene	100	Benzaldehyde	−35
Propylene	105	Acetone	20
1-Butene	95	Methanol	10
2-Butene	100	Ethanol*	25
1-Pentene	70	*n*-Propanol	45
2-Methylbut-2-ene	80		
1,3-Butadiene*	105	**Chlorinated hydrocarbons**	
Isoprene*	100	Methylene chloride	1
α-Pinene*	50	Chloroform	1
β-Pinene*	50	Methyl chloroform	0
Acetylenes			
Acetylene	15		

various VOC and NO_x inputs. It can be seen that olefins and trimethylbenzenes represent classes of VOCs with a high potential for ozone production, followed by aldehydes and ketones; on the other side of the scale lie alkyl halides and methane. Interesting is the case of benzaldehyde, which, unlike other VOCs, gives rise to ozone reduction. This happens because photolysis is unimportant, and removal of NO_x through the reaction leading to peroxybenzoyl nitrate slows down the formation of ozone. Although a general agreement is found with other POCP scales (Lowi and Carter, 1990), all of which separate VOCs in the same major classes of significance with olefins and aromatics being more important than alkanes and alcohols, the values obtained by Derwent and Jenkin seem to give more importance to alkanes and less to formaldehyde. Differences arise not only from the type of emission and scenarios considered but also from the fact that different models, one focused on long-range transport situations, the other restricted to urban scale, were used. With a procedure similar to those just described, the photochemical PAN creation potential (PPCP) of individual VOCs can also be calculated although a different reference compound (e.g. propylene) may be used (Derwent and Jenkin, 1991). Once the POCP and PPCP are obtained for a certain area, it is possible to estimate the potential of ozone and PAN production associated with a specific source or source category by simply multiplying the amount of each VOC present in the emission for the corresponding POCP and PPCP values. This information is crucial for deciding which, among the available control techniques, is the most effective one for reducing ozone, PAN or both. The use of POCP and PPCP approaches would also prevent the adoption of expensive VOC control techniques and, because they are based on mass reduction criteria would leave the ozone levels unchanged. Computer simulations performed with different control scenarios have shown that a 50% reduction of actual VOC emission in Europe would only produce a 20% decrease in the existing ozone levels if control techniques having the same efficiency for all components are used (OECD, 1990), therefore there is a significant risk that the wrong abatement strategies would simply result in a waste of money.

Although important, the formation of oxidants is just one of the aspects of photochemical pollution, as production of the species responsible for acid deposition (dry and wet) also takes place when precursors other than VOC and NO_x enter into the cycle and additional homogeneous and heterogeneous reactions are added to the scheme shown in Figure 3.6 (Penkett *et al.*, 1979a; Atkinson and Lloyd, 1982; Calvert and Stockwell, 1983; Cox and Penkett, 1983; Calvert *et al.*, 1985; Finlayson-Pitts and Pitts, 1986). OH radicals can react directly with NO_2 to form gaseous nitric acid, or can oxidize SO_2 to SO_3; by further reacting with water, this produces sulphuric acid aerosols. OH radicals are also responsible for the oxidation of reduced sulphur components (e.g. H_2S, COS and DMS produced by natural sources) to SO_2. In the presence of ammonia, both nitric and sulphuric acid are partly neutralized to form sulphate and nitrate salts. Accumulation of fine sulphate and nitrate aerosols is mainly responsible for visibility reduction observed during photochemical smog episodes. Extensive oxidation of sulphur also takes place in water droplets where hydrogen and organic peroxides formed by photochemical reactions are dissolved together with SO_2 (Calvert and Stockwell, 1983; Calvert *et al.*, 1985). When clouds pass across photochemically polluted areas with high levels of SO_2, a drastic lowering of pH combined with accumulation of sulphate and nitrate takes place in the droplets. Washout effects can further increase the acidity of wet deposition so that a pH as low as 3.5 can be measured. Acidity, together with high contents of sulphate, nitrate, organic ions (formate, acetate) and secondary pollutants, is then transferred to the surface of the earth by dry and wet (rain, fog, snow) deposition processes. Accumulation of these components in watersheds and soils incapable of buffering the acidifying effect of deposition can cause damage to aquatic ecosystems and forests (Schofield, 1976; Overrein *et al.*,

1980; Lefohn and Brocksen, 1984; Driscoll and Newton, 1985; USNAPAP, 1991). For this reason, critical loads of hydrogen, sulphate and nitrate ions (Henriksen and Brakke, 1988) have been established for sensitive regions of Europe (RIVM–ECE, 1991). In addition to natural ecosystems, acid deposition can also cause damage to ancient, carbonaceous stones by converting them to soft material (gypsum) incapable of resistance to erosion (USNAPAP, 1991). Although present knowledge of the health risks represented by acid species is incomplete, inhalation of acid aerosols (H_2SO_4, NH_4HSO_4) and nitric acid vapours is known to cause bronchitis and worsening of symptoms in asthmatic patients by altering mucociliary clearance of the lungs (USNAPAP, 1991). Exposure to sulphuric acid can increase the airway resistance, decrease the forced expiratory volume and increase the responsiveness to bronchoconstrictors. Effects in asthmatic children have been observed at concentrations of 68 $\mu g/m^3$, whereas concentrations six times higher were necessary to induce symptoms in adults (USNAPAP, 1991). As increased acidity is always associated with the presence in air of substantial ozone levels, synergistic health effects may be observed if the population is exposed to photochemically polluted airsheds. Because of the heterogeneous chemistry and larger domains covered, modelling of acid deposition is much more complex than that of ozone and PAN. Reliable predictions can only be made by using sophisticated models (such as the RADM models mentioned in section 3.2.1) requiring a large amount of input data covering large areas of the earth's surface. For this reason surrogate species must be used for describing the chemical–physical behaviour of VOCs. Although strategies aimed at reducing the effects associated with acid deposition focus mainly on sulphur and nitrogen components, VOCs play a fundamental role in the whole cycle and substantial benefits can be gained through a drastic reduction of their emission. As formation of phytotoxic components by dissolution of VOCs and their oxidation products in acid water droplets cannot be completely ruled out (Gaffney *et al.*, 1987), the importance of VOCs in acid deposition may be even greater than is presently believed and research in this field must increase.

3.3.3 VOCs as a source of stratospheric ozone depletion

Oxygen photolysis at wavelengths lower than 242 nm, followed by reaction with a neutral molecule, is the basic mechanism for ozone formation in the upper stratospheric layers. Photolysis, recombination with atomic oxygen and reactions with hydrogen and hydroxyl radicals originated by dissociation of water vapour, methane and hydrogen from the earth's surface are the most efficient mechanisms for stratospheric ozone removal. Destruction of odd oxygen (either atomic oxygen or ozone) is catalysed by the presence of neutral molecules or radicals formed in the stratosphere by irradiation or oxidation of natural and anthropogenic sources emitted at the earth's surface

(Brasseur, 1991). All reactions follow the same general scheme:

$$X + O_3 \longrightarrow XO + O_2$$
$$XO + O_3 \longrightarrow X + 2O_2$$

The balance results in a net loss of odd oxygen through recombination of ozone with atomic oxygen:

$$O + O_3 \longrightarrow 2O_2$$

or ozone conversion:

$$2O_3 \longrightarrow 3O_2$$

Species able to perform this process efficiently are nitric oxide (Crutzen, 1970) and some halogen (particularly chlorine) (Stolarski and Cicerone, 1974) and hydroxyl radicals (Finalyson-Pitts and Pitts, 1986). These are present in the stratosphere because of the long residence times (well over two years) of their precursors. The main source of NO is nitrous oxide (N_2O), with an estimated lifetime of approximately 100 years. It is emitted mainly by natural sources and is present in the troposphere at levels of 300 ppbv. Removal of atomic oxygen from the N_2O molecule by photolysis leads to molecular nitrogen, which can further react with molecular or atomic oxygen to yield NO. Methane, with a lifetime in the order of 10 years, is particularly important for OH radical formation. It is produced by 60% of anthropogenic sources, and increases in the troposphere, where it is already present at 1.7 ppmv, at levels of 1% per year. Chlorine and halogen radicals arise mainly from photolytic degradation of man-made chlorofluorocarbons (CFCs) or organic halides (Molina and Rowland, 1974). CCl_4 $CFCl_3$, CF_2Cl_2 and $C_2F_3Cl_3$ (called CFC (or Freon) 10, 11, 12 and 113 respectively) are considered the most important source gases for stratospheric chlorine atoms based on their global emissions (95, 330, 440 and 140 ktons/year in 1980, respectively), photolysis rates and atmospheric lifetimes (60–100 years) Brasseur, 1991). With the exception of CCl_4, CFCs are relatively recent components of the atmosphere as their manufacture began only in the early 1930s for use as refrigerator fluids. Subsequently, their uses expanded to include aerosol can propellants, solvents, and blowing agents for foam packaging and other products. CFCs became widely adopted as they were perceived to be non-reactive, non-toxic and generally benign. In spite of reduced production, their tropospheric concentrations increase at a rate of 5% per year (Warrick *et al.*, 1990). In the stratosphere, all radicals (Cl, ClO, OH, OH_2) and nitrogen oxides (NO, NO_2) are almost in equilibrium with more stable constituents (HNO_3, $ClONO_2$, HOCl, N_2O_5, HCl and hydrogen peroxide) which are inactive toward ozone but act as a reservoir for rapidly reacting components (Brasseur, 1991). Reservoir species are distributed all over the stratosphere and are slowly reaching the earth's surface by wet and dry deposition. Small changes in the stratospheric chemistry can rapidly deplete the ozone layer by allowing the shorter component of UV light to reach the earth's surface. Exposure to such wavelengths may result in skin cancer and mutagenic effects, as UV light can damage DNA and impair the normal genetic function of cells (Larson and Berembaum, 1988). Minor effects include erythema or reddening of epidermal cells. In 1985 enough evidence was collected to show that the ozone column over the Antarctic region had gradually decreased by as much as 40% from 1979 to 1984 (Farman *et al.*, 1985). Repeated satellite measurements showed that ozone depletion occurred at altitudes of 12–22 km and covered the entire Antarctic region from September to November (Stolarski *et al.*, 1986). The fact that depletion was not restricted to the upper stratospheric layers, combined with the observation that the polar vortex was found to contain 100 times more active chlorine species (e.g. ClO) than at mid-latitudes, and that it was substantially denitrified and dehydrated, suggested that ozone could have been destroyed through the following sequence of reactions involving formation of ClO, Cl_2O_2 and ClOO radicals (Molina *et al.*, 1987):

$$2(Cl + O_3 \longrightarrow ClO + O_2)$$

$$2ClO + M \longrightarrow Cl_2O_2$$

$$Cl_2O_2 + h\nu \longrightarrow ClOO + Cl$$

$$Cl + M \longrightarrow Cl + O_2 + M$$

$$\text{Net} \quad 2O_3 \longrightarrow 3O_2$$

Since the process does not require the presence of atomic oxygen, it can take place in the low stratosphere providing that sufficient concentrations of ClO (1 ppbv) are accumulated in the air. It is believed that such conditions can be realized in Antarctica because of polar stratospheric cloud formation during winter. Dissolution of HCl in ice crystals where $ClONO_2$ is present can produce Cl_2 and HNO_3 by heterogeneous reaction (Molina *et al.*, 1987; Wofsy *et al.*, 1988). While halogen molecules are released in the gas phases, HNO_3 is trapped in the ice crystals. Such molecular chlorine, inactive during the polar winter, rapidly photolyses when the sun returns in springtime, thereby generating enough active radicals (Cl, ClO) to deplete ozone. In spite of the uncertainty still associated with the physical and chemical processes involved in ozone depletion, there is little doubt that the springtime ozone hole over the Antarctic is mainly caused by the tropospheric release of CFCs. Although the production of CFCs has been reduced in the industrialized countries since 1976–1978 (Gamlen *et al.*, 1986) and further cuts are planned by governments which have signed the Montreal Protocol (UNEP, 1987), the future rate of CFC emissions remains extremely uncertain, and is related to the decisions that non-signatories to the Protocol will take in the years to come. If the CFC content of the atmosphere continues to grow at the present rate, there is no chance that springtime ozone depletion over the

Antarctic will disappear. Instead, the possibility exists that some depletion of the stratospheric ozone column will occur in the Arctic region whenever cold winters make possible the generation of low stratospheric clouds in the polar vortex.

3.3.4 VOCs in a changing climate

The energy associated with short-wave solar radiation reaching the earth is continuously balanced by the emission to space of terrestrial radiation produced by the earth's surface, by radiative trace gases in the atmosphere and by clouds and aerosols (Warrick *et al.*, 1990; Brasseur, 1991). Radiative emission occurs mainly in the infrared (IR) region at wavelengths higher than 4 μm. Long-wave radiation emitted by the earth's surface is partly intercepted by the atmospheric layers and re-emitted at its own temperature. The occurrence in the air of gases transparent to the incoming radiation but absorbing in the long-wave IR region (also called greenhouse gases) results in the 'capture' of heat in the troposphere thus creating a warm and hospitable environment for man. In the absence of these gases, the global temperature of the earth would have been 30°C colder than it actually is (Warrick *et al.*, 1990). However, thermal trapping of IR radiation can be regarded as beneficial as long as increases in the aerometric concentrations of greenhouse gases do not upset the natural balance existing on earth. A few degrees increase in temperature, sufficient to cause melting of polar ice and desertification of tropical and some temperate regions, would lead to dramatic ecological changes. Although long-term predictions are affected by uncertainties in time and scale, historical records obtained by analysing the composition of gases trapped in ice cores of the Antarctic seem to indicate that the 0.5–1.2°C increase in temperature observed in the last century was caused by the rapid increase in greenhouse gases in the atmosphere, related to the beginning of industrial revolution in Europe and North America (Warrick *et al.*, 1990). Based on ice core data, gases that in the last hundred years have contributed most to global warming are CO_2 and methane, whereas minor contributions are attributed to N_2O, ozone and CFCs (Tucker, 1986). Projections based on the percentage annual growth of greenhouse gases (Warrick *et al.*, 1990; Brasseur, 1991) suggest, however, that in 1994 halogenated VOCs are likely to become the second most important source of thermal trapping after CO_2. In spite of their low concentrations in air (1988 measurements indicate average levels of 0.26 and 0.44 ppbv for CFCs 11 and 12, respectively), the ability to absorb thermal IR is so strong that they already account for 20% of recent changes in radiative forcing (Tucker, 1986; Hansen *et al.*, 1988; Warrick *et al.*, 1990). Since their annual growth (5% per year) is much greater than that of carbon dioxide (0.48% per year), methane (1% per year) and nitrous oxide (0.3–0.4% per year), CFCs represent the greatest source of concern for environmental scientists investigating climatic changes. For this reason some countries are seriously thinking of reinforcing the CFC control policies contained in the Montreal Protocol (UNEP, 1987) according to which a 50% reduction of 1986 levels should have been obtained for mid-1998.

Important climatic effects may also arise from changes in the cloud albedo caused by some natural VOCs emitted in the marine atmosphere. It has been suggested that increasing amounts of dimethyl sulphide (DMS) from oceans can be converted into methane sulphonic acid (MSA) and SO_2 because of the enhanced presence of OH radicals in air following long-range transport of photochemical pollution. As gas-phase oxidation of these components ultimately leads to the production of fine sulphuric acid aerosols able to act as additional condensation nuclei for droplets, a change in the optical properties of clouds could take place (Penner, 1990). A 0.005% increase in albedo would lead to a 1.3°C decrease in the earth's surface temperature, which would counterbalance the greenhouse effect. Consequently, increasing efforts are presently being made to investigate the emission and fate of DMS in the troposphere.

3.4 VOC measurement

Understanding of the complex processes in which VOCs participate relies mainly on the availability of analytical tools capable of providing reliable information on the composition of VOCs and the levels present in air and emission sources. In this respect, the information provided by non-methane VOCs is largely insufficient to clarify what is happening in the atmosphere. Moreover, the range of time scales and domains, as well as the role of each individual component in the atmosphere, are so different that it is difficult, if not impossible, to design monitoring strategies suitable for investigating all effects related to VOC emission. This, combined with the inability of existing methods to provide real-time data on the levels existing in air and emission sources explains why different approaches have been pursued. In this section we will first discuss the strategies adopted to evaluate VOC emission and to monitor their behaviour in the atmosphere. Following this, principles and application of analytical methods commonly used for their evaluation will be described.

3.4.1 Evaluation of VOCs in emission sources

3.4.1.1 Anthropogenic sources. In presenting the approaches used for determining VOC emission from anthropogenic sources, it is useful to start with large point sources. With minor modifications these methods and techniques apply to almost any type of area source. Identification of all possible ways by which VOCs penetrate the atmospheric environment is a crucial step in starting any investigation. Collection and careful analysis of background information on the facility investigated, combined with sufficiently experienced technical staff, are essential prerequisites for the successful determinaton of VOC emission. Technical information obtained from maps showing facility location and environmental (e.g. stacks, deposits, plants) and geographic features, schematic diagrams describing cycles and processes, and official reports dealing with substances used and production levels reached are all very valuable. The analysis of the industrial process, the control devices adopted, and the storage and waste facilities makes it possible to identify critical activities and areas (e.g. stacks, storage tanks, filling and refuelling operations or valve leakage) where VOC emission is likely to take place. It also clarifies whether emission contains organic compounds potentially toxic to man, or relevant to photochemical smog pollution, global warming or stratospheric ozone depletion. When determining VOC emission rates, it is always advisable to start by measuring the total content of carbon material (ppm C or $\mu g\,C/m^3$) released per unit of time by each identified source. Indications given by total VOC monitors are important for several reasons. They allow identification, through the time dependence of the emission, of whether sources are continuous or intermittent and if VOC release is regular or irregular. Based on this information, the observation time, type of sampling, sampling duration and required number of samples can be decided. By integrating emission rates over time, the total amounts released can be calculated and compared with estimates based on production/consumption rates and process efficiency. In this way it is sometimes possible to determine the existence of unidentified sources through mass balance. For a correct evaluation of the emission rate, total VOC monitors must be connected to the sampling point through lines incorporating devices for the measurement of temperature, water content and emission velocity. Different sampling trains are required for ducted or fugitive emission. For instance, heated lines connected to filters and traps able to remove particles larger than, say, 2 μm, water and CO_2 are needed when sampling VOCs in the stack of combustion sources. By contrast plastic bags equipped with gas meters and pressure gauges must be used for evaluating fugitive emissions from valve leakage (USEPA, 1979b). Compliance with regulations designed to reduce the emission of toxic components, alkyl halides and CFCs or species characterized by a high potential for ozone production requires that speciation of VOCs is carried out in parallel with total VOC determinations. In this case several options are possible. The easiest, although not the cheapest and sensitive, is to divert part of the sample to the loop of a sampling system directly connected to a gas chromatographic (GC) unit. By switching a valve, a known volume of the emission sample is transferred to packed or capillary columns where it is analysed. Specific determinations can be performed by using GC detectors selectively responding to some VOCs or classes of VOC. Care must be taken to remove from the emission those gases (water, CO_2, SO_2,NO_x) that may cause damage to the stationary phase or generate interferences in the detection system. GC units can be directly connected to the sampling line only when no danger arises from the presence of pressure bottles supplying gases to the GC column and the detection unit, and no constraints are posed by the sampling site. If this occurs, GC and gas chromatographic–mass spectrometric (GC–MS) determinations can be carried out by collecting gas and vapours on flasks and canisters (Gholson *et al.*, 1988), or on traps filled with solid adsorbents capable of selectively retaining VOCs present in the gas stream (USEPA, 1979a; Polcyn and Hesketh, 1985). After sampling, flasks, canisters and traps are taken to the laboratory and analysed. Gas syringes can be used to transfer gaseous samples from flasks and canisters into the GC column. Such a procedure applies only to concentrated samples where 10–100 ppmv of each compound is present. For diluted samples (1–0.01 ppmv levels) enrichment on traps is needed. Traps can be either extracted with small volumes of solvents not present in the emission (usually CS_2) or thermally desorbed. In the former case, liquid injection of aliquots of the extract (1–5 μl) is performed, whereas in the latter all of the sample is transferred into the GC column by a thermal desorption unit capable of releasing VOCs by means of an inert gas. Liquid extraction applies to concentrated samples and gives quantitative results only for those constituents whose extraction efficiency is good and fugacity from solvent poor. It offers the advantage that repeated analysis of the same sample can be performed. Thermal desorption provides better sensitivity than liquid extraction as the whole amount of matter is available for the analysis. However the solid sorbent must be selected to ensure that VOCs are quantitatively retained during sampling, but released during the desorption step without undergoing thermal decomposition. To decide whether one method is preferable to another, preliminary research, aimed at evaluating retention efficiencies and sample recoveries for the individual VOCs suspected to be present in the emission, is needed (Polcyn and Hesketh, 1985). If the content of the emission is completely unknown, it is good practice to perform a parallel sampling with flasks and traps filled with different absorbents. Since the volume that can be collected in a flask or passed over a trap before compounds reach their breakthrough volume is usually quite limited (100–1000 ml), the information obtained is restricted to small portions of the total emission. This is not a problem if the release of VOCs is continuous and regular in time and composition. For intermittent and irregular sources, it is necessary to identify whether changes in the total VOC profile arise from processes that also affect composition. Based on a detailed analysis of various sources present in the facility and factors influencing their emission, a sampling protocol allowing the collection of traps when substantial changes in composition are suspected to occur must be designed. By combining information from total monitors with those provided by flasks and traps, the correct weighting factors can be estimated and applied for evaluating time-averaged emission rates for individual VOCs. Real-time determination of VOC emission is possible whenever compounds selectively respond to spectrometric or spectroscopic analysers. By connecting the sampling line to a mass spectrometer working with chemical ionization (CI–MS) (V&F, 1992), emission profiles can be determined with time resolutions of less than a second. The advantage of this method is that VOCs with a wide range of molecular weights can be measured in a single scan. The method is, however, unable to distinguish isomeric components and applies only to concentrated samples. If compounds emitted strongly absorb in the UV-visible or IR region, it is also possible to measure emission by optical methods such as differential optical absorption spectroscopy (DOAS) or Fourier transform infrared (FTIR) spectroscopy (Grant *et al.*, 1992). With these systems, VOCs can be determined in a continuous way over the entire facility or specific sections of it. Since optical systems do not need to be connected by lines to the sampling points, they can be installed outside the facility, for the purpose of auditing emission sources suspected not to meet existing regulations (Grant *et al.*, 1992). Obviously, the use of an optical system is restricted to those components whose reactivity in air is slow and emission can be derived from aerometric measurements.

Basically, there is no difference between the analytical approaches followed for determining VOC emission from large industrial plants and small stationary sources (e.g. domestic appliances, solvent evaporation from houses and filling stations). However, greater difficulties are encountered when identifying and quantifying fugitive losses which have an intermittent and irregular time pattern (e.g. household products containing solvents). Moreover, it is always difficult to establish how much indoor emission is really going into the atmosphere; this is because there are marked differences in indoor/outdoor exchange as a function of the type of house considered and the habits of the occupants. Although mathematical models of ventilation systems have recently been proposed for identifying and quantifying VOC emission from indoor sources (Molhave and Thorsen, 1991), their application to a large number of different sources is rather complicated.

In the case of mobiles sources, methods of gauging emission rates are based mainly on laboratory experiments designed to simulate reality. In particular, the mass of VOCs released by vehicular sources is very often determined by statutory tests and computed from the measured concentrations generated by the exhaust gas when a vehicle is operated on a chassy dynamometer according to certain standard driving cycles (CONCAWE, 1991). Test procedures mainly simulate driving conditions in urban traffic and on highways. Substantial differences exist between countries with respect to driving cycle (speed and distance), vehicle preconditioning and analytical equipment. The European test cycle (ECE 15), for instance, requires the vehicle to be preconditioned by driving 3500 km before testing. The vehicle is allowed to soak for at least 6 h at a test temperature of 20–30°C. It is then started and allowed to idle for 40 s. The driving cycle is repeated four times without interruption. This gives a total test cycle time of 780 s, a total distance of 4052 km and thus an average speed of 19 km/h (CONCAWE, 1991). Since the maximum speed reached is 50 km/h, this cycle is a very low duty one and is not representative of many modes of driving. For this reason, an extra urban driving cycle (EUCD) where a maximum speed of 120 km/h is reached, has been included in the European protocol. This cycle must be carried out after the standard ECE 15 test (CONCAWE, 1991). The US federal city cycle (US-75 Cycle) is divided into three phases, and has a total test time of 31 min plus 10 min for the hot soak, and a total length of 17.85 km. It has been designed to produce weighted average emissions from cold and hot start conditions and is quite representative of driving patterns in the Los Angeles area (CONCAWE, 1991). The calculation procedure weights separately the emissions measured for each phase. The highway test cycle adopted in the USA is used mainly for measuring NO_x; VOC emission is often derived from fuel consumption. In spite of these differences in cycles, all countries use similar testing facilities and analytical methods for determining VOCs in the exhaust emission. Sampling is performed by connecting the exhaust tail pipe to a dilution tunnel equipped with a heat exchanger for keeping constant the temperature of the gases and particles during the cycle. After isokinetic collection of particulate matter, the exhaust is sent to a mixing chamber or large bags where the concentrations of CO, CO_2 and NO_x, and the total VOC content are continuously measured. In the USA specific testing facilities have also been developed for the determination of fugitive emissions from engines and gasoline tanks (CONCAWE, 1991; Schuetzle *et al.*, 1991). The car is placed in a sealed housing connected to a total analyser used for controlling VOC levels. With this system it is possible to evaluate diurnal, hot-soak and running losses. The term diurnal losses identifies evaporation from the fuel tank when the engine is off; this accounts for all VOC losses resulting from normal temperature changes occurring over a 24 h period. The facility is equipped with a control system able to simulate diurnal changes in temperature. Hot-soak losses, caused by heat transfer from the engine or carburettor to the fuel tank, identify VOC emission occurring when a fully warmed up vehicle is left to stand. Running losses refer to the emission measured when the car simulates driving conditions. The sealed housing system for evaporative determination (SHED), used since 1978, has recently been included in the proposed draft to procedures to be adopted by European countries (CONCAWE, 1991). Evaporative losses are also currently measured by weighting the amount of VOCs adsorbed on carbon traps connected to the fuel system at locations where vapours may escape to the atmosphere. This system, adopted in Japan, is the only one using gravimetric methods for determining total VOC emission (CONCAWE, 1991). Techniques adopted for the speciation of VOC emission are similar to those developed for point sources, with extensive use being made of trapping systems combined with GC or GC–MS instrumentation (Schuetzle *et al.*, 1991). While dynamometer driving cycles are essential to establish uniform emission standards for regulatory purposes and for testing new technologies for VOC reduction, none of the current cycles reflects accurately the range of driving speeds and conditions experienced on the road. For this reason, on-road testing techniques have been also applied to evaluate VOC emission factors (Bailey *et al.*, 1990). By inserting miniaturized proportional constant volume samplers in the exhaust pipe (Potter and Savage, 1982), a portion of the emission is sent to a total VOC monitor or collected in Tedlar bags installed in the vehicle. VOC emission is evaluated by driving the car over a road considered to be representative of a certain area or country. Using this approach, emission rates for 18 VOCs have been measured and emission factors derived for the UK fleet (Bailey *et al.*, 1990).

3.4.1.2 Natural sources. As mentioned in previous sections of this chapter, vegetation is the most abundant natural source of VOCs and increasing efforts are devoted to improving its determination. Release to the atmosphere occurs because VOCs present in the oil cells of plants are exuded as vapour into surrounding epithelial cells or intercellular spaces and reach the leaf surface through mesophyll and epidermal cells or through cuticles (Yokouchi and Ambe, 1985). The types of hydrocarbon emitted by vegetation depend on the composition within the oil cell and may differ according to the plant family. The emission rate of each component is determined by the product of its concentration in the oil cell and the saturated vapour pressure. Since saturated vapour pressures follow a log-linear relationship with temperature, strong diurnal and seasonal variations in the flux of VOCs in the atmosphere are observed (Tingey *et al.*, 1979, 1980; Yokouchi and Ambe, 1985). Isoprene, terpenes, alcohols, carbonyl compounds and esters are the major components released by plant emission (Graedel *et al.*, 1986). Some of these can actively participate in photochemical smog pollution, being extremely reactive with OH radicals and ozone (Dimitriades, 1981; Altshuller, 1983b). Research in this field has focused mainly on the quantitative determination of isoprene and monoterpene components (e.g. α-pinene, β-pinene, δ-3-carene, limonene, myrcene) as these exhibit high potentials for ozone production and are widely emitted by evergreen and broadleaf trees. The need to collecting information on reactive organic species explains why enrichment on traps combined with GC and GC–MS determination is the preferred method for evaluating VOC emission from vegetation. Practical difficulties arise from the numerous species to be investigated, the strong dependence of their metabolism on environmental parameters and the reactivity of the VOCs emitted. To reduce uncertainty, laboratory and field experiments are performed. One way to approach the problem is to place vegetation which has been planted in pots and cultured in greenhouse for at least two years in dynamic mass balance cabinets housed in a chamber where accurate control of the temperature, relative humidity and radiation intensity is performed (Tingey *et al.*, 1980; Yokouchi and Ambe, 1985). Air, filtered to remove VOCs and air pollutants (SO_2, NO_2, ozone), is fed into the cabinet from small holes made in the floor and emerges from a grid placed on one of the vertical walls. The sampling port, usually inserted in the ceiling of the cabinet, is connected to a collection trap set in series with an aspirating pump. After sampling, the traps are thermally desorbed and individual organic components are determined by GC and GC–MS techniques. By knowing the concentration of each organic compound in the cabinet, the air volume exchanged per unit of time and the dry weight of the plant, it is possible to determine the emission in terms of μg/g dry weight (Tingey *et al.*, 1979; Yokouchi and Ambe, 1985). The total emission in forest and rural areas is thus obtained by knowing the percentage distribution of natural species, their composition and the log-linear relationship between emission rates and temperature (Altshuller, 1983b). Although these chambers are ideal for investigating the dependence of VOC emission on environmental parameters, their use is limited to small plants; additionally their behaviour does not necessarily reflect what is happening in nature. For this reason, bag enclosure techniques have been developed and applied to field measurements of biogenic emission (Zimmermann, 1979). A portion of the vegetation (usually a branch of a tree) is enclosed in a plastic bag and a sample of air is collected on a trap. The bag is then filled with hydrocarbon-free air and left over the vegetation before another sample is taken. The two samples are analysed and the concentration of individual VOCs determined. By subtracting from the amount measured in the second trap the background levels present in the first one, the net amount emitted by the piece of vegetation during the time elapsed between the two samplings is determined. The bag enclosure method is simple but can be affected by uncertainties. Damage to leaves during the bag enclosure operation and the use of zero gases (air free from VOCs, CO_2 and water) changing the amount of water and CO_2 near to the leaf surface are two possible sources of error that may affect VOC emission. To prevent such effects, tree branches have been enclosed in sealed Teflon boxes with a volume of 300 l, equipped with small fans for slowly equilibrating the air composition inside the box (Yokouchi and Ambe, 1985). Among field techniques, the most impressive and promising one is that based on large open-top chambers (Bufler and Wegmann, 1991). Used for many years by plant physiologists for assessing damages to plants arising from primary pollutants and ozone (Hileman, 1982), open-top chambers comprise large sections of plastic foil rolled around a thin, cylindrical structure made of aluminum bars to support the chamber itself. The bottom of the chamber (approximately 3 m in diameter) is placed in contact with the soil where lysimeters and other monitors can be placed to provide *in situ* information on plant metabolism, water supply and soil composition. The top of the chamber is open to the atmosphere. The plastic section near to the ground has a double wall system, with the internal part communicating with the chamber. The external wall is sealed and connected by a duct with a pumping system forcing air through the annular section. The air emerging from the holes of the internal wall washes the whole chamber and is discharged into the atmosphere at rates of 1–5 m/s. For determining VOC emission from tall trees, chambers as high as 9 m have been built and installed in pine forests (Seufert and Arndt, 1987). To ensure adequate circulation, two annular sections, each one connected to a separate pumping system, have been used. Before and during the collection of VOCs on traps, the open-top chamber is washed with air free from hydrocarbons, primary pollutants and ozone. This is obtained by inserting batteries of active charcoal filters at the inlet of the pumping system. Since open-top chambers are basically ducts where the flux of VOCs released by vegetation is forced into the atmosphere, sampling lines quite similar to those used in industrial stacks are needed for calculating emission. With such an experimental set-up, daily profiles of terpene emission from an 18-years-old Norway spruce 7 m tall were directly measured inside a German forest (Bufler and Wegmann, 1991). Until now, field determinations carried out with open-top chambers or controlled cabinets can be regarded as the most reliable tools available for measuring VOC emission from vegetation; by contrast the application of micrometeorological methods and the use of tracers is extremely critical and results obtained are difficult to interpret (Arnts *et al.*, 1982; Lamb *et al.*, 1986). The frequent observation that VOC emission from vegetation follows diurnal profiles completely opposite in trend to those shown in Figure 3.5, is highly indicative of the difficulties that would be encountered if experimental devices capable of somehow preventing rapid VOC conversion in air were not available.

The techniques described are also suitable for determining the amount of DMS released by algal fields deposited on the seashore, providing that proper collection systems are used and that flame photometry is attached to the separating column. Less problematic is the evaluation of methane, which is naturally emitted mainly by enteric fermentations in mammals, anaerobic decay of various types of organic matter occurring in sludges, swamps and rice paddies, or geochemical sources (Cullis and Hirschler, 1989). Due to its long lifetime and high concentration in the atmosphere, methane can be directly measured by injecting a small volume of air into a GC equipped with a packed column; emission can be determined by aerometric measurements of vertical gradients and micrometeorological parameters.

3.4.2 *Evaluation of VOCs in the atmosphere*

There are several reasons justifying the monitoring of VOCs in the atmosphere, each of which basically responds to the two following needs:

(1) Assessment of the exposure of the population and other vulnerable receptors to potentially toxic components released by emission sources or formed in the atmosphere.

(2) Creation of data bases to permit the analysis of long-term trends in air pollution or for other research purposes (one of the most important of which is the validation of models for dispersion, transport and deposition).

Although, in both cases, the techniques adopted are similar to those used in emission sources, a higher degree of sensitivity, and hence sophistication, is necessary for accurate determination of individual VOCs at the levels existing in air (ppb–ppt). The achievement of this task often requires that specific GC detection (e.g. electron capture, flame photometry and photoionization) must be combined both with selective capillary columns and with highly efficient enriching procedures for sample collection (cryogenic trapping on empty tubes or adsorption traps). With optical devices such as DOAS and FTIR, sufficient sensitivity can be obtained by accumulating signals on computer systems and processing them through dedicated software. When possible, automated systems allowing unattended operations are preferred. In this case, the monitoring unit must be interfaced with computers for data storage and the statistical treatment. Sensitivity requirements and unattended operations justify the frequent adoption of dedicated instrumentation for VOC determination in air. Since quality assurance is fundamental to provide meaningful data on VOC monitoring networks, special intercomparison programs must also be undertaken for validating results and procedures. This is particularly important whenever different methods, techniques or procedures are used for evaluating a given compound or a class of compounds. Although the basic needs for VOC monitoring are only two, there are several aspects that can be investigated, each of which requires a different range of compounds to be determined and a specific spatial and temporal scale in which monitoring must be performed. We will summarize here only those monitoring approaches that are strictly related to the effects described in section 3.3.

3.4.2.1 Air monitoring of potentially toxic VOCs. Since the number and type of human diseases associated with VOC emission depend both on the levels of pollution existing in air and on the number of individuals exposed, monitoring networks devoted to risk assessment are mainly designed to cover densely populated areas experiencing severe pollution. Such conditions are usually realized in large urban or suburban areas where heavy traffic is combined with substantial domestic and industrial emission. Although general indications of pollution levels can be obtained by using monitors measuring the non-methane VOC content of the air, detailed knowledge of the organic composition is fundamental for correlating the results of epidemiological studies with exposure levels whenever air quality standards are frequently exceeded. This is because toxic VOCs exhibit different effects, dose–response curves and body burdens. Using the pattern of slowly reacting

VOCs it is also possible, in some instances, to identify the sources that contribute most to changing the quality of the atmosphere through simple source–receptor models. Among the 261 components that have been identified in outdoor samples (Shah and Singh, 1988), only a portion of them can really represent a potential threat to human health because of their ubiquitous occurrence and atmospheric levels. If we refer to a previously mentioned study reporting daily average concentrations for 66 components detected in outdoor samples collected in the USA (Shah and Singh, 1988), we can estimate that, based on their toxicity, median and average daily concentrations, and frequency of observation, only 30 of them can be considered ubiquitous. This number is very close to that selected in the total exposure assessment methodology (TEAM) study (Hartwell *et al.*, 1987) carried out in California to develop and implement methodologies for measuring human exposure and body burden to toxic substances in various media. In the TEAM study, however, occurrence in water and human breath were also considered important factors when selecting target components. Since the level and composition of VOCs existing in the USA do not substantially differ from those observed in other countries, it is safe to say that the core of potentially toxic compounds for which monitoring is needed is represented by aromatic compounds, alkyl halides and aldehydes. Among aromatic compounds, the species of major importance are benzene, toluene, xylenes, styrene and their chlorinated, brominated, hydroxylated and nitrated homologues. Alkyl chlorides with one to four carbon atoms, and their corresponding unsaturated homologues are the alkyl halides on which most attention is focused. CFCs are not of importance as they are generally non-toxic and chemically inert. Formaldehyde and, to a minor extent, benzaldehyde are considered to be the relevant carbonyl components because of their health implications. In the core of organic compounds for which monitoring is needed we must also include, 1,3-butadiene and PAN as they show distinct carcinogenic, teratogenic and mutagenic features. It should be stressed that such a list of target components is somehow arbitrary, being based mainly on toxicity data, common sense, aerometric observations and control legislation applied to VOC emission. No specific air quality standards for potentially toxic components have yet been defined by the environmental agencies of the most industrialized countries. This is justified by the fact that, although outdoor exposure to toxic VOCs at levels higher than those established by the air quality standards is recognized to be responsible for the increase of diseases in humans, it is hard to decide whether such effects are dominant with respect to indoor exposure or other possible absorption routes (e.g. drinking of water, ingestion of food, use of medicaments). Until accurate multi-media assessments are made, it is impossible to establish priorities or define limits for the content of potentially toxic VOCs in ambient air. Compliance with national emission regulations for toxic compounds similar to those issued in Italy (DPR, 1990) would probably be sufficient to maintain their levels well below the threshold limit. Due to present uncertainty in this matter, the present discussion will be limited to those methods capable of providing information on the core of air toxics mentioned earlier, with the assumption that they are the important ones affecting human health. In spite of the outstanding progresses made in the selective detection of VOCs by optical methods, gas chromatography is still regarded as the most cost-effective technique for the monitoring of potentially toxic components, because of the large number of compounds that can be determined in a single run. However, to reach the sensitivity necessary for detecting pollutants at levels existing in the air enriching procedures for sample collection must be combined with highly efficient chromatographic columns and selective detection. Cryogenic focusing on empty tubes, adsorption on traps filled with solid sorbents, or a combination of both techniques are the procedures commonly adopted for enriching atmospheric samples (Liberti and Ciccioli, 1985). Suitable volumes of air (usually more than 100 ml) are passed through the enriching device for selectively retaining the components to be monitored by using aspirating pumps or pressurized systems. If cryogenic sampling is performed, condensation of water in the connecting lines and the enriching system must be avoided. Although ion-exchange membranes (Foulger and Simmonds, 1979) and dehydrating agents (USEPA, 1979a) work efficiently in preventing water condensation, none of these systems ensures that selective losses of polar compounds are not taking place during the enriching process as a result of adsorption, partition and condensation processes. Water condensation does not represent a problem if sampling is carried out at room temperature on traps filled with hydrophobic sorbents (carbons, polymers). Although water retention is low, the monolayers of water molecules continuously formed over the solid surface during sampling strongly reduce the adsorption efficiency of sorbents toward organic compounds and lead to losses in the volatile components (Ciccioli *et al.*, 1992b). Usually, good recoveries are only obtained with VOCs containing more than five carbon atoms. Whichever sampling system is used, tests must be performed to check if and to what extent organic compounds are lost during sampling. Thermal desorption, carried out under a flow rate of inert gas to prevent chemical transformation of VOCs, is the preferred technique to transfer the sample from the enriching system ot the GC column. To prevent efficiency losses due to band broadening, desorbed compounds are usually refocused at the column inlet or in a short capillary liner kept at low temperatures (approximately $-150°C$). Due to the wide range of molecular weights and polarities of potentially toxic compounds, narrow-bore columns coated with low-polar stationary phases (usually methyl or phenylmethyl silicon fluids) are preferred for analysis. Suitable temperature programmes must be selected for the complete elution and better separation of the components injected. Since none of the existing columns is capable of separating all the compounds present in air, multi-detection units comprised of GC or MS detectors need to be connected to the GC column for identification and quantitation purposes. MS is the preferred system as it allows positive identification of components by means of their mass spectra. If used in the selected ion mode (SID) it can reach sensitivities quite close to the best GC detectors (Pellizzari *et al.*, 1976). However, it is expensive and calibration is needed for each of the components to be monitored. Well-trained personnel are required for running GC–MS instruments and interpreting mass spectra, therefore it is not currently adopted in automated systems designed for unattended monitoring.

Electron capture detection (ECD) plus flame ionization detection (FID) is the basic combination needed if monitoring of toxic pollutants with multi-detection units based on GC detectors is to be performed. Since ECD responds well to halogenated and nitrated organic compounds but is quite insensitive towards other components, whereas FID is almost equally sensitive to all components, combination of the two methods allows the identification and quantification of a large number of components eluted by the column (Ciccioli *et al.*, 1976). Detectors can be connected either in parallel or in series (Ciccioli *et al.*, 1976; Possanzini *et al.*, 1981). The former solution is less sensitive as only a portion of the whole sample is sensed by each detector. Moreover, good control of the flow rate is needed to prevent changes in the amount of matter reaching the detectors during runs carried out at programmed temperature. By placing ECD before FID the maximum sensitivity for both detectors can be achieved (Possanzini *et al.*, 1981; Rudolph *et al.*, 1985). Since only a fraction is lost during electron capture processes of the strongly electrophilic components, a large amount of sample is available for FID. This combination is so advantageous that many chromatographic companies provide kits for assembling multi-detection units of this type in their standard GC equipments. Due to their low cost and easy maintenance, GC units based on ECD, FID or a combination of both are particularly suitable for automated or unattended systems (McClenny *et al.*, 1984). Photoionization detection is preferred to ECD whenever enhanced selectivity toward olefins or aromatic compounds is needed (Cox and Earp, 1982). The GC methods just described do not, however, cover the whole range of target compounds identified as potentially toxic to humans and the environment. PAN, for instance, requires a simple but dedicated GC system for its determination, as the column must be able to separate this compound from air and CFC peaks. Usually a short packed column coated with a polar phase (Carbowax 20M or 400) is used (Roberts, 1990). The column is directly connected to an ECD (Darley *et al.*, 1963) or a Luminol-Chemiluminescence detector (Burkhardt *et al.*, 1988) after the sample has been catalytically converted into NO_2. Since volumes ranging from 1 to 5 ml are sufficient for PAN determinations at tenths of pptv, direct injection of air samples can be performed. Formaldehyde is another compound for which dedicated monitoring is required. Normal GC detectors are virtually insensitive toward this component, therefore optical methods or liquid chromatographic systems are preferred for the determination of formaldehyde in air. Differential optical absorption spectroscopy (DOAS) has been found particularly suitable for detecting formaldehyde at ppbv levels (Platt *et al.*,1979) and, with the availability of commercial systems, has been used for unattended monitoring in urban areas (Brocco *et al.*, 1992). Benzene and toluene also absorb strongly in the UV region, therefore DOAS can provide information for at least three potentially toxic compounds in air. A good example of the capabilities afforded by DOAS in providing reliable information on the levels of formaldehyde, benzene and toluene in urban areas is shown in Figure 3.8. Data were collected in the centre of Rome in January 1992. From the profiles obtained, it is possible to see the regular pattern followed by all pollutants over the entire month and to clearly distinguish their day/night variations. It is worth noting that the data provided by DOAS are similar to those shown in Figure 3.5. Tunable diode laser absorption spectroscopy (TDLAS) has also proven to be a sensitive and reliable optical technique for formaldehyde determination in air (Schiff and Mackay, 1989), although the high cost of operation and complexity prevent its adoption in large monitoring networks. A cheaper alternative to optical methods is that of formaldehyde collection on impingers filled or traps coated with 2,4-dinitrophenylhydrazine (2,4-DNPH) (Lowe *et al.*, 1980). By reacting with 2,4-DNPH, formaldehyde is quantitatively converted into the corresponding dinitrophenylhydrazone derivative, which can be determined by submitting the solution to HPLC determination using a reversed-phase column connected to a UV detector. This method can be extended, although with some limitations, to all carbonyl compounds in the C2–C4 carbon range (Ciccioli *et al.*, 1987). A special problem is presented by 1,3-butadiene, which usually has a relatively low concentration in air and is difficult to separate with columns of low polarity. As will be discussed later (see section 3.5.2.1). Specific capillary columns capable of separating all alkanes, alkenes, alkines and arenes but not eluting oxygenated hydrocarbons are needed for the unambiguous detection and accurate quatification of 1,3-butadiene. By considering the complexity and amount of dedicated instrumentation needed for assessing exposure to potentially toxic VOCs, it is clear that the number of automated and unattended systems that can be installed in large polluted areas is quite limited. Furthermore, accurate siting must be made in order to extrapolate results to other portion or sectors of the area investigated. Extrapolation to areas not covered by continuous or semi-continuous monitoring is possible if data from the automated networks are integrated with those obtained by out-of-line sampling methods (Tedlar bags, canisters or adsorption traps). Although out-of-line sampling is not the best solution for evaluating temporal variations of toxic VOCs, it offers the advantage that the same analytical unit can be used for all determinations. Realistic figures can be obtained with this approach if samples are collected at sites where continuous monitoring of total VOCs less methane is performed.

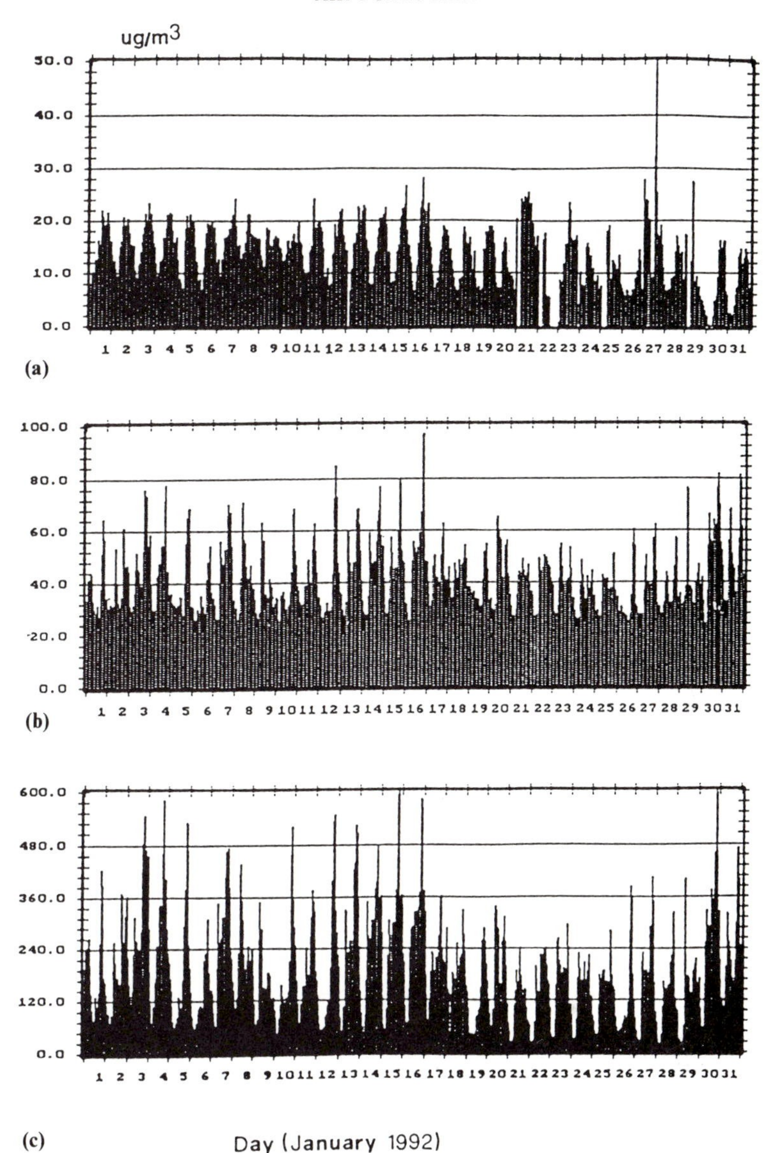

Figure 3.8 Monitoring of (a) formaldehyde, (b) benzene and (c) toluene by DOAS. Determinations were performed in the centre of Rome. Profiles refer to the month of January 1992. Redrawn and adapted from Brocco *et al.* (1992).

The availability of a central unit for out-of-line determinations can justify the use of MS as a detection system, therefore combined approaches make it possible to interpret more accurately data obtained from unattended systems based on FID and/or ECD. By knowing the range of potentially toxic compounds in a given area, the analytical performances of automated systems can be optimized to provide the best performances. When possible attempts should be made to integrate and share information with networks which, although designed for other purposes (e.g. validation of photochemical models), are also detecting species relevant for risk assessment. In many countries, different agencies (e.g. health and welfare, environmental, agricultural) and authorities (international, national, local) are dealing with aspects related to VOC pollution and many of them own monitoring networks. Consequently, the creation of joint committees for integrating networks and treating composite data would be of help in reducing unnecessary duplication of expensive instrumentation.

3.4.2.2 Air monitoring of VOCs relevant to photochemical pollution and acid deposition. Air quality determinations, validation of models, budgeting of atmospheric cycles or evaluation of deposition velocities and emission rates are just some of the items for which speciation of photochemically reactive VOCs is required. For the validation of models aimed at predicting photochemical pollution and acid deposition, detailed information both on precursors (VOCs, NO_x), and on products (ozone, PAN, H_2O_2, HNO_3, nitrate and sulphate in particles) pertaining to the photochemical cycle must be provided for each grid of the domain investigated. By comparing predictions with the observations made in the various grids it is possible to verify whether a model provides satisfactory results or needs to be somehow modified. As the area covered by a grid can be quite large and the grids into which the domain is split numerous, a complex network comprising monitoring stations located in urban, suburban, rural and forest areas is often required. Meaningful results are obtained only if the selected sites are truly representative of the average pollution levels existing in each grid (or a certain number of them), and the data provided by the monitoring stations are highly homogeneous in type and quality. Although the total mass of non-methane VOCs is useful to alert communities to the fact that oxidant levels exceed the air quality standards, it gives a poor representation of the reactivity existing in the atmosphere. Therefore, speciation of VOCs is increasingly regarded as the only means of improving existing models, through a careful check of the chemistry used for describing the various steps leading to the formation of photochemical oxidants and acidic species (Hov *et al.*, 1986, 1987; Seinfeld, 1988; Derwent, 1990). Although complete knowledge of all organic compounds in air is highly desirable, it can be fully exploited only if models able to treat extensively the chemical and physical transformations are tested. Model limitations and the number of stations and amount of instrumentation required for a robust validation programme explain why VOC monitoring is often restricted to a limited number of target compounds requiring relatively simple and, possibly, automated systems for their determination. The core of target compounds is usually comprised of a well-balanced mix of paraffins, olefins, arenes, carbonyl compounds and PAN. These species are representative both of the reactivity classes used by existing models (Lurmann *et al.*, 1988; Middleton *et al.*, 1990), and of the organics emitted by natural and anthropogenic sources or formed by their photochemical degradation. The exact number and type of VOCs to be monitored can be quite variable, being a compromise between what is desired by the modellers and what is feasible with the available instrumentation, personnel and funding. To enable the validation of photochemical models existing in Europe and to check the influence on VOC emission and concentration pattern that reduction programmes and new legislation would have on the member countries, 31 organic components were identified as priority species to be measured by the CEC working group on VOCs set up by the DG XI at the Joint Research Centre in Ispra (Kotzias and Hjorth, 1991). Specific guidelines on the duration and frequency of their sampling as a function of the type of site to be investigated were prepared, and intercomparison programmes aimed at validating methods and procedures were undertaken (Kotzias and Hjorth, 1991). The target compounds selected by the CEC working group are listed in Table 3.2. Although they are representative of the reactivity classes used by models, and of VOC emission existing in Europe, some highly reactive natural compounds (e.g. α-pinene, δ-3-carene) and important reaction products (e.g. acetaldehyde, acetone, acrolein, propionaldehyde and PAN) are missing. If these are included in the list of priority VOC to be monitored, the minimum bed for validating both Eulerian and

Table 3.2 Priority VOCs to be measured according to the indication given by the VOC working group set up by the CEC (DG XI). Adapted from Kotzias and Hjorth (1991).

Methane	Isoprene
Ethane	*n*-Hexane
Ethylene	1-Hexene
Acetylene	*n*-Heptane
Propane	*n*-Octane
Propene	Isoctane
n-Butane	Benzene
Isobutane	Toluene
trans-2-Butene	$(m+p)$-Xylene
1-Butene	*o*-Xylene
cis-2-Butene	Ethylbenzene
1,3-Butadiene	1,2,4-Trimethylbenzene
n-Pentane	1,2,3-Trimethylbenzene
Isopentane	1,3,5-Trimethylbenzene
1-Pentene	Formaldehyde
2-Pentene	

VOCs suggested by the author to be included in the monitoring protocol for validating models are: α-pinene; δ-3-carene; acetone; PAN; acetaldehyde; propionaldehyde; acrolein; and benzaldehyde.

Lagrangian models used for predicting photochemical pollution and acid deposition deposition can be identified. While carbonyl compounds and PAN can be monitored using the techniques described in section 3.4.2.1, some differences between the sampling and analytical procedures exist due to the chemical nature of target compounds and the wide range of carbon numbers to be investigated. Since all components to be monitored are basically non-polar, ion-exchange membranes and dehydrating agents can safely be used to remove water from the air stream during cryogenic sampling on empty tubes or adsorption traps (Schimdbauer and Oheme, 1986; Bloemen *et al.*, 1990). Water removal is important not only to avoid plugging of the connecting lines and enriching devices during the sampling process, but also to prevent deactivation of the capillary columns necessary for separating in a single GC run all alkanes, alkenes, alkines and arenes in the C2–C9 carbon range. The columns most suited to performing such separation are those coated with aluminum oxide and KCl (Schneider *et al.*, 1978). Deposition of large amounts of water on their internal walls would dramatically change their adsorption properties giving rise to severe peak tailing, changes in retention and dramatic losses in efficiency and selectivity. The almost irreversible adsorption of all polar and moderately polar components, together with the mild temperatures required for the elution of very volatile constituents, have also contributed to the success of aluminum oxide/KCl capillary columns in the determination of atmospheric VOCs. These unique features enable identification of all eluted compounds by means of their retention indices, therefore FID can be used for both detection and quantification purposes (Intersociety Committee (IC), 1990). By fully exploiting these principles, an automated unit allowing the unattended monitoring of the target VOCs identified as relevant to photochemical smog pollution and acid deposition by the CEC working group has been developed in Europe (Bloemen *et al.*, 1990). Water is efficiently removed by passing the air stream through a specially designed unit comprised of two Nafion membranes. VOCs, retained on a three-stage carbon adsorption trap maintained at $-10^{\circ}C$, are transferred into a cryofocusing unit by thermal desorption. By raising the temperature of the fused-silica liner from -150 to $200^{\circ}C$, compounds are injected into the aluminum oxide/KCl capillary column, where they are separated and detected by FID. A mini-computer is used to control the various operations and to store chromatographic runs. Dedicated software is provided for integrating and quantifying peaks. An example of the GC profile obtained with this system is shown in Figure 3.9. Although this is currently the most advanced apparatus for the monitoring of non-polar hydrocarbons, the operational costs arising from the substantial consumption of liquid nitrogen (approximately 100 l/week) to cool the sampling and cryofocusing units (Miliazza and Dooper, 1992) discourages its adoption by large monitoring networks. Consequently, mixed approaches whereby data provided by a few automated systems are integrated with information

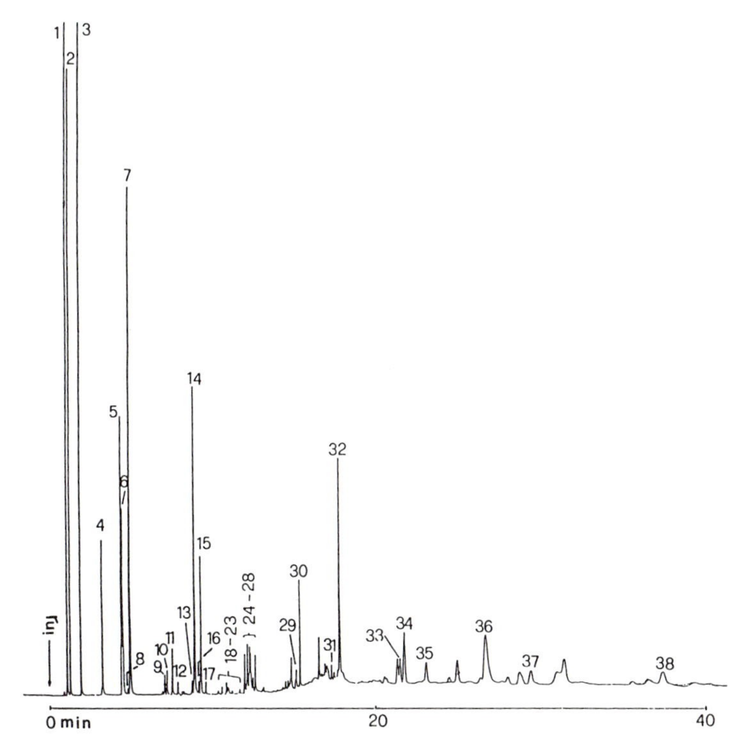

Figure 3.9 Chromatographic profile of VOCs relevant to photochemical smog formation obtained by an automated and unattended system operated in Bilthoven. Separation of all alkanes, alkenes, alkines and arenes in the C2–C9 carbon range was accomplished by capillary chromatography on an aluminum oxide/KCl column. Peak identification; 1, ethane; 2, ethene; 3, propane; 4, propene; 5, acetylene; 6, isobutane; 7, *n*-butane; 8, propadiene; 9, *trans*-2-butene; 10, 1-butene; 11, isobutene; 12, *cis*-2-butene; 13, cyclopentane; 14, 2-methylpentane; 15, *n*-pentane; 16, methylacetylene; 17, 1,3-butadiene; 18, 3-methyl-1-butene; 19, *trans*-2-pentene; 20, 2-methyl-2-butene; 21, 1-pentene; 22, 2-methyl-1-butene; 23, *cis*-2-pentene; 24, methylcyclopentane; 25, cyclohexane; 26, 2-methylpentane; 27, 3-methylpentane; 28, *n*-hexane; 29, *n*-heptane; 30, benzene; 31, *n*-octane; 32, toluene; 33, ethylbenzene; 34, 1,3-/1,4-dimethylbenzene; 35, 1,2-dimethylbenzene; 36, 1,3,5-trimethylbenzene; 37, 1,2,4-trimethylbenzene; 38, 1,2,3-trimethylbenzene. (Kindly provided by H.J.Th. Bloemen, RIVM, The Netherlands).

obtained by sampling VOCs on electropolished canisters and traps (IC, 1990) are still required. As each of these out-of-line methods (or their combination) present certain practical limitations, selection is made on a case-by-base approach. Canisters offer the advantages that they are easy to use in the field, basically inert towards VOCs and can be directly interfaced with the automated GC system just described. However, dark reactions (e.g. ozonolysis) and adsorption of reactive gases and particles on the walls may result in unpredictable losses of reactive VOCs if sampling is carried out in highly polluted atmospheres. Although reduced storage is the best way to minimize possible changes in the VOC composition, it is difficult to predict whether and to what extent transformation occurs. The use of canisters also requires the availability of enriching systems for analysis. Tedlar bags have intentionally not been mentioned as they suffer the same limitations as canisters, are less inert toward VOCs and are more difficult to clean (Jayanty, 1989). Adsorption traps, which preserve better the sample quality, are more difficult to operate in the field, as dehydrating agents and cryogenic units must be inserted in the sampling line to prevent losses of low-boiling hydrocarbons (C2–C4) during the enrichment process. In addition, care must be taken to prevent contamination during cleaning, transport and storage operations. As a result, out-of-line sampling is preferred to on-line systems only when simultaneous collection of different air parcels is to be performed (e.g. vertical gradients) or severe constraints are posed by the sampling platform (vehicles, aircraft, flying balloons) (Rudolph *et al.*, 1981; Kanakidou *et al.*, 1988). Although the information obtained using methods based on aluminum oxide/KCl capillary columns is fundamental for evaluating model performances, it is not always sufficient to decide whether possible deviations of predictions from observations should be attributed to a poor chemical description of the model or to incomplete information on the composition of the VOC in air (Nollet and Dechaux, 1991). In this case, methods allowing the determination of compounds not listed in Table 3.2 must be included in the monitoring protocol. To prevent losses of polar components arising from the use of dehydrating systems, sample enrichment is carried out at room temperature and compounds are retained on traps filled with hydrophobic sorbents (Ciccioli *et al.*, 1976; Knoeppel *et al.*, 1980; Arnts, 1985; Yokouchi *et al.*, 1990; Ciccioli *et al.*, 1992b). In this way, only C5–C14 VOCs can be quantitatively collected and fully recovered by thermal desorption. Analysis is performed on the same type of capillary columns used for separating toxic VOCs. Because of the large number of compounds eluted by the column and the incomplete separation of many critical species, MS combined with an accurate knowledge of the VOC elution sequence is mandatory for unambiguous identification (Ciccioli *et al.*, 1976, 1986, 1992a; Arnts, 1985). Uncertainties in VOC quantification associated with the different fragmentation patterns exhibited by eluted compounds can be reduced if the same parcel of air is simultaneously collected on two different traps, which are separately analysed by GC–FID and GC–MS. With this approach, fully or partly resolved components can be quantified through their GC–FID signals and MS exclusively applied to those overlapping constituents for which selected ion detection is needed (Ciccioli *et al.*, 1992b). An example of the GC–MS profile obtained by analysing an urban sample collected on traps filled with carbon adsorbents is shown in Figure 3.10. The numbers refer to the compounds listed in Table 3.3, where retention indices and concentrations are also listed. Coeluted compounds (e.g. 1,2,4-trimethylbenzene and 2-(2-ethoxyethoxy)-ethanol) were quantified by selected ion detection. While the number and types of compounds (chlorinated solvents, acids, alcohols, carbonyl compounds, monoterpenes,

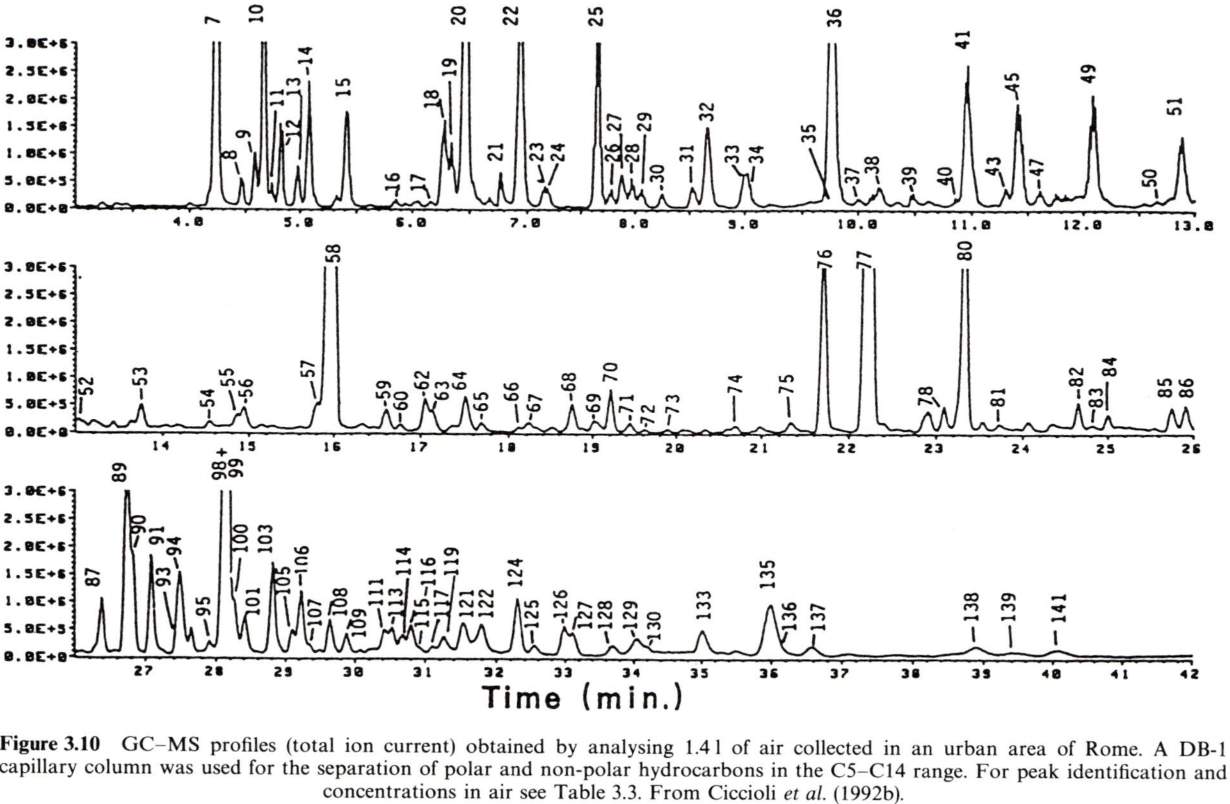

Figure 3.10 GC–MS profiles (total ion current) obtained by analysing 1.4 l of air collected in an urban area of Rome. A DB-1 capillary column was used for the separation of polar and non-polar hydrocarbons in the C5–C14 range. For peak identification and concentrations in air see Table 3.3. From Ciccioli *et al.* (1992b).

Table 3.3 Typical VOC (C5–C14 carbon range) concentrations ($\mu g/m^3$) measured in urban, suburban and pine forest sites of the Italian peninsula. Data refer to areas characterized by different VOC emissions. Rome (urban), Milan (urban) and Montelibretti (suburban) are representative of grids (50 × 50 km) with annual emissions of more than 40 tons per year per km^2. Taranto (urban) is representative of a grid where the emission was between 10 and 40 tons per year per km^2. The Monti Cimini Park (forest) belongs to a grid where the emission was between 1 and 4 tons per year per km^2. From Ciccioli *et al.* (1992a).

Compound	Retention index	Concentration				
		Rome	Milan	Taranto	Monti Cimini	Monte-libretti
7 2-Methylbutane	464.5	107.1	71.5	111.7	1.04	1.20
8 1-Pentene	483.2	2.9	2.5	13.9	0.09	0.20
9 2-Methyl,1-butene	493.0	3.6	3.3	10.3	0.18	0.03
10 *n*-C5	500.0	31.4	39.5	74.5	0.83	0.65
11 Isoprene	503.3	0.3	0.3	2.9	1.33	0.13
12 *trans*-2-Pentene	505.5	7.4	4.1	10.6	0.07	0.04
13 *cis*-2-Pentene	510.8	3.7	1.7	6.4	—	0.02
14 2-Methyl,2-butene	514.9	10.8	6.1	15.1	0.13	0.94
15 2,2-Dimethylbutane	528.1	11.7	6.2	11.5	0.07	—
16 Cyclopentene	545.0	1.5	0.5	1.6	—	—
17 Ethylcyclopropane	550.1	1.1	0.8	1.7	—	0.05
18 2,3-Dimethylbutane	556.4	3.6	2.1	13.5	—	—
19 2-Methyl,2-methoxypropane	558.5	9.7	6.4	15.2	—	0.14
20 2-Methylpentane	563.5	40.6	28.0	57.4	0.41	0.48
21 2-Butanone	575.7	0.3	—	8.5	—	—
22 3-Methylpentane	579.7	24.5	16.9	45.7	0.74	0.28
23 1-Hexene	585.2	1.4	3.1	—	—	—
24 1-Methylpentene	586.6	1.2	—	10.3	0.21	—
25 *n*-C6	600.0	16.8	16.6	44.9	0.61	0.43
26 *trans*-2-Hexene	601.9	1.3	—	2.0	—	—
27 *cis*-2-Hexene	604.3	2.1	1.0	3.9	0.07	—
28 2-Methyl,2-pentene	606.8	2.5	1.6	2.0	—	—
29 3-Methyl,2-pentene	609.7	1.5	1.4	—	—	—
30 3-Hexene	613.5	1.0	1.0	1.1	—	—
31 3-Methyl,1-pentene	623.5	0.4	0.5	1.5	—	—
32 Methylcyclopentane	625.6	7.6	6.4	12.7	0.18	0.14

33 2,4-Dimethylpentane	629.7	2.9	1.9	6.8	—	—
34 1,1,1-Trichloroethane	633.6	0.8	3.2	—	—	—
35 Acetic acid	646.5	—	0.1	15.9	0.22	—
36 Benzene	648.8	39.0	39.1	43.9	4.39	1.84
37 Carbon tetrachloride	654.4	1.5	0.8	—	0.11	0.08
38 Cyclohexane	657.9	2.1	5.2	3.7	—	0.07
39 1-Butanol	661.2	0.2	0.2	—	—	—
40 4-Methyl,2-hexene	663.2	0.3	0.1	—	—	—
41 2-Methylhexane	667.3	11.0	6.4	24.0	0.07	0.11
42 2,3-Dimethylpentane	668.8	4.0	2.6	2.5	—	0.04
43 1,2-Dichloropropane	673.6	0.5	0.4	—	—	—
44 Pentanal	675.5	—	—	—	1.27	0.18
45 3-Methylhexane	675.9	11.7	7.7	16.6	—	0.10
46 1,2-Dimethylpentane	677.4	1.9	1.2	—	—	—
47 1,3-Dimethylcyclopentane	682.1	1.4	1.0	—	—	—
48 Isooctane	685.8	2.9	2.6	—	—	0.08
49 2,2,4-Trimethylpentane	689.4	12.8	8.6	16.7	0.46	0.14
50 3-Heptene	697.9	1.2	0.5	—	—	—
51 *n*-C7	700.0	8.5	6.1	11.7	0.24	0.18
52 4-Methyl,2-hexene	704.7	0.8	0.4	—	—	—
53 Methylcyclohexane	721.0	2.6	3.4	3.2	—	0.06
54 Ethylcyclopentane	727.0	0.1	—	—	—	—
55 2,5-Dimethylhexane	733.0	2.2	1.5	4.7	—	0.02
56 2,4-Dimethylhexane	735.1	2.9	1.9	1.7	—	0.03
57 2,3,4-Trimethylpentane	751.4	4.4	2.5	4.9	—	0.03
58 Toluene	755.8	96.9	84.4	102.5	1.99	1.05
59 2,3-Dimethylhexane	761.9	3.1	28.5	—	—	0.02
60 4-Methylcyclohexanone	763.4	0.9	0.7	—	—	—
61 Cyclopentanone	767.3	3.9	2.9	—	0.09	0.07
62 4-Methylheptane	768.9	1.5	1.1	2.7	—	0.02
63 1,4-Dimethylcyclohexane	770.7	0.8	0.7	6.4	—	—
64 2-Methylheptane	774.7	4.5	3.4	5.1	0.07	0.04
65 Hexanal	778.5	1.9	2.6	3.3	3.49	0.82
66 2-Methyl,1-heptene	785.4	0.8	1.0	—	—	—
67 2,3-Dimethylcyclopentanone	787.1	0.2	1.3	—	—	—
68 Tetrachloroethene	794.5	0.6	0.4	3.5	—	—

Table 3.3 (*contd.*)

Compound	Retention index	Concentration				
		Rome	Milan	Taranto	Monti Cimini	Monte-libretti
69 1,2-Dimethylcyclohexane	797.1	1.4	2.8	—	—	—
70 *n*-C8	800.0	3.9	8.4	5.6	0.52	0.18
71 Isooctadiene	804.1	1.0	0.5	—	—	—
72 3-Methyl,2-heptene	807.6	0.7	0.3	—	—	—
73 2-Octene	812.1	0.4	0.9	—	—	—
74 Ethylcyclohexane	824.9	0.5	0.9	—	—	0.03
75 3-Methyloctane	838.3	1.1	0.8	—	—	—
76 Ethylbenzene	851.7	16.9	14.3	23.4	0.24	0.20
77 $(m+p)$-Xylene	860.2	58.9	45.7	79.3	0.63	0.42
78 Vinylbenzene	878.0	2.3	1.4	9.7	0.07	0.04
79 Heptanal	880.9	—	1.4	4.9	9.67	0.87
80 *o*-Xylene	883.2	24.2	18.4	31.4	—	0.30
81 Isononene	889.0	0.4	2.2	9.8	—	—
82 *n*-C9	900.0	1.9	5.0	16.5	0.32	0.15
83 Methyloctene	909.6	0.4	0.7	—	—	—
84 Isopropylbenzene	915.9	1.1	1.3	3.0	—	0.02
85 Benzaldehyde	937.9	4.0	5.1	11.4	—	—
86 α-Pinene	937.9	—	—	—	7.37	0.47
87 *n*-Propylbenzene	946.5	4.3	3.3	9.2	0.10	0.05
88 Camphene	948.4	—	—	—	0.20	0.08
89 1-Methyl,3-ethylbenzene	953.3	15.2	9.7	1.2	0.28	0.11
90 1-Methyl,4-ethylbenzene	955.7	7.6	5.4	28.0	0.07	0.06
91 1,3,5-Trimethylbenzene	961.5	5.4	3.7	12.0	0.07	0.05
92 Phenol	963.6	0.6	4.9	—	—	—
93 6-Methyl,5-hepten-2-one	965.7	6.1	—	—	9.66	1.56
94 1-Methyl,2-ethylbenzene	972.2	6.1	5.9	17.3	0.24	0.13
95 Isodecane	979.4	0.4	1.0	—	—	0.12
96 β-Pinene	980.1	—	—	—	0.45	0.04

No.	Compound						
97	Octanal	983.4	1.6	1.9	2.4	8.38	2.91
98	1,2,4-Trimethylbenzene	986.8	14.0	6.0	23.2	—	0.14
99	2-(2-Ethoxy,ethoxy)ethanol		15.5	6.6	—	—	0.15
100	1-(2-Methoxy,1-methylethoxy)-propan-2-ol		1.1	—	—	—	—
101	1-(1-Methyl,2-(2-propenyloxy)-ethoxy)-propan-2-ol		1.5	—	—	—	—
102	Isobutylbenzene	994.4	0.1	—	—	0.17	—
103	1-(2-Methoxy,propoxy)propan-2-ol		5.5	—	—	—	—
104	*sec*-Butylbenzene	997.1	0.6	—	—	—	0.03
105	*n*-C10	1000.0	0.9	4.8	5.1	0.12	0.11
106	1,2,3-Trimethylbenzene	1015.6	3.5	3.0	3.0	—	0.03
107	*p*-Cymene	1018.3	0.3	1.1	—	0.15	0.03
108	1-Methyl,3-isopropylbenzene	1025.2	0.1	0.3	—	—	—
109	Indane	1028.8	3.5	3.7	—	—	—
110	Limonene	1029.0	—	—	—	0.49	0.14
111	1H-Indene	1035.5	1.1	1.6	—	—	—
112	β-Phellandrene	1036.3	—	—	—	0.3	0.06
113	1-Methyl,3-propylbenzene	1039.1	0.1	—	—	—	—
114	1,3-Diethylbenzene	1042.3	0.8	0.7	—	—	—
115	*n*-Butylbenzene	1044.9	1.4	1.3	—	—	—
116	1-Methyl,4-propylbenzene	1047.1	0.6	1.0	—	—	0.06
117	1,4-Dimethyl,2-ethylbenzene	1051.0	2.2	2.8	—	—	—
118	2-Nonen,1-ol	1051.2	—	—	—	1.06	0.66
119	1-Methyl,2-propylbenzene	1055.6	0.3	1.7	—	—	—
120	γ-Terpinene	1056.1	—	—	—	0.08	0.02
121	1,3-Dimethyl,2-ethylbenzene	1061.8	0.5	1.5	3.3	—	0.05
122	1,2-Dimethyl,4-ethylbenzene	1064.2	0.4	1.6	3.1	—	—
123	Dimethylethylbenzene	1079.2	0.8	1.1	—	—	0.03
124	Nonanal	1085.5	1.8	3.1	7.3	9.51	8.85
125	*n*-C11	1100.0	0.5	1.9	0.9	0.20	0.09
126	1,2,4,5-Tetramethylbenzene	1112.8	0.4	1.0	3.9	—	0.02
127	1,2,3,5-Tetramethylbenzene	1116.3	0.5	0.9	1.6	—	0.03
128	2,3-Dihydro,5-methyl-1H-Indene	1147.0	0.3	1.8	—	—	—
129	2,3-Dihydro,2-methyl-1H-Indene	1152.4	0.5	2.0	—	—	—
130	Iso-C4-benzene	1157.7	0.2	5.0	—	—	0.05

Table 3.3 (*contd.*)

Compound	Retention index	Concentration				
		Rome	Milan	Taranto	Monti Cimini	Monte-libretti
131 4-Terpinenol	1160.3	—	—	—	0.31	0.09
132 Menthol	1174.7	—	—	—	0.1	0.02
133 Naphthalene	1180.0	0.3	8.3	—	—	0.08
134 α-Terpinol	1183.7	—	—	—	0.13	0.05
135 Decanal	1187.8	0.5	2.3	6.3	8.20	7.24
136 2-Phenoxyethanol	1193.6	0.3	—	—	—	—
137 *n*-C12	1200.0	0.3	—	3.5	0.44	0.06
138 2-Methylnaphthalene	1272.6	0.02	0.4	—	—	—
139 1-Methylnaphthalene	1276.3	0.02	0.3	—	—	—
140 Geraniol	1296.5	—	—	—	0.17	0.02
141 *n*-C13	1300.0	0.1	4.1	—	0.35	0.11
142 *n*-C14	1400.0	—	2.9	—	0.36	0.05
Total		826	731	1260	91	39

arenes) found in this sample provide a clear picture of the complexity of VOC distribution in air, comparison of the chromatograms shown in Figures 3.9 and 3.10 illustrates the different types of information provided by the two methods adopted for VOC monitoring. Only through their combination is it possible to determine how and to what extent emitted compounds are converted in the atmosphere, and the nature of their products. The data sets listed in Table 3.3, which are relative to urban, suburban and forest samples, are useful to determine the typical amounts and compositions of VOCs found in stations which, although belonging to the same monitoring network, are located in grids where different emission occurs. Although sophisticated double-column instrumentation has been developed for the determination of all VOCs emitted or formed in the air (Rudolph *et al.*, 1985), its capability to positively identify components is limited by the fact that the separation of the very volatile fraction is carried out on packed columns, with peak identity derived from retention and identification based on multi-detection units comprising of three GC detectors (namely, photoionization, electron-capture and flame ionization) set in series. Regardless of the conceptual approach followed, all methods mentioned (and their numerous modifications reported in the literature) are useful not only for validating models but also for deriving, through aerometric measurements, specific information on VOC emission, reactivity and transport. A special case is that of airborne measurement where, due to the low concentrations existing at high altitude and platform limitations, out-of-line sampling on electro-polished canisters or adsorption traps is preferred for collecting VOCs, and enriching procedures (mainly deep freezing) are required for PAN determination (Meyrahn *et al.*, 1984). High-volume collection is required for carbonyl compounds only if HPLC methods are adopted for their evaluation (Schmidt and Lowe, 1981). Simplified instrumentation is also preferred when monitoring campaigns are to be performed on ships at sea. In this case, however, the space and facilities available on-board often make possible *in situ* cryogenic enrichment of air samples and analysis by GC. Among the methods adapted for on-board use, particularly interesting examples are those developed for PAN (Vierkorn-Rudolph *et al.*, 1985; Muller *et al.*, 1990) and DMS (Pashalidis *et al.*, 1990) determinations. In both cases, cryogenic trapping combined with GC selective detection was used to meet the sensitivity requirements dictated by the low background levels existing at sea. While on-line sampling techniques combined with GC–ECD facilitated the evaluation of PAN at pptv levels, out-of-line collection was preferred for DMS. This was performed on cartridges filled with Tenax adsorbent kept at $-20\,^{\circ}C$. After thermal desorption, DMS was analysed by GC and selectively detected by flame photometry. Both methods were used on the Polarstern ship during the 1988 cruise across the Atlantic ocean.

3.4.2.3 Monitoring of VOCs relevant to stratospheric ozone depletion and thermal trapping. Knowledge of secular trends followed by the gases responsible for global warming and stratospheric ozone depletion is the basic information needed to forecast long-term climatic changes occurring as a result of increased man-made emissions. Knowledge of the annual percentage growth of these gases, established over several decades of monitoring, enables the efficacy of control strategies adopted to be checked, correlations between existing atmospheric levels and observed effects to be established, and models to be validated. Background monitoring is required for methane, CFCs and chlorinated solvents. These compounds are characterized by long atmospheric life times and are thus subject to long-range transport before being slowly equilibrated in the atmosphere by diffusion. In particular, CFC-11 is so persistent that it is commonly used as an ideal tracer for tracking long-range transport of anthropogenically polluted air masses travelling in the ABL (Lovelock *et al.*, 1973) or for detecting intrusion of stratospheric gases (e.g. ozone and PAN) into the low troposphere via diabatic descent (Brice *et al.*, 1984). To obtain values which are truly representative of the background levels of methane and CFCs in the Northern and Southern hemisphere, monitoring stations must be installed in remote areas not directly influenced by anthropogenic or strong natural sources. The importance of collecting background data on CFCs and methane has been recognized since the early 1970s when the United Nations Environment Program (UNEP) and the World Meteorological Organization (WMO) encouraged the establishment of a global network comprising background air pollution stations equally distributed among the two hemispheres. For many years Barrow (Alaska), Mauna Loa (Hawaii), Samoa, Cape Grim (Tasmania) and the South Pole were the five principal stations of this global network. More recently, other stations have been set up to provide complete global coverage. Data collected by these stations from the mid-1970s to the mid-1980s have provided clear evidence for the constant increase in gases responsible for global warming and stratospheric ozone depletion in both hemispheres (Tucker, 1986; Warrick *et al.*, 1990). The 100% increase in CFC-11 and methyl chloroform observed in ten years at Cape Grim, a station representative of the Southern hemisphere, is highly indicative of the rapid growth of CFCs in regions where consumption of the ozone column has been observed (Tucker, 1986). Until now, gas chromatography has been the preferred method for ground-based measurents of both methane and CFCs in remote stations. While determination of methane is accomplished with the same methods as those adopted for investigating emission, dedicated instrumentation is required for CFCs. The simplest one consists of a six-way valve, an aspirating pump, a calibrated loop and a packed column connected to an ECD. Air (5–7 ml) sampled into the loop by the pump is directly injected into the column by switching the valve. The compounds eluted are identified by their retention time (Lovelock *et al.*, 1973; Lillian *et al.*, 1976; Penkett *et al.*, 1979b). Instrumentation of this type can run unattended for long peri ds of time, provided proper timing of sampling, GC analysis and data acquisition is made. In remote areas, however, the practical applications of this method are restricted to those compounds (mainly CFC-11, methyl chloroform and CCl_4) exhibiting elevated background concentrations in air, high reactivity toward thermal electrons (i.e. electrophores capable of capturing one electron for each molecule) and reduced retention (narrow peaks result in a larger amount of matter entering the ECD per unit of time and hence greater response). Only through enrichment procedures enabling the injection of larger amounts of sample (Grimsrud and Rasmussen, 1975) is it possible to extend monitoring to CFC-12, CFC-113 and CFC-114, and to cover the six CFCs and chlorinated solvents indicated in the Montreal Protocol (UNEP, 1987) as the most potentially dangerous for the ozone layer. This has been accomplished by on-line enrichment of 50 ml of air into cooled microtraps filled with graphitic carbons (Crescentini *et al.*, 1991). The air sample, collected in a calibrated loop by an aspirating pump, is first transferred into the trap and maintained at $-50°C$ to retain CFCs. After the enrichment process is completed, the trap is connected to a packed column where compounds are injected by flash-heating. ECD is used as the sensing device. All six compounds are completely separated, and identification is based on peak retention. The system is fully automated and suitable for unattended monitoring in remote stations. An example of the GC profile recorded when the system was in operating at the Italian Station located in Antarctica (Terra Nova Bay) is shown in Figure 3.11. An advantage of cryogenic enrichment on empty tubes (Grimsrud and Rasmussen, 1975) or cold traps (Bruner *et al.*, 1981) compared with on-line-systems is that large volumes of air can be used for CFC determination. As a result, sufficient quantities of material are available for analysis, and MS working in the selected ion detection mode can be used for positive identification of CFCs present at trace levels (Grimsrud and Rasmussen, 1975; Bruner *et al.*, 1981). The results shown in Figure 3.12

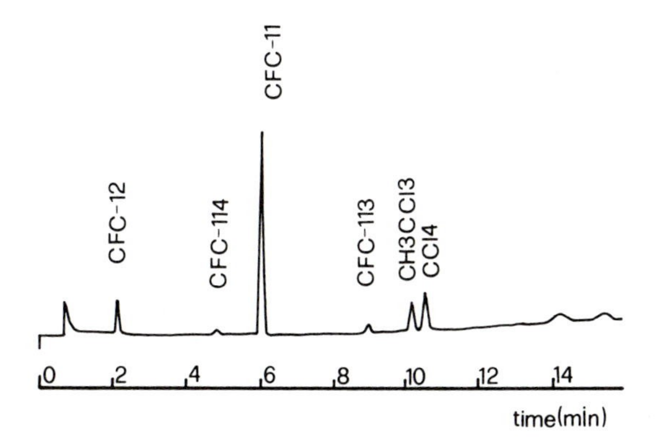

Figure 3.11 Typical GC–ECD profile of CFCs recorded by automated instruments installed in the Italian Station in Antarctica (Terra Nova Bay) in 1989–1990. Redrawn and adapted from Crescentini *et al.* (1991).

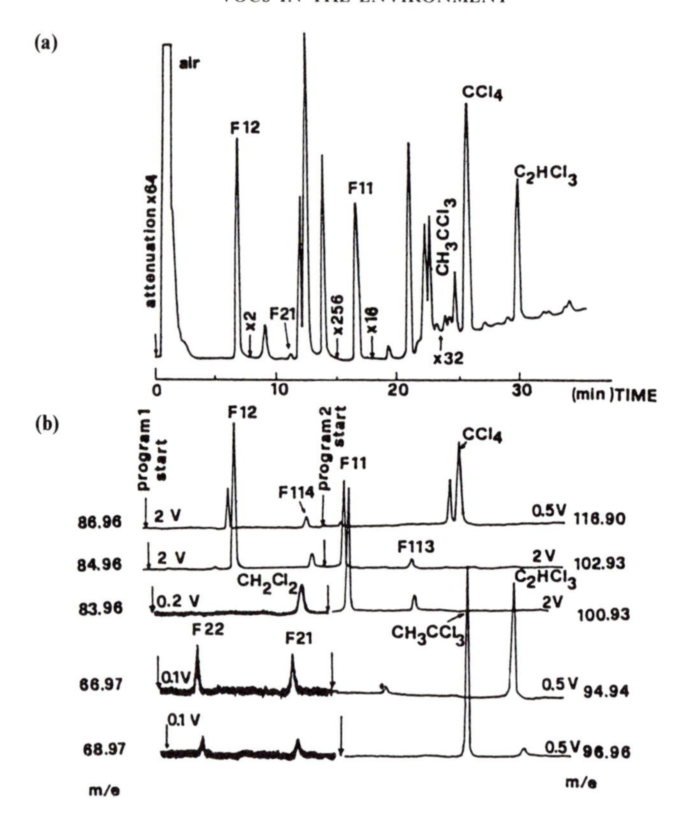

Figure 3.12 Comparison of (a) the GC–ECD and (b) the GC–MS profiles obtained during the analysis of CFCs sampled in a semi-rural area of Montelibretti. From Bruner *et al.* (1981).

illustrate well the unique capabilities afforded by GC–MS. Only through its use was it possible to unambiguously identify CFC-21 and avoid interference from coelution of species which, due to their high abundance in air, respond to the ECD. If samples are collected in evacuated canisters installed on balloons or flying aircraft, the methods described can also be used to evaluate the vertical distribution of CFCs through the atmosphere (Schmidt *et al.*, 1990). An attractive alternative for measuring real-time amounts of CFCs present in the total atmospheric column, and for following seasonal variations, is to use high-resolution, ground-based FTIR instruments, with the sun as the IR source (Adrian *et al.*, 1990). By investigating the spectral range between 750 and 1140 cm^{-1} the total amounts of methane, CFC-11, CFC-12, N_2O and ozone present in the atmospheric column can be determined simultaneously. A tracker is needed to line up the FTIR spectrometer with the sun so that spectra can be taken at favourable angles (0.5–23°). Although this method is rather sophisticated (a radiative transfer model is needed to analyse solar spectra), results obtained are of great importance when studying reactions occurring in stratospheric regions.

3.5 Methods for VOC determination

3.5.1 Total and non-methane VOC content

The simplest way to determine the total VOC content of air and emission sources is to pass filtered air sampled by a sealed PTFE-coated diaphragm pump directly to an FID (Lawrence Berkeley Laboratory (LBL), 1973; IC, 1990). VOCs present in the air stream are combusted over a hydrogen flame and the carbon ions formed are collected on a cylindrical electrode to which a positive potential is applied. Since the number of carbon ions produced by combustion differs slightly from compound to compound (LBL, 1973) the total VOC content must be expressed in terms of units relative to the hydrocarbon used for calibrating the detector (e.g. ppmv equivalent of methane or hexane). Semi-continuous determinations can be obtained if the air is sampled in a calibrated loop and transferred to the sensing device by a flow of inert gas (N_2 or He). Total monitors based on FID are rather simple and are sensitive enough to detect VOCs down to 0.1 ppmv methane equivalent. When emission is monitored, connecting lines must be heated to prevent water condensation, and flow control devices should be used to tune the amount of sample entering the FID. Non-dispersive IR spectroscopy is another technique suitable for evaluating the total VOC content of emission sources (LBL, 1973). It is based on the broadband absorption of VOCs at wavelengths in the region of a few micrometres. IR radiation is directed through two separate adsorption cells. One, filled with non-absorbing gases, operates as reference cell, the other is the measure cell. By passing the sample containing VOCs through the measure cell, IR radiation is absorbed. The difference in heat energy between the reference cell and the measure cell is detected by the change in capacitance caused by the distension of a diaphragm separating the two parts of the detection system receiving radiation from the cells. In this case, the signal recorded must also be referred to a specific hydrocarbon (usually hexane) or a mixture of hydrocarbons. VOC detectors of this type are subject to interference from water and CO_2 present in combustion sources. They also gave a disproportionate response if the VOCs contained in the sample are different from those used for calibrating the instrument. Dispersive IR spectrometers employing a mirror chopper and concave gratings to reflect the appropriate IR radiation into several different detectors offer an advantage over non-dispersive instruments in that CO and CO_2 can be detected along with the VOCs (LBL, 1973). Another principle exploited for the measurement of total VOCs in emission sources is catalytic oxidation (LBL, 1973; USEPA, 1979a). If air containing VOCs is passed through a cell where organic species are oxidized by a catalyst (a vanadium–alumina bed or Hopcalite) a change in temperature occurs with respect to a cell where no oxidation takes place. By measuring the difference in temperature between the two cells with thermocouples, and by amplifying the signal with the aid of an electronic circuit, a current proportional to the amount of VOCs present in the gas stream is obtained (LBL, 1973). Carbon monoxide and water must be removed, both to avoid deactivation of the catalyst, and to prevent inaccurate readings. Although instruments based on IR and catalytic oxidation detection are less sensitive and more prone to interference than those exploiting FID, they are often preferred for safety reasons.

Unfortunately, knowledge of the total VOC content present in atmospheric samples is of little practical interest, since the largest portion of carbon material dispersed in air is represented by methane, which is essentially non-toxic and relatively unreactive. For this reason, monitors capable of determining the non-methane VOC content in air have been developed and operated since the late 1960s. In the last thirty years a large number of different principles have been exploited to derive this aerometric index (Sexton *et al.*, 1982; Liberti and Ciccioli, 1985), but methods based on GC separation of methane from other VOCs remain the most common and reliable. The simplest way to obtain the non-methane VOC content is to subtract from the total VOC content (obtained with continuous monitors) the amount of methane determined by injecting a known volume of air directly into a GC column filled with highly selective adsorbents (usually Porapak Q) and connected to an FID. The non-methane content can also be obtained by backflushing VOCs from the GC column into an FID after methane is completely eluted by the column (Sexton *et al.*, 1982). In some instruments, backflushed non-methane VOCs are first oxidized to CO_2 and then reduced to CH_4 prior to being quantified by FID (USEPA, 1979b). The advantage of this oxidation/reduction step is that the response is proportional to the number of carbon atoms present in the sample. With this type of instrument, the non-methane VOC content in air is given as ppmv of carbon (ppm C), defined as the product of the number of carbon atoms present in an organic molecule and its concentration in ppmv. If the column is selective enough to separate methane from CO and CO_2, the reduction step can also be exploited to convert inorganic carbon components into methane so as to detect them by FID (USEPA, 1979b).

3.5.2 Determination of individual VOCs

Direct evaluation of the radiation absorbed or reflected, separation on chromatographic columns, and selective detection and titration with specific reagents are the methods used to evaluate individual VOCs in air and emission samples. This section will focus mainly on chromatographic techniques as they are the only ones capable of providing the full spectrum of compounds present in air and emission sources. Discussion of spectroscopic techniques will be restricted to the working principles of DOAS, FTIR and tunable diode laser absorption spectroscopy (TDLAS). Instrumental designs, fields of application and limitations are discussed in chapter 6 and have been reviewed elsewhere (Ciccioli and Cecinato, 1992; Grant *et al.*, 1992). Wet chemical techniques will not be mentioned as they have been in use for several years. Adequate descriptions of these methods can be found in the manual issued by the Intersociety Committee of experts, representing eight major American professional societies concerned with environmental measurements (IC, 1990).

3.5.2.1 Chromatographic methods. Chromatography, in its various forms, exploits the ability of a solid sorbent or a liquid coating to retain molecules with different vapour pressures, molecular sizes, solubilities and heats of absorption. Separation is obtained by injecting gaseous or liquid mixtures into a column filled with a solid material, the surface of which may be coated with a thin film of the liquid phase. A liquid or a gas is passed through the column until a dynamic equilibrium of the components between the stationary and the mobile phase is reached. The different affinities of the compounds present in the mixture for the stationary phase result in differential retention within the column such that individual components reach the column outlet at different times. Depending on the selectivity and efficiency of the column used, complete separation of the injected mixture can be obtained. The efficiency depends on the geometrical features of the column (column diameter and length, particle size of the solid support), while the selectivity is related to the nature of the stationary phase. In gas chromatography, the role of the mobile phase is negligible compared to that of the stationary phase. In liquid chromatography, both phases are important in determining the separation process. Detection of eluted compounds is obtained by connecting the column outlet to a physical or chemical device capable of sensing the molecules emerging from the column in the presence of the mobile phase. Chromatography cannot be used for continuous monitoring of VOCs and also requires the use of injection or collection devices to transfer fixed volumes of air into the column. For successful identification of the compounds present in the mixture, they must be separated efficiently and the amount of matter eluted by the column must be sufficient to be sensed by the detection system used. The larger the volume injected into the column, the greater the sensitivity. The maximum volume that can be received by a column without affecting its performance depends on the internal diameter of the column, therefore a limited amount of matter can be directly analysed. By considering the maximum volume that can be injected into GC packed columns with internal diameter ranging from 4 to 6 mm to be 10 ml, and the minimum amount detected by FID to be in the nanogram range, only mixtures containing components at levels higher than approximately 0.2 ppmv can be quantified by direct injection. Minimum detectable amounts by flame-photometry, photoionization detection and MS are equal or even lower than by FID, therefore the application field of direct GC analysers is restricted mainly to determination of VOCs in emission samples and of methane present in air. Among GC detectors, only ECD is likely to reach the required sensitivity for direct determinations in air; this is because ECD is capable of detecting strong electrophores at sub-picogram levels and is virtually blind to the remaining organic species. Due to the limited number of compounds to be separated, short packed columns (0.5–2 m) with an efficiency lower than 1000 theoretical plates per metre can successfully be used for the evaluation of PAN (Darley *et al.*, 1963; Sandalls *et al.*, 1974), CFCs and low-boiling halogenated solvents (Lillian *et al.*, 1976; Penkett *et al.*, 1979b) whenever levels in air exceed 0.05–0.1 ppbv. Achievement of such high performances requires, however, careful optimization of all relevant parameters, including the ECD working conditions and the stationary phase used for separation. PAN is a good example for illustrating the difficulties that can be encountered when developing semi-continuous monitors based on direct analysis of air samples. First, the column must be operated between 20 and 30°C to prevent PAN decomposition (Sandalls *et al.*, 1974; Roumelis and Glavas, 1989). At this temperature it must separate PAN from its upper homologues (peroxypropionil (PPN) and peroxybutyl nitrate (PBN)), air, CFCs, chlorinated solvents and water. Elution of PAN must be carried out within 2–3 min to concentrate the eluted matter in a sharp peak and exploit efficiently the sensitivity of the ECD. Column bleeding must be negligible so that low background currents are measured and enhanced signal-to-noise ratios are obtained. Only a few types of columns are capable of meeting these stringent requirements (Roberts, 1990). A further problem is represented by the selection of the most suitable ECD operating conditions. There are two basic ways of working with ECD: (1) constant current; and (2) constant frequency. The former provides a wide range of operation (from approximately 100 pptv to 400 ppbv), while the latter gives a better signal-to-noise ratio and hence better sensitivity (Ciccioli and Cecinato, 1992). If ECD is operated under constant frequency the linear range is reduced from approximately 30 pptv to approximately 30 ppbv. Consequently, different operating conditions may be required depending on the levels of PAN occurring or expected to occur at the site where the instrument is installed. These difficulties explain why dedicated instrumentation for PAN monitoring has been produced and commercialized only recently by some European and Canadian companies (Ciccioli and Cecinato, 1992). Only air volumes ranging from approximately 0.25 (urban areas) to 2 l (rural, remote areas) contain enough matter to be detected by FID, therefore photoionization detection and MS enrichment procedures, capable of concentrating the total VOC content in volumes of the order of tenths of μl, have been developed for atmospheric investigations. Reduction of the injected volume to such a small size is justified by the need to exploit fully the efficiency and selectivity of capillary columns which, because of their small internal diameter (0.25–0.35 mm), are incapable of tolerating injection samples of volumes larger than a fraction of their dead volume. With column lengths varying from 30 to 60 m, a number of theoretical plates one to two orders of magnitude higher than those available with the best packed columns can easily be reached, and a large number of compounds separated in a single run. The development of capillary columns internally coated with aluminum oxide/KCl has made possible the separation of all alkanes, alkenes, alkines and arenes in the C1–C9 range (Schneider *et al.*, 1978). The recent availability of chemically bonded phases (mainly methylphenyl silicons linked to the silica surface) has allowed the elution from the same GC column of organics exhibiting different polarities (McClenny *et al.*, 1984; Yokouchi *et al.*, 1990; Schuetzle *et al.*, 1991; Ciccioli *et al.*, 1992b). The low surface activity of fused silica, combined with the high thermal stability of chemically bonded phases (i.e. low column bleeding) and the low flow rate of narrow-bore columns, have made capillary chromatography the ideal separation system for MS detection. The column outlet can be directly connected to the ion source, such that all the matter is available for ionization by electron impact or by chemical ionization (Arnts, 1985; Yokouchi *et al.*, 1990; Schuetzle *et al.*, 1991; Ciccioli *et al.*, 1992b). Although the capacity of narrow-bore columns is much smaller than that of packed columns, the higher concentration of matter released at the column outlet makes it possible to reach sensitivities quite close to those of larger columns. The versatility of capillary chromatography is such that columns internally coated with thick films (1.2 μm) of methylphenyl silicon fluids bonded to the silica surface can be used to investigate the whole range of VOCs present in air and emission sources in a single run (Matuska *et al.*, 1986). Good separation of the low-boiling fraction (C2–C5) can be only achieved at rather low temperatures (-40°C), therefore this method cannot be considered truly competitive with that based on aluminum/KCl columns. Moreover, it has not yet been proven that columns of this type are actually able to separate all components in the C2–C10 range. However, the selectivity and sensitivity provided by capillary chromatography are considered so important that the increased expense associated with the use of sophisticated enriching/desorption units to provide sufficient amounts of VOC for GC determination and to reduce band broadening at the column inlet is a small price to pay. A schematic diagram of the main operational steps required for an enrichment/desorption unit designed for VOC determinations by capillary GC are shown in Figure 3.13. The instrumental design shown in this figure is of general application, as it can be included in a GC–FID system dedicated to automated and unattended monitoring of VOCs (on-line sampling) (McClenny *et al.*, 1984; Bloemen *et al.*, 1990; Schuetzle *et al.*, 1991) or used in central GC–FID or GC–MS units where samples collected in canisters or Tedlar bags are to be analysed (Rudolph *et al.*, 1985; Schmidbauer and Oheme, 1986; Kanakidou *et al.*,

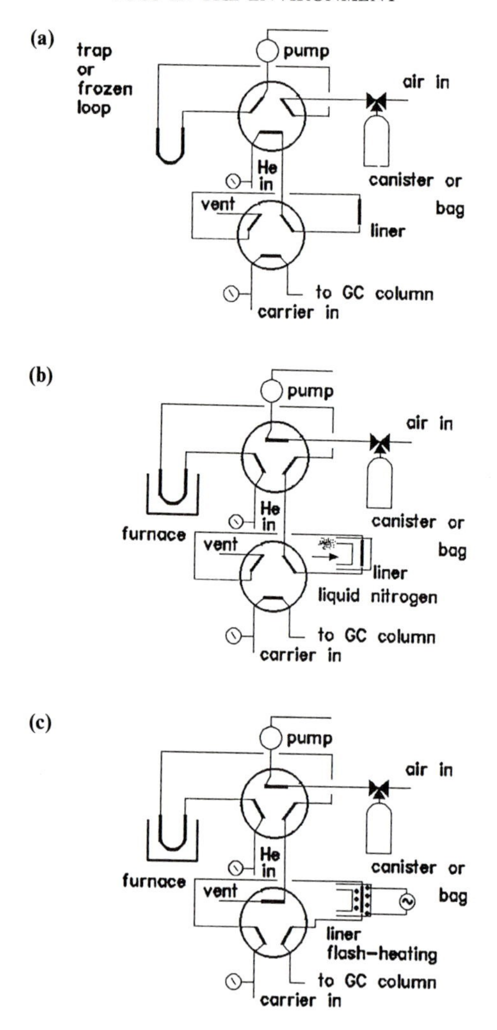

Figure 3.13 Schematic diagram of a desorption/cryofocusing unit for the determination of VOCs by capillary GC. (a) Injection; (b) desorption and cryofocusing; (c) sampling.

1988). With small modifications to the connecting lines and valves, a similar arrangement can be exploited for desorbing VOCs previously collected on cartridges filled with solid sorbents (Knoeppel *et al.*, 1980; Arnts, 1985; Yokouchy *et al.*, 1990; Ciccioli *et al.*, 1992b). If cryogenic enrichment on traps (Bloemen *et al.*, 1990) or empty tubes (McClenny *et al.*, 1984), is performed and/or aluminum oxide/KCl capillary columns are used for separating VOCs (Schmidbauer and Oheme, 1986; Kanakidou *et al.*, 1988; Bloemen *et al.*, 1990), ion exchange membranes and dehydrating agents should be added to the set-up shown in Figure 3.13. These must be inserted between the sampling point and the enrichment devices, during the collection step. Whether cryogenic trapping on empty tubes or collection on cartridges filled with solid sorbents is preferable is still undecided as no field studies demonstrating the advantages and disadvantages associated with the adoption of either enrichment system with respect to the other have been performed. If VOC determination is restricted to the compounds listed in Table 3.2, similar performances are likely to be obtained. Differences are likely to occur, however, if a whole range of VOCs is to be investigated; any device used for removing water leads inevitably to losses of the polar or moderately polar fraction by adsorption and dissolution processes (Ciccioli *et al.*, 1992b). Adsorbents play such an important role in VOC collection, that it is useful to discuss the preferred materials for concentrating air or emission samples. Since the mid-1970s a large number of solid sorbents has been tested and used for collecting VOCs. The list includes porous polymers, such as Tenax (Pellizzari *et al.*, 1976; Knoeppel *et al.*, 1980; Arnts, 1985; Yokouchi *et al.*, 1986; Clement *et al.*, 1990), Chromosorb 102 (Louw *et al.*, 1977) and Porapak Q (IC, 1990), graphitic carbons with different surface areas (Bruner *et al.*, 1974; Ciccioli *et al.*, 1976; Bloemen *et al.*, 1990; Ciccioli *et al.*, 1992b) and carbon molecular sieves (Rudolph *et al.*, 1981; Bloemen *et al.*, 1990, Ciccioli *et al.*, 1992a). Figure 3.14 shows the adsorption efficiency of some of these materials toward selected olefinic and aromatic components. Data were obtained by injecting the same VOC mixture into traps filled with different materials and by measuring the amount of each compound recovered when volumes of air varying from 2 to 10 l were passed through the various traps (Ciccioli *et al.*, 1984). Experiments were carried out at room temperature and under controlled humidity conditions. Cartridges filled with the same volume of solid sorbent were used. From the results obtained, it can be concluded that Carbopack B (a graphitic carbon with a specific surface area of approximately 90 m^2/g) is probably the most versatile material for collecting VOCs as it covers a wide range of carbon number, and enriched compounds can be recovered either by liquid extraction or by thermal desorption. It is also thermally stable up to 2000°C, chemically inert and highly hydrophobic, the basic structure being comprised of crystalline graphite. Stability to chemical attack is particularly important as the appearance of peaks in the chromatographic profile due to degradation of the polymeric matrix have often been observed after exposure of Tenax to ozone and other atmospheric pollutants (Knoeppel *et al.*, 1980; Ciccioli *et al.*, 1984; Pellizzari *et al.*, 1984). In spite of these limitations, Tenax still remains the most commonly used material for sampling VOCs in air.

No matter how efficient an adsorbing material is, some losses of VOCs always occur, therefore traps filled with different materials set in series have been suggested to extend the range of VOCs that can be retained efficiently and recovered (Rudolph *et al.*, 1981; Bloemen *et al.*, 1990; Ciccioli *et al.*, 1992b). The sequence is selected in such a way that, during sampling, the

Olefins and Terpenes
TENAX GC (Thermal Desorption)
100
50
0
Hex A Hept B Oct C Pin Car Terp Lim

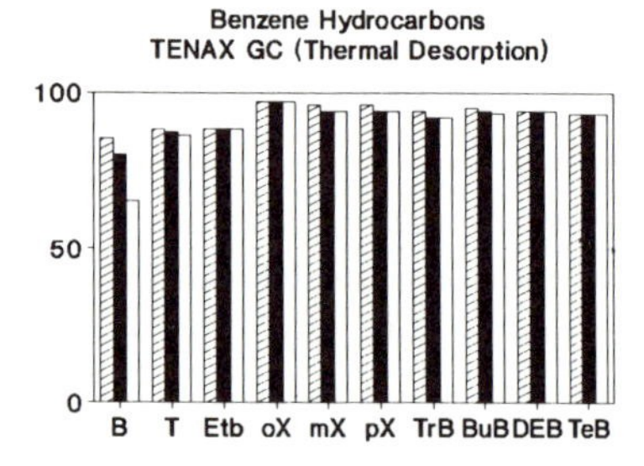

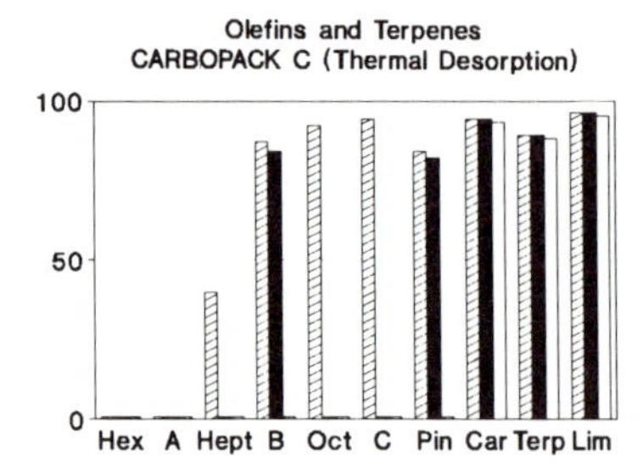

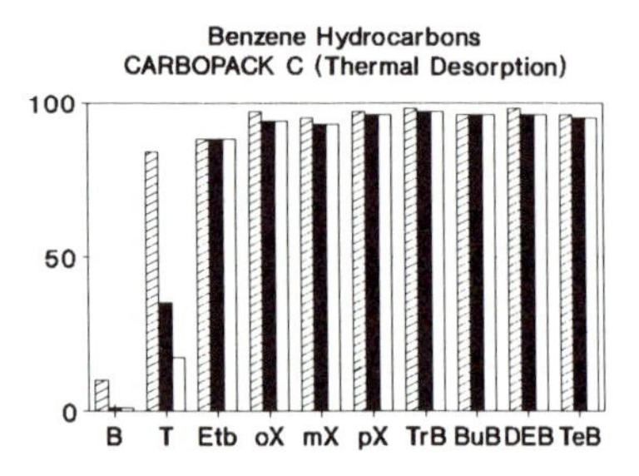

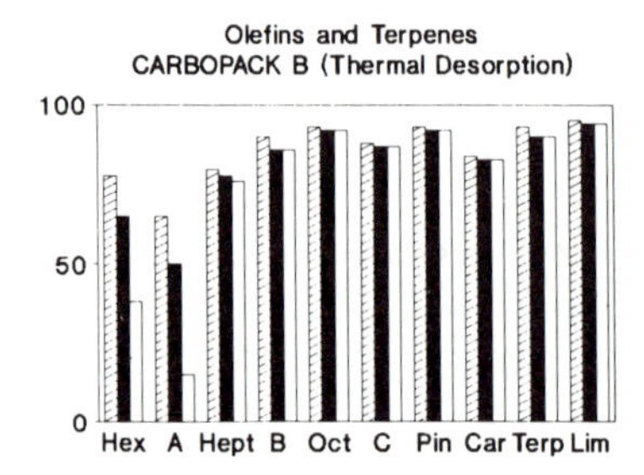

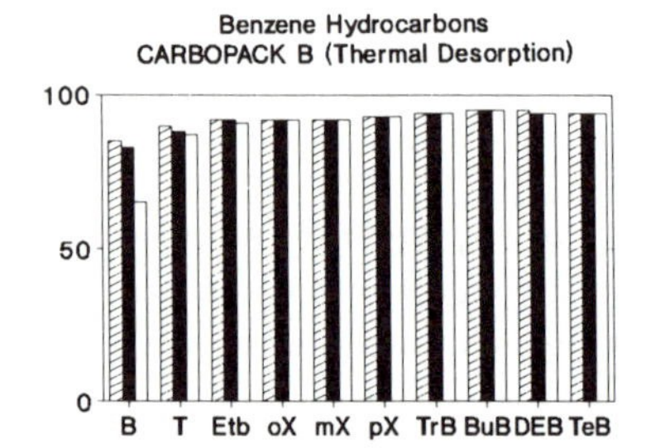

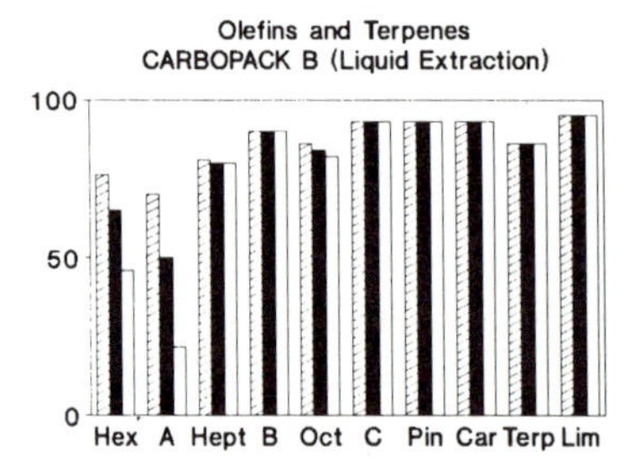

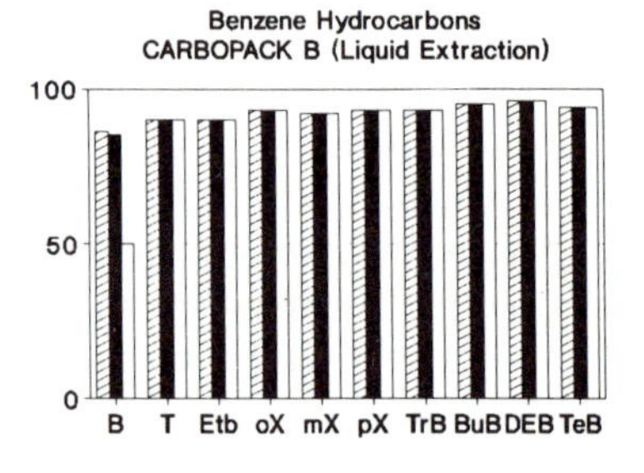

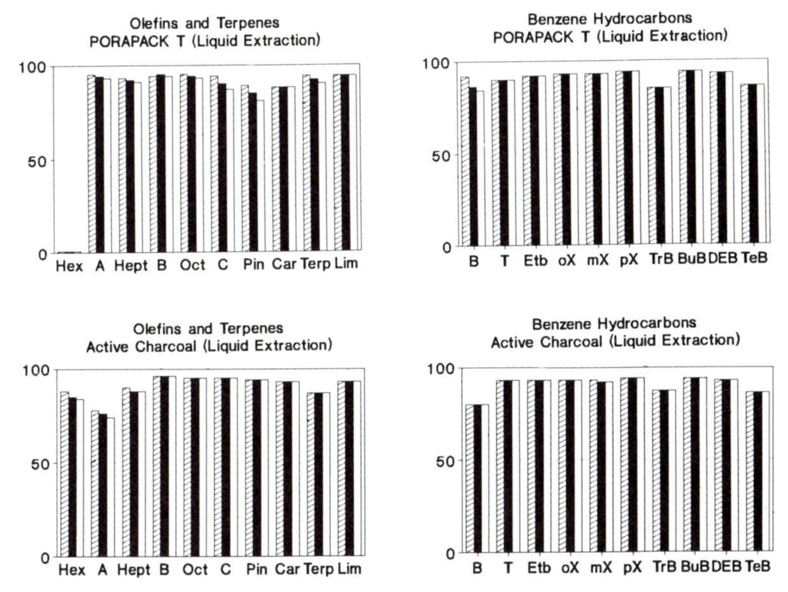

Figure 3.14 Adsorption efficiency (%) of various materials used for the enrichment of VOCs following the injection of 2, 5 and 10 l of air through cartridges (15 mm × 4 mm) filled with the same volume of adsorbent (40–60 mesh). Olefins: Hex; Hexene; A, bicycloheptadiene; Hept, Heptene; B, vinylcyclohexene; Oct, octene; C, *cis-cis* octadiene; Pin, α-pinene; Car, δ-3-carene; Terp, α-terpinene; Lim, limonene. Arenes: B, benzene; T, toluene; Etb, ethylbenzene; oX, *o*-xylene; mX, *m*-xylene; pX, *p*-xylene; TrB, 1,3,5-trimethylbenzene; BuB, *n*-butylbenzene; DEB, 1,3-diethylbenzene; TeB, 1,2,4,5-tetramethylbenzene. Redrawn and adapted from Ciccioli *et al.* (1984).

'lightest' adsorbent is placed at the trap inlet whereas the 'strongest' one is set at the end. Because of this specific set-up, backflushing procedures are required for the thermal desorption of traps filled with different solid sorbents (see Figure 3.13). Although their adoption does not provide solutions to all of the problems associated with the collection of VOCs in air and emission sources, the principle is promising as a high number of combinations are still to be proposed and tested. Interesting collection devices, application of which has until now been restricted to the evaluation of volatile aldehydes (Possanzini *et al.*, 1987) and organic acids (Puxbaum *et al.*, 1988) in air, are diffusion systems based on denuders. Information on their working mechanism and applications have recently been reviewed (Ciccioli and Cecinato, 1992), therefore they will not be discussed further here, except to mention the considerable possibilities they offer to discriminate gases from particles, a problem not yet resolved either by collection methods based on cryogenic trapping or by adsorption on cartridges filled with solid sorbents. Diffusion methods based on denuders have so far only been used in connection with

HPLC after liquid extraction of the coating used for collecting gaseous VOCs, no attempts have been made to exploit thermal desorption for sample recovery.

3.5.2.2 Differential optical absorption spectroscopy (DOAS). DOAS is based on the measurement of the intensity of absorption bands characteristic of a given VOC recorded in the UV visible region (Platt *et al.*, 1979). Through this measure, the column density of the compound along the light path is derived by knowing its differential absorption cross-section. The light source, which can be an Xe or an incandescent quartz-iodine lamp, is placed from a few hundreds to thousands of metres from the receiving point. The light beam crossing the air is received on a telescope focusing the light into the entrance slit of a spectrograph (Platt and Perner, 1980). The grating disperses the light so that different wavelengths are projected across the exit slit. A small segment of the spectrum is rapidly scanned by a slotted disk scanning device, consisting of a series of slits etched radially into a thin metal disk rotating into the focal plane. The slotted disk uses only one slit as the exit slit at any one time. The light passing through the exit slit is amplified and the signals recorded are sent to a computer where they are stored in separate channels. By accumulating scans on several hundreds of channels, each one corresponding to intervals of approximately 0.2 nm, the absorption spectrum on a region of 20 to 40 nm is reconstructed. The contribution of each VOC to the spectrum is obtained by sequential subtraction. Detection limits depend on the cross-section of the compound to be detected, the light path used and the number of scans accumulated. Integration in space and short-time resolution render DOAS extremely useful for continuous and unattended monitoring. The main limitation is the dependence of the signal on visibility conditions and on the constancy of light emission. For some years, the application of DOAS to VOC monitoring has been restricted to formaldehyde. Only recently has it been extended to benzene, toluene and xylenes (Brocco *et al.*, 1992; Grant *et al.*, 1992) by scanning the region between 230 and 270 nm. Although the results shown in Figure 3.8 seem to be extremely encouraging, evaluation of the accuracy of DOAS for determining benzene, toluene and xylenes is complicated by the intrinsic difficulties in correlating single-point data with long-path measurements. Based on the limited number of organic species that can be determined, other priorities, such as simultaneous and real-time determination of other important primary and secondary pollutants (ozone, SO_2, NO_2, NO, ammonia and nitrous acid) can justify the acquisition of a DOAS instrument for atmospheric monitoring. Although evaluation of the emission sources and compliance with existing regulations, detection of accidental releases and integrated monitoring of selected 'criteria' pollutants in urban areas (with the aim of establishing when, to what extent and with what frequency air quality standards are exceeded) are important items justifying the use of DOAS, the real advantage is the unique possibility of detecting the radical species (OH and NO_3) which are mainly responsible for photochemical reactivity in air during daytime and at night.

3.5.2.3 Fourier transform infrared spectroscopy (FTIR). Long-path FTIR, used in atmospheric investigations since the middle of the 1970s (Finalyson-Pitts and Pitts, 1986), has played a fundamental role in the knowledge of atmospheric chemistry. Through its use it has been possible to evaluate the concentration air of species (e.g. HNO_3) relevant to photochemical pollution and acid deposition (Tuazon *et al.*, 1981). The instrument used comprises a Michelson interferometer, an IR light source, a multi-reflection White cell, two detectors and a computer system for deconvolution of sinusoidal signals by Fourier transform. When IR light crosses the interferometer the detector receives a signal (called an interferogram), which is the result of constructive and destructive interferences produced by all wavelengths. When air flows through the chamber, compounds absorb the light and the interferogram is altered. Deconvolution of the interferogram by Fourier transform and comparison with the empty cell leads to generation of a classical IR spectrum by computer. Constant resolution over the entire range, use of the whole source energy for generating the spectrum and simultaneous sampling of different wavelengths are major advantages of FTIR compared with scanning devices (Finalyson-Pitts and Pitts, 1986). PAN, formaldehyde, formic acid, ethylene and non-methane paraffinic hydrocarbons are some of the organic species that have been monitored by FTIR (Harnst *et al.*, 1982). As detection limits fall in tenths of ppbv the range, monitoring of certain species, such as PAN, can be performed only in rather polluted areas. Important limitations of this type of instrument are complexity, high cost and difficulties in transportation (the multi-reflection cell can be as long as 25 m). More recently, an arrangement very similar to that used by DOAS has been proposed for collecting spectra by FTIR. The IR source placed near the receiving point is directed toward a retroreflector placed at a distance of approximately 200 m from the FTIR (Grant *et al.*, 1992). Although a large number of organic compounds absorb in the IR region, the minimum detectable concentrations of open-path FTIR are quite high compared with the levels existing in air, and possible applications are restricted to the evaluation of VOC emission or alterting the population in cases of accidental releases (Grant *et al.*, 1992). As mentioned previously (section 3.4.2.3), sensitive determination of CFCs was obtained by FTIR using the sun as a radiation source (Adrian *et al.*, 1990).

3.5.2.4 Tunable diode laser absorption spectroscopy (TDLAS). The operating principle of TDLAS is based on the ability of certain lead-salt semiconductors to act as IR laser sources when a *p–n* junction is formed in the crystal and electrical current is applied to the diode. By changing the temperature of the semi-conductor, the laser can be tuned over several hundred wave numbers. This procedure requires good temperature control because the laser works in the range 10–80 K. As linewidths of tunable diode lasers are rather narrow, absorbancies due to a single rotation line can be measured. In addition to the diode laser, the instrument requires an electronically controlled cryocooler fed with liquid He, a current control, a lock-in amplifier, a White cell where the sample is sucked by an aspirating pump, and a detection system. The instrument is usually operated in frequency modulation mode at an amplitude 2.2 times larger than the peak-to-peak modulation amplitude. The detector output is also analysed at twice the modulation frequency. By scanning the modulated radiation over an absorption line, a derivative-like signal is recorded. The working principle, calibration procedures and instrumental set-up for tropospheric monitoring were reviewed by Shiff *et al.* (1983). With tunable diode lasers it is possible to scan regions where IR absorption of some important organic compounds occurs (Hinkley *et al.*, 1976); however, to date formaldehyde is the only VOC for which monitoring has been performed. Intercomparison with DOAS and FTIR has shown the accuracy of TDLAS in determining formaldehyde in the low ppbv range (Schiff and Mackay, 1989). Results obtained using wet methods based on 2,4-DNPH seem to agree with those obtained using TDLAS in the 50–100 pptv range (Harris *et al.*, 1990). It is worth mentioning that good results were collected on-board a boat cruising over the Atlantic ocean. While the reliability and extremely high sensitivity of commercially available TDLAS instruments for the monitoring of formaldehyde and some important 'criteria' (mainly NO and NO_2) and secondary pollutants (HNO_3, NH_3 and H_2O_2) is undisputed, the range of TDLAS applications is still largely unexplored and, at the present time, it is difficult to envisage whether and to what cases it could be extended.

3.6 Conclusions

The large number of initiative undertaken worldwide seem to indicate that the time has come for a critical revision of the concepts that, enacted twenty years ago, inspired VOC legislation. There is an urgent need for a better definition of air quality standards in terms of individual VOCs in order to take into account both short-term effects arising from exposure to primary and secondary pollutants, and long-term effects associated with the disruption of secular equilibria. A list of VOCs potentially harmful to man and the environment should be prepared based on toxicity, potential for ozone and acidic species production, and capability both of depleting ozone and of increasing thermal trapping. Clear identification of the species to be monitored in the atmosphere and emission sources would also allow better definition of the strategies to be adopted for determination and, consequently, would stimulate research aimed at developing analytical tools more accurate and precise than those existing today. The great proliferation of methods intended for VOC determination, together with the larger number of modelling studies for predicting effects, have indicated the path to follow. However, a degree of international harmonization, both in the methods used and in the strategies for VOC control, must be reached if the transboundary implications associated with VOC emission are to be avoided. Only by making the VOC problem an international issue on which a consensus must be reached, will it be possible to preserve our environment for future generations.

References

Adrian, G.P., Blumenstock, T., Fisher, H., Gerhardt, L., Gulde, T., Oelhaf, H., Thomas, P. and Trieschmann, O. (1990) Column amounts of trace gases derived from ground based IR-spectroscopic measurements in the North Polar winter. In Pyle, J.A. and Harris, N.R.P. (eds) *Proceedings of the First European Workshop on Polar Stratospheric Ozone*, Schliersee, 1990, pp. 25–28. *CEC Air Pollution Research Report No. 34.* Brussels.

Altshuller, A.P. (1983a) Measurements of the poducts of atmospheric photochemical reactions in laboratory studies and in ambient air-relationships between ozone and other products. *Atmos. Environ.* **17** 2383–2427.

Altshuller, A.P. (1983b) Natural volatile organic substances and their effect on air quality in the United States. *Atmos. Environ.* **17** 2131–2165.

American Society for Testing Material (ASTM) (1981) Standard recommended practises for referencing supra-threshold odor intensity. *E-544-75 Annual Book of Standards.* Philadelphia, PA.

Arnts, R.R. (1985) Precolumn sample enrichment device for analysis of ambient volatile organics by GC–MS. *J. Chromatogr.* **329** 399–405.

Arnts, R.R., Peterson, W.B., Seila, R.L. and Gay, B.W. Jr. (1982) Estimates of alpha-pinene emissions from a loblolly pine forest using an atmospheric diffusion model. *Atmos. Environ.* **16** 2127–2137.

Atkinson, R. (1986) Kinetics and mechanisms of the gas-phase reaction of the hydroxyl radical with organic compounds under atmospheric conditions. *Chem. Rev.* **88** 69–201.

Atkinson, R. and Carter, W.P.L. (1984) Kinetics and mechanisms of the gas-phase reactions of ozone with organic compounds under atmospheric conditions. *Chem. Rev.* **84** 69–201.

Atkinson, R. and Lloyd, A.C. (1982) An updated chemical mechanism for hydrocarbons/NO_x/SO_2 photoxidation suitable for inclusion in atmospheric simulation model,. *Atmos. Environ.* **16** 1341–1355.

Bailey, J.C., Schmidl, B. and Williams, M.L. (1990) Speciated hydrocarbon emissions from vehicles operated over the normal speed range on the road. *Atmos. Environ.* **24A** 43–52.

Barnes, I., Bonsang, B., Brauers, T., Carlier, P., Cox, R.A., Dorn, H.P., Jenkin, M.E., Le Bras, G. and Platt, U. (1991) Laboratory and field studies of oxidation processes occurring in the atmospheric marine boundary layer (OCEANO-NOX CEC Project). *CEC Air Pollution Research Report No. 35.* Brussels.

Bates, T.S., Cline, J.D., Gammon, R.H. and Kell-Hansen, S.R. (1987) Regional and seasonal variations in the flux of oceanic dimethylsulfide to the atmosphere. *J. Geophys. Res.* **92** 2930–2938.

Beilke, S. and Gravenhorst, G. (1980) Cycles of pollutants in the troposphere. In Versino, B. and Ott, H. (eds) *Physico-Chemical Behaviour of Atmospheric Pollutants, Proceedings of the First European Symposium*, Ispra, Italy, 16–18 October 1979, pp. 331–353. EUR 6621 DE, EN, FR, CEC. Brussels.

Bloemen, H.J. Th., Bos H.P. and Dooper, H.P.M. (1990) Measuring VOC for the study of atmospheric processes. *Int. Laboratory* **20** 23–26.

Bouscarin, R. (1988) The CORINAIR Project—An EC air pollutant emission inventory. Paper presented at the Riso International Conference on Environmental Models: Emission Consequences, Roskilde, 24–25 May.

Brasseur, G. (1991) Atmospheric and climate. In Fantechi, R., Maracchi, G. and Almeida-Texeira, M.E. (eds) *Climatic Change and Impacts: A general Introduction. CEC-Directorate General, Science, Research and Development Report No. EUR 11943*. Brussels.

Brice, K.A., Penkett, S.A., Atkins, D.A.F., Sandalls, F.J., Bamber, D.J., Tuck, A.F. and Vaugham, G. (1984) Atmospheric measurements of peroxyacetyl nitrate (PAN) in rural south-east England: seasonal variations, winter photochemistry and long-range transport. *Atmos. Environ.* **18** 2691–2702.

Brocco, D., Fratarcangeli, R., Lepore, L. and Ventrone, I. (1992) Monitoraggio degli inquinanti gassosi primari e secondari mediante Spettroscopia d'Assorbimento Ottico differenziale (DOAS). Atti delle 'Giornate di Studio Europeo per l'Ambiente—Inquinamento Atmosferico-Citta' ed ambiente'. *SEP Pollution 1992*—29 Marzo – 2 Aprile, pp. 477–488. PADOVAFIERE Editrice, Padova.

Brodzinsky, R. and Singh, H.B. (1980) Volatile organic chemicals in the atmosphere: an assessment of available data. *EPA Report No. EPA-600/3-83-027a*. Research Triangle Park, NC.

Bruner, F., Ciccioli, P. and Di Nardo, F., (1974) Use of graphitised carbon black in environmental analysis. *J. Chromatogr.* **99** 661–672.

Bruner, F., Crescentini, G., Mangani, F., Brancaleoni, E., Cappiello, A. and Ciccioli, P. (1981) Determination of halocarbons in air by gas chromatography–high resolution mass spectrometry. *Analyt. Chem.* **53** 798–801.

Bufler, U. and Wegmann, K. (1991) Diurnal variation of monoterpene concentrations in open-top chambers and in the Wlexheim forest air. *FRG Atmos. Environ.* **24A** 251–256.

Burkhardt, M.R., Maniga, N.I., Stedman, D.H. and Paur, R.J. (1988) Gas chromatographic method for measuring nitrogen dioxide and peroxyacetyl nitrate in air without compressed gas cylinders. *Analyt. Chem.* **60** 816–819.

Calvert, J.C. and Stockwell, W.R. (1983) Acid generation in the troposphere by gas-phase chemistry. *Environ. Sci. Technol.* **17** 428A–433A.

Calvert, J.C., Lazrus, A.J., Kok, G.L., Heikes, B.G., Walega, J.G., Lind, J. and Cantrell, C.A. (1985) Chemical mechanisms of acid generation in the troposphere. *Nature* **317** 27–35.

Carlier, P. and Mouvier, G. (1988) Initiation à la physicochimie de la basse troposphère. *Pollut. Atmosph.* **117** 12–23.

Carter, W.P.L. and Atkinson, R. (1988) Computer modelling study of incremental hydrocarbon reactivity. *Environ. Sci. Technol.* **23** 864–880.

Chang, J.S., Brost, R.A., Isaksen, I.S.A., Madronich, S., Middleton, P., Stockwell, W.R. and Walcek, C.J. (1987) A three-dimensional Eulerian acid deposition model: physical concepts and formulations. *J. Geophys. Res.* **92** 14681–14700.

Chock, D.P. and Heuss, J.M. (1987) Urban ozone and its precursors. *Environ. Sci. Technol.* **21** 1146–1153.

Ciccioli, P. and Cecinato, A. (1992) Advanced methods for the evaluation of atmospheric pollutants relevant to photochemical smog pollution and acid deposition. In Nriagu, J.O. (ed) *Gaseous Pollutants: Characterization and Cycling* John Wiley and Sons, Inc., New York.

Ciccioli, P., Bertoni, G., Brancaleoni, E., Fratarcangeli, R. and Bruner, F. (1976) Evaluation of organic pollutants in the open air and atmospheres in industrial sites using graphitized carbon black traps and GC–MS analysis with specific detectors. *J. Chromatogr.* **126** 757–770.

Ciccioli, P., Brancaleoni, E., Possanzini, M., Brachetti, A. and Di Palo, C. (1984) Sampling, identification and quantitative determination of biogenic and anthropogenic hydrocarbons in forest areas. In Versino, B. and Angeletti, G. (eds) *Proceedings of the Third European Symposium on the Physico-Chemical Behaviour of Atmospheric Pollutants*, Varese, 1984, pp. 62–73. D. Reidel, Dordrecht.

Ciccioli, P., Brancaleoni, E., Di Palo, V., Liberti, M. and Di Palo, C. (1986) Misura delle alterazioni della qualità dell'aria da fenomeni di smog fotochimico. *Acqua ed Aria* **7** 675–683.

Ciccioli, P., Draisci, R., Cecinato, A. and Liberti, A. (1987) Sampling of aldehydes and carbonyl compounds in air and their determination by liquid chromatographic techniques. In Angeletti, G. and Restelli, G. (eds) *Proceedings of the Fourth European Symposium on the Physico-Chemical Behaviour of Atmospheric Pollutants*, Stresa, 1986, D. Reidel, pp. 131–141. Dordrecht.

Ciccioli, P., Brancaleoni, E., Cecinato, A. and Brachetti, A. (1989) Diurnal and seasonal variations of peroxyacetyl nitrate (PAN) in a semirural area of Central Italy. In Brasser, L.J. and Mulder, W.C. (eds) *Proceedings of the VIII Clean Air Congress*, The Hauge, 1989, Vol. 3, pp. 497–502. Elsevier, Amsterdam.

Ciccioli, P., Cecinato, A., Brancaleoni, E. and Frattoni, M. (1992a) Identification and quantitative evaluation of C4–C14 volatile organic compounds in some urban, suburban and forest sites in Italy. *Fres. Envir. Bull.* **1** 73–78.

Ciccioli, P., Cecinato, A., Brancaleoni, E. and Frattoni, M. (1992b) Use of carbon adsorption traps combined with high resolution gas chromatography–mass spectrometry for the analysis of polar and non-polar C4–C14 hydrocarbons involved in photochemical smog formation. *J. High Resolut. Chromatogr. Chromatogr. Commun.* **15** 75–84.

Clement, B., Riba, M.L., Leduc, R., Haziza, M. and Torres, L. (1990) Concentration of monoterpenes in a maple forest in Quebec. *Atmos. Environ.* **24A** 2513–2516.

Committee on the Challenges of Modern Society (CCMS)–NATO (1974) Air Quality Criteria for Photochemical Oxidants and Related Hydrocarbons. *CCMS-NATO Report No. 29.* Brussels.

CONCAWE (1991) Motor vehicle emission regulations and fuel specifications—1991 update. *CONCAWE Report No. 3/91.* Brussels.

Cox, R.A. (1979) Rates, reactivity and mechanism of homogeneous atmospheric oxidation reactions. In Versino, B. and Ott, H. (eds) *Physico-Chemical Behaviour of Atmospheric Pollutants, Proceedings of the First European Symposium,* Ispra, Italy, 16–18 October 1979, pp. 91–109. EUR 6621 DE, EN, FR, CEC. Brussels.

Cox, R.A. and Penkett, S.A. (1983) Formation of atmospheric acidity. In Beilke S. and Elshout A. (eds) *Proceedings of a CEC Workshop on Acid Deposition,* Berlin (Reichstag), pp. 56–81. D. Reidel, Dordrecht.

Cox, R.D. and Earp, R.F. (1982) Determination of trace level organics in ambient air by high-resolution gas chromatography with simultaneous photoionization and flame ionization detection. *Analyt. Chem.* **54** 2265–2270.

Crescentini, G., Maione, M. and Bruner, F. (1991) Measurements of tropospheric concentration of halocarbons in Antarctica. *Annali di Chimica* **81** 491–501.

Crutzen, P.J. (1970) The influence of nitrogen oxides on the atmospheric ozone content. *Quart. J. Roy. Met. Soc.* **96** 320–325.

Cullis, C.F. and Hirschler, M.M. (1989) Man's emission of carbon monoxide and hydocarbons into the atmosphere. *Atmos. Environ.* **23** 1195–1203.

Dann, T., Bunbaco, M., Chiu, C., Dlouhy, J., Li, K., Mar, J. and Wang, D. (1989) Measurement of potentially toxic airborne contaminants in Canadian urban air. In Brasser, L.J. and Mulder, W.C. (eds) *Proceedings of the VIII Clean Air Congress,* The Hague, September 1989, Vol. 3, pp. 165–170. Elsevier, Amsterdam.

Darley, E.F., Kettner, K.A. and Stephens, E.R. (1963) Analysis of peroxyacyl nitrates by gas chromatography with electron capture detection. *Anal. Chem.* **35** 389–391.

De Bortoli, M., Knoeppel, H., Peccio, E., Pail, A., Rogora, L., Schauenburg, H., Schlitt, H. and Vissers, H. (1985) Measurements of indoor air quality and comparison with ambient air. A study of 15 homes in Northern Italy. *CEC Report EUR 9656 EN.* Luxembourg.

Decreto del Presidente della Repubblica Italiana (DPR) (1983) Limiti massimi di accettabilità delle concentrazioni e di esposizione relativi ad inquinanti dell'aria in ambienti esterni. *Gazzetta Ufficiale della Repubblica Italiana* **145** 3–17.

Decreto del Presidente della Repubblica Italiana (DPR) (1990) Linee guida per il contenimento delle emissioni inquinanti degli impianti industriali e la fissazione dei valori minimi di emissione. *Gazzetta Ufficiale della Repubblica Italiana* **176** 5–94.

Demerjan, K.L., Kerr, J.A. and Calvert, J.G. (1974) The mechanism of photochemical smog formation. In Pitts, J.N. Jr. and Metcalf, R.L. (eds) *Advances in Environmental Science and Technology* Vol. 4, pp. 1–263. John Wiley and Sons, Inc., New York.

Derwent, R.G. (1990) Evaluation of a number of chemical mechanisms for their application in models describing the formation of photochemical ozone in Europe. *Atmos. Environ.* **24A** 2615–2624.

Derwent, R.G. (1991) Photochemical ozone creation potentials for over 150 individual organic compounds. Paper presented at the informal UN-ECE national expert meeting on VOC held in Berlin (April, 1991).

Derwent, R.G. and Jenkin, M.E. (1991) Hydrocarbon and the long-range transport of ozone and PAN across Europe. *Atmos. Environ.* **25A** 1661–1678.

Dimitriades, B. (1981) The role of natural organics in photochemical pollution. *J. Air Pollut. Control. Assoc.* **32** 229–235.

Driscoll, C.T. and Newton, R.M. (1985) Chemical characteristics of Adirondack lakes. *Environ. Sci. Technol.* **19** 1018–1024.

Farman, J.C., Gardiner, B.G. and Shanklin, J.D. (1985) Large losses of total ozone in Antarctica reveal ClO_x/NO_x interaction. *Nature* **315** 207–210.

Finlayson-Pitts, B.J. and Pitts, J.N. Jr. (1977) The chemical basis of air quality: kinetics and mechanisms of photochemical air pollution and application to control strategies. In Pitts, J.N. Jr., Metcalf, R.L. and Lloyd, A.C. (eds) *Advances in Environmental Science and Technology*, Vol. 7, pp. 75–162. John Wiley and Sons, Inc., New York.

Finlayson-Pitts, B.J. and Pitts, J.N. Jr. (1986) *Atmospheric Chemistry: Fundamentals and Experimental Techniques*. John Wiley and Sons, Inc., New York.

Foulger, B.E. and Simmonds, P.G. (1979) Drier for field use in the determination of trace atmospheric gases. *Analyt. Chem.* **51** 1089–1090.

Gaffney, J.S., Streit, G.E., Spale, W.D. and Hall, J.H. (1987) Beyond acid rain: do soluble oxidants and toxins interact with SO_2 and NO_x to increase ecosystem effects? *Environ. Sci. Technol.* **21** 519–524.

Gamlen, P.H., Lane, B.C., Midgely, P.M. and Steed, J.M. (1986) The production and release to the atmosphere of CCl_3F and CCl_2F_2 (chlorofluorocarbons CFC 11 and CFC 12). *Atmos. Environ.* **20** 1077–1085.

Gaudioso, D., Vaccaro, R., Brini, S., Cirillo, M.C. and Trozzi, C. (1991) Le emissioni di composti volatili in Italia. *Ingegneria Ambientale* **20** 244–251.

Gholson, A.R., Storm, J.F., Jayanty, R.K.M., Fuerst, R.G., Logan, T.J. and Midgett, R.M. (1988) Evaluation of canisters for measuring emissions of volatile organic air pollutants from hazardous waste incineration. *JAPCA* **39** 1210–1217.

Goldsmith, J.R. and Friberg, L.T. (1987) Effects of air pollution in human health. In Stern, A.C. (ed.) *Air Pollution*. Academic Press, New York.

Graedel, T.E., Hawkins, D.T. and Claxton, L.D. (1986) *Atmospheric Chemical Compounds—Sources, Occurrence and Bioassay*. Academic Press, Orlando.

Grant, W.B., Kagann, H.R. and McClenny, W.A. (1992) Optical remote measurement of toxic gases. *J. Air Waste Manage. Assoc.* **42** 18–30.

Grennfelt, P., Saltbones, J. and Schjoldager, J. (1988) Oxidant data collection in OECD Europe 1985–1987 (OXIDATE). Report on ozone, nitrogen dioxide and peroxyacetyl nitrate—October 1985–March, April, September 1986. *NILU Report OR 31/88*. Norwegian Institute for Air Research, Lillestrom.

Grimsrud, E.P. and Rasmussen, R.A. (1975) The analysis of chlorofluorocarbons in the troposphere by gas chromatography–mass spectrometry. *Atmos. Environ.* **9** 1010–1013.

Guicherit, R. and Van Dop, H. (1977) Photochemical production of ozone in Western Europe (1971–1975) and its relation to meteorology. *Atmos. Environ.* **11** 145–155.

Haagen-Smit, A.J. and Fox, M.M. (1956) Ozone formation in photochemical oxidation of organic substances. *Indust. Eng. Chem.* **48** 1484–1487.

Hansen, J.I., Fung, A., Lacis, D., Rind, S., Lebedeff, R., Ruedy, G., Russell, C. and Stone, P. (1988) Global climate changes as forecast by Goddard Institute for space studies three-dimensional model. *J. Geophys. Res.* **93** 9341–9364.

Harnst, P.L., Wong, N.W. and Bragin, J. (1982) A long path infrared study of Los Angeles smog. *Atmos. Environ.* **16** 969–971.

Harris, G.W., Klemp, D., Zenker, T., Burrows, J.P., Mattieu, B. and Jacob, P. (1990) Polarstern 1988: measurements of trace gases using tunable diode lasers and intercomparison with other methods. In Restelli, G. and Angeletti, G. (eds) *Proceedings of the Fifth European Symposium on the Physico-Chemical Behaviour of Atmospheric Pollutants*, Varese, 1989, pp. 644–650. Kluwer Academic, Dordrecht.

Hartwell, T.D., Pellizzari, E.D., Perritt, L.R., Whitmore, R.W., Zelson, H.S., Sheldon, L.S., Sparacino, C.M. and Wallace, L. (1987) Results from the total exposure methodology (TEAM) study in selected communities in Northern and Southern California. *Atmos. Environ.* **21** 1995–2004.

Henriksen, A. and Brakke, D.F. (1988) Sulfate deposition to surface waters. *Environ. Sci. Technol.* **22** 8–14.

Hileman, B. (1982) Crop losses from air pollutants. *Environ. Sci. Technol.* **16** 495A–499A.

Hinkley, E.D., Ku, R.T., Nill, K.W. and Butler, J.F. (1976) Long-path monitoring: advanced instrumentation with a tunable diode laser. *Appl. Opt.* **15** 1653–1658.

Hov, O., Hesstvedt, E. and Isaksen, I.S.A. (1978) Long range transport of tropospheric ozone. *Nature* **273** 341–344.

Hov, O., Becker, K.H., Builtjes, P., Cox, R.A. and Kley, D. (1986) Evaluation of photooxidants–precursor relationship in Europe, 1986. *Air Pollution Research Report 1, CEC, AP/60/87*. Brussels.

Hov, O., Allegrini, I., Beilke, S., Cox, R.A. Eliassen, A., Elshout, A.J., Gravenorst, G., Penkett, S.A. and Stern, R. (1987) Evaluation of atmospheric processes leading to acid deposition in Europe, 1987. *Air Pollution Research Report 10, CEC Report Eur 11441*. Brussels.

Intersociety Committee APCA, ACS, AChE, APWA, ASME, AOAC, HPS, ISA (IC) (1990) *Methods of Air Sampling and Analysis. 100 Carbon Compounds*. Lodge, J.P. Jr. (ed.). Lewis Publishers, Inc., Chelsea, MI.

Jayanty, R.K. (1989) Evaluation of sampling and analytical methods for monitoring toxic organics in air. *Atmos. Environ.* **23** 777–782.

Kanakidou, M., Bonsang, B. and Lambert, G. (1988) Light hydrocarbons vertical profiles and fluxes in a French rural area. *Atmos. Environ.* **23** 921–927.

Knoeppel, H., Versino, B., Schlitt, H., Peil, A., Schauenburg, H. and Vissers, H. (1980) Organics in air, sampling and identification. In Versino, B. and Ott, H. (eds) *Physico-Chemical Behaviour of Atmospheric Pollutants, Proceedings of the First European Symposium*, Ispra, Italy, 16–18 October 1979, pp. 25–40. EUR 6621 DE, EN, FR, CEC. Brussels.

Kotzias, D. and Hjorth, J. (1991) Sampling and analysis of selected volatile organic compounds (VOC) relevant for the formation of photochemical oxidants. *Pollut. Atmosph.*, numèro special Juillet, 1991 **131** 209–216.

Lamb, B., Westberg, H. and Quarles, T. (1986) Isoprene emission fluxes determined by an atmospheric tracer technique. *Atmos. Environ.* **20** 1–8.

Lamb, B., Guenther, A., Gay, D. and Westberg, H. (1987) A national inventory of biogenic hydrocarbon emissions. *Atmos. Environ.* **21** 1695–1705.

Larson, R.A. and Berembaum, M.R. (1988) Environmental phototoxicity. *Environ. Sci. Technol.* **22** 354–360.

Lawrence Berkeley Laboratory Survey (LBL) (1973) *Survey on Instrumentation for Environmental Monitoring—LBL-1, Vol. 1. Hydrocarbon Monitoring Instrumentation*. Lawrence Berkeley Laboratories, Berkeley, CA.

Lefohn, A.S. and Brocksen, R.W. (1984) Acid rain effects research—A status report. *JAPCA* **34** 1005–1013.

Levy, H., Mahlman, J.D. and Moxim, W.J. (1985) Tropospheric ozone: the role of transport. *J. Geophys. Res.* **90** 3753–3772.

Liberti, A. and Ciccioli, P. (1985) Chromatography for the evaluation of the atmospheric environment. *The Science of Chromatography*. Journal of Chromatography Library, Vol. 32. Bruner, F. (ed.). Elsevier, Amsterdam.

Lillian, D., Singh, H.B. and Appleby, A. (1976) Gas chromatographic analysis of ambient halogenated compounds. *JAPCA* **26** 141–142.

Liu, M.K., Morris, L.E. and Killus, J.P. (1984) Development and application of a regional oxidant model to the northeast United States. *Atmos. Environ.* **18** 1145–1186.

Lovelock, J.E., Maggs, R.J. and Wade, R.J. (1973) Halogenated hydrocarbons in and over the Atlantic. *Nature* **241** 194–196.

Louw, C.W., Richards, J.F. and Faure, P.K. (1977) The determination of volatile organic compounds in city air by gas chromatography combined with standard addition, selective subtraction infrared spectrometry and mass spectrometry. *Atmos. Environ.* **11** 703–717.

Lowe, D.C., Schmidt, U. and Ehhalt, D.H. (1980) A new technique for measuring tropospheric formaldehyde (CH_2O). *Geophys. Res. Lett.* **7** 825–828.

Lowi, A. and Carter, W.P.L. (1990) A method for evaluating the atmospheric ozone impact of actual vehicle emission. *SAE Report No. 900710*. Society Automotive Engineers, Warrendale, PA.

Lubkert, B. and De Tilly, S. (1989) The OECD–MAP emission inventory for SO_2, NO_x and VOC in Western Europe. *Atmos. Environ.* **23** 3–15.

Lubkert, B. and Zierock, K.H. (1989) European emission inventories—a proposal of worksharing. *Atmos. Environ.* **23** 37–48.

Lurmann, F.W., Carter, W.P.L. and Coyner Lori, A. (1988) A surrogate species chemical reaction mechanism for urban scale air quality simulation models. Vol. I—adaptation of the mechanism. *Final Report on EPA Contract No. 60-02-4104.* Atmospheric Sciences Research Laboratory, Research Triangle Park, NC.

Matuska, P., Kova, M. and Seiler, W. (1986) A high resolution GC-analysis method for the determination of C2–C10 hydrocarbons in air samples. *J. High Resolut. Chromatogr. Chromatogr. Commun.* **9** 577–583.

McClenny, W.A. Pleil, J.D., Oliver, K.D. and Holdren, M.W. (1984) Automated cryogenic preconcentration and gas chromatographic determination of volatile organic compounds in air. *Analyt. Chem.* **56** 2947–2951.

Meyrahn, H., Hahn, J., Helas, G., Warneck, P. and Penkett, S.A. (1984) Cryogenic sampling and analysis of peroxyacetyl nitrate in the atmosphere. In Versino, B. and Angeletti, G. (eds) *Proceedings of the Third European Symposium on the Physico-Chemical Behaviour of Atmospheric Pollutants*, Varese, 1984, pp. 38–43. D. Reidel, Dordrecht.

Middleton, P., Stockwell, W.R. and Carter, W.P.L. (1990) Aggregation and analysis of volatile organic compound emissions in regional models. *Atmos. Environ.* **24A** 1107–1133.

Miliazza, A. and Dooper, R. (1992) Determinazione dei composti organici volatili (COV) in aria mediante analizzatore automatico. Atti delle 'Giornate di Studio Europeo per l'Ambiente—Inquinamento Atmosferico-Città ed ambiente'. *SEP Pollution 1992—29 Marzo–2 Aprile*, pp. 495–509. PADOVAFIERE Editrice, Padova.

Molhave, L. and Thorsen, M. (1991) A model for investigations of ventilation systems as sources for volatile organic compounds in indoor climate. *Atmos. Environ.* **25A** 241–251.

Molina, M.J. and Rowland, F.S. (1974) Stratospheric sink for chlorofluoromethanes: Chlorine atom-catalyzed destruction of ozone. *Nature* **249** 810–812.

Molina, M.J., Tso, T.L., Molina, L.T. and Wang, Y.F.C. (1987) Antarctic stratospheric chemistry and chlorine nitrate, hydrogen chloride: Release of active chlorine in Antarctic stratosphere. *Science* **238** 1253–1257.

Mueller, W.J. and Stickney, P.B. (1970) A survey and economic assessment of air pollution on elastomers. Battelle Memorial Institute-Columbus Laboratories; National Air Pollution Control Administration Contract No. CPA-22-69-146. Columbus, OH.

Muller, K.P., Rudolph, J. and Wohlfart, K. (1990) Measurements of peroxyacetyl nitrate in the marine atmosphere. In Restelli, G. and Angeletti, G. (eds) *Proceedings of the Fifth European Symposium on the Physico-Chemical Behaviour of Atmospheric Pollutants*, Varese, Italy, 25–28 September, 1989, pp. 705–710. Kluwer Academic, Dordrecht.

Nollet, V. and Dechaux, J.C. (1991) Etude par simulation numérique de l'intérêt d'un dosage exhaustif des composés organiques volatils atmosphériques. *Pollut. Atmosph.* **130** 466–478.

Organization for Economic Cooperation and Development (OECD) (1975) *Photochemical Oxidant Air Pollution.* A report of the Air Management sector group on the problem of photochemical oxidants and their precursors in the atmosphere. OECD, Paris.

Organization for Economic Cooperation and Development (OECD) (1977) *The OECD Programme on Long Range Transport of Air Pollutants.* OECD, Paris.

Organization for Economic Cooperation and Development (OECD) (1990) *Control Strategies for Photochemical Oxidants across Europe.* OECD, Paris.

Overrein, L.N., Seip, H.M. and Tollan, A. (1980) *Acid-Precipitation—Effects on Forest and Fish.* Final Report of the SNSF Project, 1972–1980. RECLAMO, Oslo, Norway.

Pashalidis, S., Carlier, P. and Mouvier, G. (1990) DMS distribution study over the Atlantic ocean during the Polarstern crossing from Bremerhaven to Rio Grande do Sul. In Restelli, G. and Angeletti, G. (eds) *Physico-Chemical Behaviour of Atmospheric Pollutants—Air Pollution Research Report*, pp. 681–685. Kluwer Academic, Dordrecht.

Pellizzari, E.D., Bunch, J.E., Berkley, R.E. and McRae, J. (1976) Determination of trace hazardous organic vapor pollutants in ambient atmospheres by gas chromatography–mass spectrometry–computer. *Analyt. Chem.* **48** 803–807.

Pellizzari, E.D., Demian, B. and Krost, K. (1984) Sampling of organic compounds in the presence of reactive inorganic gases. *Analyt. Chem.* **56** 793–798.

Penkett, S.A., Jones, B.M.R., Brice, K.A. and Eggleton, A.E.J. (1979a) The importance of atmospheric ozone and hydrogen peroxide in oxidizing sulfur dioxide in cloud and rainwater. *Atmos. Environ.* **13** 123–137.

Penkett, S.A., Brice, K.A., Derwent, R.G. and Eggleton, A.E.J. (1979b) Measurements of CCl_3F and CCl_4 at Harwell over the period January 1974–November 1977. *Atmos. Environ.* **13** 1011–1019.

Penner, J.E. (1990) Cloud albedo, greenhouse effects, atmospheric chemistry and climate change. *J. Air Waste Manage. Assoc.* **40** 456–461.

Perrin, M.L., Fauconnier, J.M., Guillet, P. and MacLeod, P. (1986) Développement d'un olfactometre correspondant à la norme française. In Hatmann, H.F. (ed.) *Proceedings of the Seventh World Clean Air Congress*, Sidney, 1986, Vol. 2, pp. 120–126. Clean Air Society of Australia and New Zealand.

Pitts, J.N. Jr., Biermann, H.W., Tuazon, E.C., Green, M., Long, W.D. and Winer, A.M. (1989) Time-resolved identification and measurements of indoor air pollutants by spectroscopic techniques: gaseous nitrous acid, methanol, formaldehyde and formic acid. *JAPCA* **39** 1344–1347.

Platt, U. and Perner, D. (1980) Direct measurements of atmospheric HCHO, HNO_2, O_3, NO_2 and SO_2 by differential optical absorption in the near UV. *J. Geophys. Res.* **85** 7453–7458.

Platt, U., Perner, D. and Patz, H.W. (1979) Simultaneous measurement of atmospheric HCHO, O_3 and NO_2 by differential optical absorption. *J. Geophys. Res.* **84** 6329–6335.

Polcyn, A.J. and Hesketh, H.E. (1985) A review of current sampling and analytical methods for assessing toxic and hazardous organic emissions from stationary sources. *JAPCA* **35** 54–60.

Possanzini, M. (1981) La valutazione dell'inquinamento fotochimico nell'area di Roma. Collana del Programma Finalizzato Promozione della Qualità dell'Ambiente. *Consiglio Nazionale delle Ricerche Report No. AQ/3/16.* Rome.

Possanzini, M., Ciccioli, P., Brancaleoni, E., Tappa, R. and Brachetti, A. (1981) Gas chromatographic detection of hydrocarbons in the atmosphere by using specific GC detectors and mass-spectrometry in the selected ion monitoring mode. In Versino, B. and Ott, H. (eds) *Proceedings of the Second European Symposium on the Physico-Chemical Behaviour of Atmospheric Pollutants*, Varese, pp. 76–81. D. Reidel, Dordrecht.

Possanzini, M., Ciccioli, P., Di Palo, V. and Draisci, R. (1987) Determination of low boiling aldehydes in air and exhaust gases by using annular denuders combined with HPLC techniques. *Chromatographia* **23** 829–834.

Potter, C.J. and Savage, C.A. (1982) The evaluation of the Warren Spring Laboratory Vehicle exhaust gas proportional sampler. *Warren Spring Laboratory Report No. LR 417 (AP).* Stevenage, UK.

Puxbaum, H., Rosemberg, C., Gregoeri, M., Lanzerstorfer, C., Ober, E. and Winiwarter, W. (1988) Atmospheric concentrations of formic and acetic acid and related compounds in Eastern and Northern Austria. *Atmos. Environ.* **22** 2481–2850.

Rijsinstitut Voor Volksgesondheid en Milieuhygiene (RIVM)–ECE (1991) Mapping critical loads for Europe. *CEC Technical Report No. 1. RIVM Report No. 259101001.* Bilthoven.

Roberts, J.M. (1990) The atmospheric chemistry of organic nitrates. *Atmos. Environ.* **24A** 243–287.

Roumelis, N. and Glavas, S. (1989) Decomposition of peroxyacetyl nitrate and peroxypropionyl nitrate during gas chromatographic determination with a wide bore capillary and two packed columns. *Analyt. Chem.* **61** 2731–2734.

Rudolph, J., Ehhalt, O.H., Khedim, A. and Jebsen, C. (1981) Determination of C2–C5 hydrocarbons in the atmosphere at low parts per 10 to the 9 to high parts of 10 to the 12 levels. *J. Chromatogr.* **247** 301–310.

Rudolph, J., Khedim, A. and Johnen, F.J. (1985) Gas chromatographic techniques for the measurement of halocarbons and hydrocarbons. In *Proceedings of the COST 611 Workshop on Pollutants Cycles, Transport Modelling and Field Experiments*, Bilthoven, 23–25 September, 1985. *CEC Report No. AP/51/85, XII/ENV/59/85.* Brussels.

Sandalls, F.J., Penkett, S.A. and Jones, B.M.R. (1974) Prepration of peroxyacetylnitrate (PAN) and its determination in the atmosphere. *AERE Report No. R7807.* HMSO, Harwell, UK.

Schiff, H.I. and Mackay, G.I. (1989) Tunable diode laser absorption spectrometry as reference method for tropospheric measurements. In Grisar, R., Schmidtke, G., Tacke, M. and Restelli, G. (eds) *Proceedings of the International Symposium on Monitoring of Gaseous Pollutants by Tunable Diode Lasers*, Freiburg, 1988, pp. 36–45. Kluwer Academic, Dordrecht.

Schiff, H.I., Hastie, D.R., Mackay, G., Iguchi, T. and Ridley, B.A. (1983) Tunable diode laser systems for measuring trace gases in tropospheric air. *Environ. Sci. Technol.* **17** 352A–364A.

Schmidbauer, N. and Oehme, M. (1986) Improvement of a cryogenic preconcentration unit for C2–C6 hydrocarbons in ambient air at ppt levels. *J. High Resolut. Chromatogr. Chromatogr. Commun.* **9** 502–505.

Schmidt, U. and Lowe, D.C. (1981) Vertical profiles of formaldehyde in the troposphere. In Versino, B. and Angeletti, G. (eds) *Proceedings of the Third European Symposium on the Physico-Chemical Behaviour of Atmospheric Pollutants*, Varese, 1984, pp. 377–386. D. Reidel, Dordrecht.

Schmidt, U., Bauer, R., Brod, M., Khedim, A., Klein, E. and Kulessa, G. (1990) In situ profile observation of long-lived trace gases in the lower Arctic stratosphere during winter. In *Pyle, J.A. and Harris, N.R.P. (eds) Proceedings of the First European Workshop on Polar Stratospheric Ozone*, Schliersee, 1990, pp. 29–32. *CEC Air Pollution Research Report No. 34*. Brussels.

Schneider, W., Frohne, J.C. and Bruderreck, H. (1978) Determination of hydrocarbons in the parts per 10 to the 9 range using glass capillary columns coated with aluminum oxide. *J. Chromatogr.* **155** 311–327.

Schofield, C.L. (1976) Effects of acid precipitation on fish. *Ambio* **5** 228–230.

Schroeder, W.H. and Lane, A.D. (1988) The fate of toxic airborne pollutants. *Environ. Sci. Technol.* **22** 240–246.

Schuetzle, D., Jensen, T.E., Nagy, D., Prostak, A. and Hochhauser, A. (1991) Analytical chemistry and auto emissions. *Analyt. Chem.* **63** 1149A–1159A.

Seinfeld, J.H. (1988) Ozone air quality models—A critical review. *JAPCA* **38** 616–645.

Seufert, G. and Arndt, U. (1987) Vergleichende immissionsokologische Versuche and Fichten in Waldschadensgebiet by Ausschluss gasformiger Schadstoffe in Open-top Kammern. *KFK*-PEF **12** 12–23.

Sexton, F.W., Michie, R.M. jr., McElroy, F.F. and Thompson, V.L. (1982) A comparative evaluation of seven automated ambient non-methane organic compound analyzers. *EPA Report No. EPA-600/4-02-046*. Research Triangle Park, NC.

Sexton, K. and Hayward, S.B. (1987) Source apportionment of indoor air pollution. *Atmos. Environ.* **21** 407–418.

Shah, J.J. and Singh, H.B. (1988) Distribution of volatile organic chemicals in outdoor and indoor air. *Environ. Sci. Technol.* **22** 1381–1388.

Shareef, G.S., Butler, W., Bravo, L.A. and Stockton, M.B. (1988) Air emission species manual. Vol. 1. Volatile organic compounds species profiles. *EPA Report No. EPA-450/2-88-003a*. Research Triangle Park, NC.

Smith, F.B. and Hunt, R.D. (1978) Meteorological aspects of the transport and pollution over long distances. *Atmos. Environ.* **12** 461–478.

Stahal, Q.R. (1969) Preliminary air pollution survey on ethylene. US Department of Health, Education and Welfare. Public Health Service. *Publication No. APTD 69-35*. Raleigh, NC.

Stern, R.M. and Builtjes, P. (1986) Long-range transport modeling of the formation, transport and deposition of photochemical oxidants and acidifying pollutants. In Angeletti, G. and Restelli, G. (eds) *Proceedings of the Fourth European Symposium on the Physico-Chemical Behaviour of Atmospheric Pollutants*, Stresa, Italy, 23–25 September 1986, pp. 656–666. D. Reidel, Dordrecht.

Stolarski, R.S. and Cicerone, R.J. (1974) Stratospheric chlorine: a possible sink for ozone. *Can. J. Chem.* **52** 1610–1615.

Stolarski, R.S., Krueger, A.J., Schoeberl, M.R., McPeters, R.D., Newman, P.A. and Alpert, J.C. (1986). Nimbus 7 satellite measurements of the springtime Antarctic ozone decrease. *Nature* **322** 808–811.

Sutton, O.G. (1971) *The Challenge of the Atmosphere*. Translated and published in Italy by A. Mondadori, Milano.

Taylor, O.C. (1969) Importance of peroxyacetyl nitrate (PAN) as a phytotoxic air pollutant. *J. Air Pollut. Control Assoc.* **19** 347–351.

Temple, P.J. and Taylor, O.C. (1983) World-wide ambient measurements of peroxyacetyl nitrate (PAN) and implications for plant injury. *Atmos. Environ.* **17** 1583–1587.

Tilton, B.E. (1989) Health effects of tropospheric ozone. *Environ. Sci. Technol.* **23** 257–263.

Tingey, D.T. and Taylor, G.E. jr. (1982) Variation in plant response to ozone: a conceptual model of physiological events. In Unsworth, M.H. and Ormod, D.P. (eds) *Effects of Gaseous*

Air Pollution in Agriculture and Horticulture. Butterworth Scientific, London.
Tingey, D.T., Manning, M., Ratsch, H.C., Burns, W.F., Grothaus, L.C. and Field, W. (1978) Monoterpene emission rates from slash pine. *EPA Report No. EPA 904/9-78-013.* Research Triangle Park, NC.
Tingey, D.T., Manning, M. Ratsch, H.C., Burns, W.F. Grothaus, L.C. and Field, R.W. (1980) Influence of light and temperature on monoterpene emission rates from slash pine. *Plant Physiol.* **65** 797–801.
Tuazon, E.C., Winer, A.M. and Pitts, J.N. Jr. (1981) Trace pollutant concentrations in a multiday smog episode in the California South Coast Air Basin by long path length Fourier transform infrared spectrometry. *Environ. Sci. Technol.* **15** 1232–1237.
Tucker, G.B. (1986) Trace gas trends in the Southern hemisphere. In Hartmann, H.F. (ed.) *Proceedings of the Seventh World Clean Air Congress,* Sidney, 1986, Vol. 1, pp. 4–16. Clean Air Society of Australia and New Zealand.
United Nations–Economic Commission For Europe (UN–ECE) (1991) Protocol for the 1979 convention on long-range transboundary air pollution concerning the control of emission of volatile organic compounds or their transboundary fluxes. *EB.AIR/R.54,* November, 1991. Geneva.
United Nations Environmental Programme (UNEP) (1987) *Montreal Protocol on Substances that Deplete the Ozone Layer. Final Act. 1987.* Geneva.
US Department of Health, Education and Welfare (USDHEW) (1970) Air quality criteria for photochemical oxidants. National Air Pollution Control Administration, *NAPCA Report No. AP-63.* Washington, DC.
US Environmental Protection Agency (USEPA) (1977) Procedures for the preparation of emission inventories for volatile organic compounds—Volume I. Office of Air and Waste Management. *EPA Report No. EPA-450/2-77-027.* Research Triangle Park, NC.
US Environmental Protection Agency (USEPA) (1979a) Procedures for the preparation of emission inventories for volatile organic compounds—Volume II: Emission inventory requirements for photochemical air quality simulation models. Office of Air Quality Planning and Standards. *EPA Report No. EPA-450/4-79-018.* Research Triangle Park, NC.
US Environmental Protection Agency (USEPA) (1979b) Measurement of volatile organic compounds. Environmental Standards and Engineering Division. *EPA Report No. EPA-450/2-78-041.* Research Triangle Park, NC.
US Environmental Protection Agency (USEPA) (1980) Volatile organic compounds (VOC) species data manual. Office of Air Quality Planning and Standards. *EPA Report No. EPA-450/4-80-015.* Research Triangle Park, NC.
US Environmental Protection Agency (USEPA) (1984) National Air Quality and Emission Trend Report—1983. US Environmental Protection Agency—Office of Air Quality Planning and Standards. *EPA Report No. EPA-450/4-84-029.* Research Triangle Park, NC.
US Environmental Protection Agency (USEPA) (1986) Air Quality Criteria for Photochemical Oxidants. Volumes III to V. US Environmental Protection Agency, Environmental Criteria and Assessment Office. *EPA Reports No. EPA-600/8-84/020 d, e and f.* Research Triangle Park, NC.
US Environmental Protection Agency (USEPA) (1987) National Air Toxics Information Clearing House: Qualitative and Quantitative Carcinogenic Risk Assessment. *EPA Report No. EPA-450/4-88-014.* Research Triangle Park, NC.
US National Acid Precipitation Assessment Program (USNAPAP) (1991) *Acidic Deposition: State of Science and Technology. A Summary Report of US-NAPAP.* Irving, P.M. (ed.). Washington DC.
V & F, Analyse und Messtechnik GesmbH (1992) *Manual of the CI-MS 500 Real Time Gas Analyzer.* Released by V & F GesmbH, Absam, Tirol, Austria.
Veldt, C. and Bakkum, A. (1987) Atmospheric Emission Inventory Project CORINAR—emission factors. *TNO Report to the Commission of the European Communities DG XI, CORINE, Ref. No. 87-127.* Gravenhage.
Vierkorn-Rudolph, B., Rudolph, J. and Diederich, S. (1985) Determination of peroxyacetyl nitrate (PAN) in unpolluted areas. *Int. J. Environ. Anal. Chem.* **20** 131–148.
Volz, A., Ehhalt, D.H. and Derwent, R.G. (1981) Measurement of 14-CO: a method for determining OH concentrations in the troposphere. *J. Geophys. Res.* **86** 5163–5171.

Wallace, L.A., Pellizzari, E., Leaderer, B., Zelon, H. and Sheldon, L. (1978) Emissions of volatile organic compounds from building materials and consumer products. *Atoms. Environ.* **21** 385–393.

Warrick, R.A., Barrow, E.M. and Wigley, T.M. (1990) The greenhouse effect and its implications for the European Community. *CEC Publication No. EUR 12707 EN*. Luxembourg.

Wofsy, S.C., Molina, M.J., Salawitch, R.J., Fox, L.E. and McElroy, M.B. (1988) Interactions between HCl, NO_x and H_2O ice in the Antarctic stratosphere: implications for ozone. *J. Geophys. Res.* **93** 2442–2450.

Wolff, G.T., Kelly, N.A. and Ferman, M.A. (1982) Source regions of summertime ozone and haze episodes in the eastern United States. *Water, Air, Soil Pollut.* **18** 65–81.

World Health Organization (WHO) (1986) *Air Quality Guidelines, Review Draft*. Regional Office of Europe, Copenhagen.

Yocom, J.E. (1982) Indoor–outdoor air quality realtionships: a critical review. *JAPCA* **32** 500–520.

Yokouchi, Y. and Ambe, Y. (1985) A study on the behavior of monoterpenes in the atmosphere. *Research Report from the National Institute for Environmental Studies of Japan. Report No. R-76-85*. The National Institute for Environmental Studies, Chemistry and Physics Division, Tsukuba.

Yokouchi, Y., Ambe, Y. and Maeda, T. (1986) Automated analysis of C3–C13 hydrocarbons in the atmosphere by capillary gas chromatography with a cryogenic preconcentration. *Analyt. Sci.* **2** 571–575.

Yokouchi, Y., Mukai, H., Nakajima, K. and Ambe, Y. (1990) Semi-volatile aldehydes as predominant organic gases in remote areas. *Atmos. Environ.* **24A** 439–442.

Zimmerman, D., Tax, W., Smith, M., Demmy, J. and Battye, R. (1988) Anthropogenic emission data for the 1985 NAPAP inventory. *EPA Report No. EPA-600/7-88-022*. Research Triangle Park, NC.

Zimmerman, P.R. (1977) Procedures for conducting hydrocarbon emission inventories of biogenic sources and some results of recent investigations. Paper presented at the 1977 Environmental Protection Agency Emission Inventory/Factor Workshop, Raleigh, NC, September 13–15.

Zimmerman, P.R. (1979) Testing for hydrocarbons emissions from vegetation leaf litter and aquatic surfaces, and development of a methodology for compiling biogenic emission inventories. *EPA Report No. EPA-450/4-4-79-004*. Research Triangle Park, NC.

4 VOCs and water pollution

D. KOTZIAS and C. SPARTÀ

4.1 Introduction

The occurrence of organic compounds in water is a matter of increasing concern to the industry, ecologists and general public from the point of view of possible health hazards presented to both human and animal life. It has been well established that an amazing number of organic substances (pesticides, detergents, solvents) are released into the sea, rivers and lakes. Industrial (e.g. paint industry, metal industry) and domestic effluents as well as agricultural run-off are the main sources for these compounds. The availability of sophisticated analytical methods sensitive enough to determine extremely low concentrations of pollutants enables us to realize the extent and the nature of environmental contamination in natural water.

Among the organic compounds released, the volatile organic compounds (VOCs) represent a group of pollutants with similar and very specific chemical and physicochemical characteristics, as they are mainly lipophilic and of low reactivity. Main sources for the discharge of VOCs into the waters are the dry-cleaning industry and accidental events during transportation of crude oil. With drinking water, some volatile chlorinated hydrocarbons are produced during chlorination processes.

For pesticides, detergents and some inorganic water pollutants enough data are available with regard to their ecotoxicological behaviour, e.g. effects on ecosystems, uptake by living organisms, and fate and adsorption onto sediments. However, for VOCs, few data exist on the behaviour in water.

Despite their high evaporation rates VOCs have been detected in natural waters. This is probably due to the high volume of production followed by large emissions into the environment. The persistance of VOCs differs according to environmental conditions. For example, for 1,2-dichloroethane the half-life in the atmosphere is 10–190 days, whereas in soil it is expressed in months. In water the half-life can be very short due to evaporation; in ground water, however, where no degradation or evaporation occurs, the half-life is very long.

Evaluation of water pollution caused by VOCs, as well as the study of discharges of certain—in particular halogenated—compounds into the aquatic environment, and the best available technology for the reduction of water pollution from such discharges are of major concern for the public authorities.

In this chapter an attempt has been made to systematize the information available on occurrence and fate (regulation and legislation), sampling and analysis of VOCs in most types of water and effluents. Emphasis is given to the analysis of individual compounds rather than to methods describing sum parameters such as total organic halide (TOX) or purgeable organic halide. (POX). Substances of particular environmental significance, e.g. halogenated hydrocarbons are discussed in detail.

4.2 Definition

According to the definition used by WHO and the USEPA volatile organic compounds (VOCs) are substances which contain carbon atoms and which have a minimum vapour pressure of 0.13 kPa at standard temperature and pressure (293 K, 101 kPa). They exclude CO, CO_2, organometallic compounds and organic acids. Physicochemical data for the VOCs frequently found as water micropollutants are shown in Table 4.1.

4.3 Distribution in air, water and soil

Generally, distribution of chemicals between air, water and soil is governed by the partition coefficients for air/water and soil/water e.g. Henry's law constant, the *n*-octanol/water partition coefficient (K_{ow}) and the water/soil or sediment partition coefficient (K_{oc}) at equilibrium. Consequently, equilibrium distribution helps to assess potential concentrations of chemicals in the different environmental compartments and can be regarded as an important parameter to determine the behaviour patterns of chemical compounds in the environment.

4.3.1 *Henry's law constant* (H)

Henry's law constant is a physical property of a chemical, characterizing its partitioning between the two phases in an air/water binary system at equilibrium. Determination of Henry's law constant is based on the equation:

$$P = HC$$

where P is the partial pressure of the gas, H is Henry's law constant and C is the concentration of the dissolved gas. Henry's law constant is a function of temperature for a particular gas–solvent system.

Henry's law constants are required to predict the behaviour of organic compounds in the environment, e.g. to evaluate the performance of air-stripping processes and to describe movement of volatile pollutants from/to the air/water. Chemicals with low Henry's law constants will tend to accumu-

Table 4.1 Physiocochemical data for the VOCs frequently found as water pollutants.

Compound	Boiling point[1] (°C)	Vapour pressure at 25°C (kPa)	Solubility in water at 25°C (g/l)	Henry's law constant at 25°C (kPa.m^3/mol)
Benzene	80.1	12.7	1.780	0.55
Bromomethane	3.56	183.9	–	–
Carbon tetrachloride	76.7	15.06	1.160	2.0
Chlorobenzene	132	1.581	0.471	0.35
Choloroform	61.7	25.60	7.900	0.38
1,1-Dibromoethane	113	–	–	–
1,2-Dibromoethane	131.7	–	–	–
1,2-Dichlorobenzene	179	0.196	0.145	0.19
1,3-Dichlorobenzene	172	0.307	0.123	0.36
1,4-Dichlorobenzene	173.4	0.902	0.083	0.16
1,1-Dichlorethane	57.3	30.10	5.100	0.58
1,2-Dichlorethane	83.5	10.93	8.700	0.11
1,1-Dichloroethylene	31.6	79.73	0.400	13.32
Dichloromethane	40.5	58.40	19.400	0.26
Ethylbenzene	136.2	1.27	0.152	0.80
Hexanal	131	–	–	–
Isopropylbenzene	152.4	0.611	0.050	0.13
Tetrachloroethylene	121.1	2.48	0.140	2.30
Toluene	110.8	3.80	0.515	0.67
1,1,1-Trichloroethane	74	16.53	0.720	2.8
1,1,2-Trichloroethane	113.5	4.04	4.420	0.12
1,1,2-Trichloroethylene	86.7	9.87	1.100	1.18
1,1,1-Trichloro-2,2,2-trifluoroethane	46	–	–	–
1,1,2-Trichloro-1,2,2-trifluoroethane	47.6	–	–	–
1,2,3-Trimethylbenzene	176.1	0.202	0.075	0.32
1,2,4-Trimethylbenzene	169.4	0.271	0.057	0.59
1,3,5-Trimethylbenzene	164.7	0.328	0.097	0.60
Vinylchloride	– 13.9	–	2.700	2.35
o-Xylene	144.4	0.882	0.175	0.50
m-Xylene	139.1	1.10	0.162	0.70
p-Xylene	138.4	1.17	0.185	0.71

[1] Boiling point is given at atmospheric pressure (760 mmHg)

late in the aqueous phase, while those with high values will move more into the gas phase. Therefore, knowledge of Henry's law constants is very important in assessing environmental risks caused by chemicals. Unfortunately, only limited data exist on reliable values of Henry's law constants for chemicals with a certain environmental significance. Mackay and Shiu (1981) have critically reviewed the Henry's law constant for approximately 160 chemicals and reported on approximately 40 experimentally measured data. In addition, they have presented a comprehensive review of common methods for measuring Henry's constants. According to Mackay and Shiu (1981) there are three basic methods; (i) use of vapour pressure and solubility data, (ii) direct measurement of air and aqueous concentrations in a system at equilibrium, and (iii) measurement of relative changes in concentration within one phase, while effecting a near-equilibrium exchange with the other phase. However, because of the large number of chemicals in use (more than 60 000) and the lack of experimentally determined data, there is a growing interest to apply QSAR (quantitative structure activity relationship) methods to establish the physico-chemical properties of chemicals, e.g. for estimating Henry's law constants (Nirmalakhandan and Speece, 1998).

4.3.2 *n-Octanol/water partition coefficient* (K_{ow})

This is the ratio of the concentration of dissolved substances in a two-phase system of immiscible liquids at equilibrium. The partition coefficient between *n*-octanol and water gives an indication of the accumulation behaviour of a chemical. As the quotient of two concentrations, the distribution ratio C(*n*-octanol)/C(water) is a dimensionless constant.

For VOCs in general, the partition coefficient K_{ow} is rather low, ranging from 15 to 500 (K_{ow}s for the non-volatile compounds hexachlorobenzene, DDT and fluoranthene are 1.5×10^6, 1.5×10^6 and 1.6×10^5, respectively), which can be explained by the good solubility of these compounds in water (Scheele, 1980). Characterization of a VOC according to its K_{ow} has, therefore, a limited value only.

4.3.3 *Water/soil or sediment partition coefficient* (K_{oc})

Sorption of organic compounds by soil or sediment follows closely the patterns found for the partitioning of chemicals between water and *n*-octanol. It is expressed as the dimensionless ratio C(soil):C(water). Sorption appears to be a distribution process rather than a classical interaction with a surface. In these cases distribution occurs between water and the organic or liquid fraction of the soil or sediment. Few data on K_{oc} values exist; for example benzene, toluene and trichloroethylene K_{oc} values are estimated to be very low—1.7, 2.8 and 2 respectively. Values for non-volatile compounds, e.g. DDT and DDE, range from 660 to 2000 (Scheele, 1980). Again, characterization of VOCs according to their distribution between water/soil or sediment has limited value.

4.4 Occurrence

4.4.1 *Seawater*

Volatile organic compounds (VOCs) have been found in seawater samples throughout the world. Gschwend *et al.* (1980) reported on the presence of alkylaromatics in the upwelling area off Peru. Concentrations ranged from 2 ng/l to 60 ng/l. Sauer (1980) reported on the occurrence of volatile compounds in the Gulf of Mexico and the Caribbean Sea. In open-ocean surface waters of the Gulf of Mexico VOC concentrations were about 60 ng/l, while in the heavily polluted Lousiana coastal area values reached about 500 ng/l. Caribbean surface samples were found to have low concentrations of VOCs, only 30 ng/l. Sauer stressed that aromatic VOCs accounted for 60–85% of the total VOCs in surface waters, while cycloalkanes ranged from 60 to 100 ng/l in polluted areas (20% of the total VOCs). Subsurface concentrations in samples of polluted areas collected from depths of 50 m were only 35–40 ng/l below surface concentrations. Open-ocean subsurface samples had concentrations of approximately 30 ng/l at 30–50 m depths, comparable to those of Caribbean surface water.

Low molecular weight (C1–C4) and volatile hydrocarbons (C5–C14) were detected in water samples around a subsurface well blow-out in the Gulf of Mexico (Brooks *et al.*, 1981). Concentrations as high as 400 μg/l in the surface seawater near the wellhead were measured. The VOCs detected in the vicinity of the wellhead were mostly benzene and alkylated benzenes. Rapid dilution reduced the VOC levels to 63 μg/l and 4 μg/l at 6 and 12 miles, respectively, from the wellhead.

Contamination of Atlantic and Baltic Sea surface waters with C3 alkylated benzenes has been reported in the work of Derenbach (1985). Large-scale transport in the atmosphere and in water masses appears to be the dominating factor in the VOC distribution in these areas. Average concentrations for the most abundant C3 benzenes—1,3,5-trimethylbenzene, 2-ethyltoluene, 1,2,4-trimethylbenzene and 3,4-ethyltoluene—were found to range from 1.8 ng/l to 8.1 ng/l (Baltic Sea) and from 1.4 ng/l to 4.2 ng/l (Central Atlantic).

Few data exist for the occurrence of VOCs in the Mediterranean Sea. Zsolnay (1979) found, in selected Mediterranean areas, VOC concentrations (mainly *n*-alkanes, aromatics, petroleum hydrocarbons) which ranged from 6.9 to 25.8 μg/l. More recently Burns and Saliot (1986) mentioned in their work that over three quarters of a million tonnes of oil were estimated to be introduced annually into the Mediterranean Sea from land-based and open-sea discharges.

Seasonal variations of VOC-concentrations were reported by Gschwend *et al.* (1982) at a coastal site near Vineyard Sound, Massachusetts, USA. Alkylaromatics, alkanes and dimethylpolysulphides were monitored over a 15 month period. Concentrations of, for example, 1,2,4-trimethylbenzene and 1,3,5-trimethylbenzene ranged from 2.2 to 2.5 ng/l and 0.8 to 11 ng/l, respectively. Mean concentrations of the alkylaromatics found compare well with average values for the C3 benzenes measured in the Baltic Sea and Central Atlantic. Concentration of dimethylsulphide ranged from 2.7 to 60 ng/l. Dimethylsulphide is produced by cultures of freshwater bacteria and blue-green or green algae, and may also be formed by the oxidation coupling of methyl mercaptan, which is a common product from microorganisms and may originate in reducing environments such as sediments of coastal ponds and marshes (Gschwend *et al.*, 1982 and references therein).

Halogenated VOCs were also detected in the marine environment (Pearson and McConnell, 1975; Gschwend *et al.*, 1982; Biziuk *et al.*, 1991). The most abundant halogenated VOCs identified were trichloroethylene, perchloroethylene, tribromomethane and chloroform. Concentrations reached values up to 3.6 μg/l for trichloroethylene, 2.6 μg/l for tetrachloroethylene and 1 μg/l for chloroform in samples from Liverpool Bay (Pearson and McConnell, 1975). Gschwend *et al.* (1982) reported mean concentrations of tetrachloroethylene in seawater of 1.3 ng/l. Occurrence of VOCs in estuarine waters (Southampton) has been well documented in the work of Bianchi *et al.* (1989). Concentrations of three volatile aromatic hydrocarbons (benzene, toluene, ethylbenzene) monitored within a period of 12 months reached values in excess of 100 μg/l. Organochlorines, e.g. trichloroethylene, were also found in concentrations ranged from 5 to 75 μg/l (see also Table 4.2).

4.4.2 River, lake, ground and tap water

Rivers generally constitute a significant transport route for VOCs into the marine environment. However, little work has been done and little information is available on the distribution of VOCs in rivers. The occurrence of VOCs in two rivers in Spain has been recently reported by Gomez-Belinchon *et al.* (1991). They found a variety of alkylated benzenes at concentrations reaching values of 110 μg/l and 16 μg/l in surface water of the Bèsos river and Llobregat river, respectively. Individual compounds, e.g. toluene, *m*- and *p*-xylene and ethylbenzene, reached values up to 22 μg/l. These concentrations are extremely high in comparison with those reported for the Loire river (100–200 ng/l) or the Brazos river (McDonald *et al.*, 1988). McDonald *et al.* (1988) reported that VOC concentrations (aromatics, alkanes) in surface water from the Brazos river (located in the northwestern part of the Gulf of Mexico) and across the river/seawater mixing zone ranged from 30 ng/l to 18 μg/l. In the same study a variety of halogenated compounds including chlorinated and brominated hydrocarbons were identified in the river water system. Dichloromethane, carbon tetrachloride, chloroform, tetrachloroethylene, bromoform and dibromochloromethane were measured at concentrations rarely exceeding 75 ng/l.

Grob and Grob (1974) reported on the presence of selected VOCs in different water samples originated from the Zurich city area in Switzerland. For benzene, concentrations ranged from 18 ng/l (spring water) to 45 ng/l (ground water), while in Lake Zurich 28 ng/l and 22 ng/l in surface water and in 30 m deep water, respectively, were measured. Tap water samples, even after several purification procedures, contained 36 ng/l of benzene. Trichloroethylene and tetrachloroethylene were the most abundant halocarbons, with concentrations reaching up to 340 ng/l (lake) or 2 μg/l (ground and tap water).

The occurrence of halogenated volatile compounds in drinking water and waste water was reported by Glaze and Henderson (1975), Morris and McKay

Table 4.2 VOCs in seawater, coastal water and river water

Location	Concentration	Dominant compounds
Open-ocean Gulf of Mexico, surface water (1)	Total VLH* Approx. 60 ng/l Cycloalkanes < 1 ng/l Alkanes Approx. 15 ng/l	Aromatic hydrocarbons were 60–85% of the total VLH in surface waters
Open-ocean Gulf of Mexico, 30–50 m depths (1)	Total VLH Approx. 30 ng/l	
Open-ocean Caribbean Sea, surface waters (1)	Total VLH Approx. 30 ng/l	
Louisiana shelf and coastal waters, surface waters (1)	Total VLH reached Approx. 500 ng/l Cycloalkanes Approx. 60–100 ng/l Alkanes ranged up to 40 ng/l	
Louisiana shelf and coastal waters, 50 m depths (1)	Total VLH Approx. 35–40 ng/l	
Coastal Gulf of Mexico waters (2)	Total VOC 50–1500 ng/l	Aromatic hydrocarbons, but *n*-alkanes, cycloalkanes, aldehydes and halocarbons were also present
Near Mississippi river delta (3)	Total VOC Approx. 500 ng/l	
Peru upwelling region, surface samples (3)	Total VOC Approx. 100 ng/kg Individual compound concentrations varied from <2 to approx. 60 ng/kg	Among the alkylated benzenes were abundant C2 alkylated benzenes (< 10 ng/kg); among the *n*-alkanes normal pentadecane was prominent
Peru upwelling region, deep water samples (3)	Total VOC Approx. 20 ng/kg	
At coastal site, Vineyard Sound, Massachusetts (4)	Total VOC 200–500 ng/l	Aromatic hydrocarons, normal aldehydes, *n*-alkanes

(*continued*)

Table 4.2 (*continued*)

Location	Concentration	Dominant compounds
Brazos river and across the river/seawater mixing zone (5)	Total VOC 30–18 200 ng/l Individual compound concentrations were generally < 10 ng/l	Among the volatile organic compounds benzene and toluene were dominant among the *n*-alkanes, those in the C6–C10 range were dominant in many areas
Atlantic sea (6)	C3 Alkylated benzenes 11.2 ng/kg	
Baltic sea (6)	C3 Alkylated benzenes 18 ng/kg	
Loire river (7)	Alkylbenzenes 100–200 ng/l Alkanes 63 ng/l Alkylbenzenes 99 ng/l Alkanes 52–78 ng/l	
Mediterranean Sea (8)	Total VOC 6.9–25.8 μg/l	Petroleum hydrocarbons, *n*-alkanes, aromatics
Central North Sea (9)	Chlorobenzene 1 ng/l Ethylbenzene 30 ng/l Isopropyl benzene < 1 ng/l Tetrachloroethylene < 1 ng/l Xylene < 10 ng/l	
Two polluted rivers in (Barcelona (Besós) and Llobregat) (10)	*Besòs* Alkylbenzes 111 g/l Alkanes 100 ng/l Alkynaphthalenes 6900 ng/l Chlorinated VOC 1,2,4-Trichlorobenzene 8100 ng/l	The major VOCs in river and seawater were essentially C1–C5 alkylbenzenes, predominantly the xylenes; C8–C20 *n*-alkanes predominated at C10. Another important group was the chlorobenzenes

	Llobregat Alkylbenzenes 16 000 ng/l Alkanes 14 000 ng/l Alkylnaphthalenes 1000 ng/l 1,2,4-Trichlorobenzene 1200 ng/l
Coastal waters of Barcelona	Total VOC 79 ng/l Alkanes 45 ng/l
Other coastal sites situated away from river influence: Pineda beach (near a petrochemical complex) and Vilanova Sitges coasts	*Pineda beach* Total VOC 280 ng/l Alkanes 76 ng/l Chlorinated VOC Absent *p*-Chlorobenzene 120 ng/l *Vilanova Sitges* Total VOC 340 ng/l Alkanes 59 ng/l Chlorinated VOC Absent
Marine waters	Alkylbenzenes 270 ng/l Alkanes 71 ng/l Alkylnaphthalenes 15 ng/l

*VLH (C6–C14 hydrocarbons)
(1) Sauer (1980); (2) Sauer (1981); (3) Gschwend *et al.* (1980); (4) Gschwend *et al.* (1982); (5) McDonald *et al.* (1988); (6) Derenbach (1985); (7) Saliot *et al.* (1984); (8) Zsolnay (1979); (9) Hurford *et al.* (1990); (10) Gomez-Belinchon *et al.* (1991)

Table 4.3 VOCs mostly presented in industrial waste waters (modified from *J. Environ. Sci. Technol.* (1979) **13** 416–423. EST special report priority pollutants by L.H. Keith and W.A. Telliard).

Percent[a] of samples	Number[b] of industrial categories	Compound
29.1	25	Benzene
29.3	28	Toluene
40.2	28	Chloroform
34.2	25	Methylene chloride
10.2	19	Tetrachloroethylene
10.5	21	Trichloroethylene

[a]The percent of samples respresents the number of times this compound was found in all samples divided by the total.
[b]A total of 32 industrial categories and subcategories were analysed for organics.

(1975), Smith (1975) and others. Chlorination of drinking and waste water results in the formation of chloroform and other halogenated compounds. Bellar *et al.* (1974) found that, after chlorination, effluents from sewage treatment plants have greater chloroform concentrations than before chlorination—12.1 μg/l and 7 μg/l, respectively. Similar findings were reported for tri- and tetrachloroethylene. Biziuk *et al.* (1991) recently estimated volatile organic halogen (VOX) in water samples of different origin, i.e. tap water, river water and seawater. Concentrations, expressed in mg Cl/l, ranged from 13 mg/l to 65 mg Cl/l, with the highest concentration being present in tap water samples (see also Table 4.2).

4.5 Sources

Volatile organic compounds are present in fuels, and in the fumes from combustion processes. They are also often used for industrial applications, e.g. as solvents. A list of VOCs and the relative frequency of their occurrence in industrial waste waters is shown in Table 4.3. Benzene, toluene, 1,1,1-trichloroethane, chloroform, methylene chloride, trichloro- and tetrachloroethylene were mostly found in industrial waste waters from a large number of industrial categories.

Thus, the occurrence of VOCs in the marine environment, and in rivers and lakes can be explained as the result of enhanced anthropogenic activities and direct discharge into the various environmental compartments. However, the major input of VOCs into seawater, for example, is caused by the various accidental events during oil transport and dumping (Table 4.4).

In 1979 in the Gulf of Mexico the world's largest oil spill at wellhead

Table 4.4 Major accidental events since 1967 with estimates of the amount of oil spilled (tonnes).

Torrey Canyon (March 1967)	British coast	119 000
Amoco Cadiz (1978)	French coast	213 000
IXTOC (1979)	Gulf of Mexico	530 000
Exxon Valdez (March 1989)	Prince William Sound (Alaska)	50 000
Haven (1991)	Ligurian coast	10 000

IXTOC caused a 530 000 t crude oil release into this area. More recently (March 24, 1989) the Exxon Valdez oil spill was the largest oil spill ever recorded in the United States and one of the largest in the world. Oil from the tanker, which ran aground on Bligh Reef in Prince William Sound in Alaska, soaked and killed plants and animals over 2000 miles of shoreline and 500 miles away. Some 30 000 dead birds of 90 species were retrieved within the first four months following the spill. About 11.2 million gallons (50 000 t) were released into the sea, which is well known as an ecologically extremely sensitive area (Schnoor, 1991).

Halogenated compounds in seawater and river water are mainly introduced through direct discharge (dry-cleaning industry), while chlorination of municipal water supplies and waste water leads to the formation of halogenated VOCs (e.g. chloroform). The atmosphere can also act as an additional source of VOCs. Winds containing polluted coastal and industrial air result in atmospheric concentrations sufficient for air–sea exchange to surface waters. According to Sauer (1980), 1 ppbv concentration of an aromatic in air could yield a single component equilibrium concentration of 20 ng/l in surface water. At 50 ppbv air concentration of the most important aromatic compounds, e.g. benzene, toluene and the xylenes, equilibrium concentrations in marine water could reach levels of 800–1 000 ng/l, that is about 0.5% of the pollutant concentration.

Thus, winds containing polluted coastal and industrial air masses result in atmospheric concentrations sufficient for air–sea exchange to surface waters.

4.6 Behaviour and fate of VOCs in aquatic systems

4.6.1 Volatilization

Volatilization of VOCs is a major process for their removal in surface waters. This process is influenced by various factors which are directly connected with the physicochemical properties of the individual compounds, as well as the fluid mechanics of the air–water interphase. Water solubility, vapour pressure and the Henry's Law constant determine widely the rate of VOC evaporation. In lakes, atmospheric turbulence and wind speed dominate the mass transfer from the water to the lower atmosphere, while in river water interaction of the current with the river bottom governs the vertical transport of VOCs. Predicting volatilization is important for estimating residence time in waters.

Although enough physicochemical data are available, difficulties still exist for predicting volatilization of VOCs. Changes in the conditions of the lower atmosphere in a certain area result in changes in volatilization rates. Furthermore, the presence of other surface-active substances in water could modify the properties of the air–water interphase, increasing resistance to volatilization (Zoeteman *et al.*, 1980).

In ground, water volatilization of VOCs plays a minor role.

4.6.2 Biodegradation

Microorganisms are potentially capable of degrading xenobiotics (Alexander, 1981, 1985). In the last ten years, investigations concerning biodegradation of VOCs have increased, particularly on halogenated VOCs, among them chlorinated C1 and C2 compounds.

The halogenated alkanes, at least those compounds with more then three halogen substituents per carbon atom, can be metabolised by multiple, competing pathways in microbial systems. Competing pathways operate within a single organism and it is possible that different microbial populations carry out competing transformations.

Compounds such as trichloroethylene, tetrachlorethylene, 1,1,1-trichloroethane and chloroform (suspected teratogens) are resistant to biodegradation in aerobic subsurface environments, so they persist in polluted groundwaters. Under aerobic conditions, in laboratory experiments, their transformation is often incomplete and may lead to accumulation of equally harmful metabolites, such as dichloroethylene and vinylchloride.

In the removal of pollutants, the applicability of microorganisms, investigated for their ability to degrade these pollutants without leading to harmful metabolites, may be restricted by the long treatment periods and/or high cell concentrations required, and/or by the inappropriate growth conditions for these microorganisms in a certain environment.

The biotransformation of trichloroethylene in vinylchloride by anaerobic bacteria could represent a serious problem for drinking water supplies. Research has been made on microorganisms that can degrade trichloroethylene (TCE). Wackett *et al.* (1989) found that five *Mycobacterium* strains—*Mycobacterium convulutum*, *M. rhodochrous w-21*, *M. rhodochrous w-25*, *M. rhodochrous w-24*, and *M. vaccae* JOB5—grown on propane as the sole source of carbon and energy, degraded TCE. *M. vaccae* JOB5 degraded TCE more rapidly and to a greater extent (in 24 h at a starting concentration of

20 mM, it removed up to 99% of the TCE), and was also observed to degrade vinylchloride, 1,1-dichloroethylene and 1,2-dichloroethylene.

Oldenhuis *et al.* (1989) studied the degradation of TCE by the methanotrophic bacterium *Methylosynus trichosporium* OB3b. This bacterium degraded TCE in a strictly co-metabolic process, and almost total dechlorination occurred. Dichloromethane, chloroform, 1,1-dichloroethane and 1,2-dichloroethane were completely degraded.

Egli *et al.* (1987) found that transformations of tetrachloromethane by *Acetobacterium woodii* may lead to carbon dioxide or to chloroform.

Criddle *et al.* (1990a) investigated the fate of carbon tetrachloride (CT) in cultures of *Escherichia coli* K12 in four different electron acceptor conditions. They found that use of oxygen and nitrate as electron acceptors prevented CT metabolism, but at low oxygen levels (1%) transformation to CO_2 did occur. The same authors in another work (1990) found that a denitrifying *Pseudomonas* sp. (strain KC) isolated from ground water aquifer solids, transformed carbon tetrachloride under denitrification conditions, giving CO_2 as major end product. Little or no chloroform was produced.

Little is known about the influence of one organic compound on the removal of another. Pettigrew *et al.* (1991) demonstrated the simultaneous degradation of chlorobenzene and toluene by *Pseudomonas* sp. strain J56, which grows on a wide range of chloro- and methylaromatic substrates. This simultaneous degradation is impossible for most bacteria because catabolic pathways are incompatible.

Evans *et al.* (1919) used various sources of inocula to establish enrichment cultures of denitrifying bacteria on benzene, toluene and xylenes in the absence of molecular oxygen. They observed complete depletion of toluene and partial depletion of *o*-xylene within three months of incubation. No loss of benzene, *p*-xylene or *m*-xylene was observed; *m*-xylene, however, was degraded by a subculture incubated on *m*-xylene alone.

Substrate interactions of benzene, toluene and *p*-xylene during microbial degradation by indigenous mixed cultures in sandy aquifer material, and by two pure cultures isolated from the same site (*Pseudomonas* sp. strain CFS-215 and *Arthrobacter* sp. strain HCB), were investigated by Alvarez and Vogel (1991).

At concentrations typical of a moderately polluted estuary (0.2–4 μg/l) and under experimental conditions simulating winter, spring and summer, Wakeham *et al.* (1983) found that, in experimental marine ecosystems, biodegration is also important for aromatic hydrocarbons in summer, while for all the investigated classes, volatilization appears to be the major removal process. For toluene, for example, half-lives of 16 days, 1.5 days and 13 days were found for spring, summer and winter, respectively. Trichloro- and tetrachloroethylene show similar half-lives, which range from 25 days to 15 days.

Jensen and Rosenberg (1975), in experiments with both open and closed systems involving seawater, found that evaporative losses of perchloroethylene, for example, are greater than losses by degradation.

4.6.3 Adsorption

In 1982, Knap and Williams reported on laboratory studies to determine the fate of petroleum hydrocarbons, originating from refinery effluents, in estuarine systems. Adsorption of the investigated hydrocarbons, as a mixture, to estuarine sediment was found to be most important removal mechanism. Experiments with low (1.4 mg/l hydrocarbon to 1 g sediment) and high (3.2 mg/l hydrocarbon to 1 g sediment) concentrations of hydrocarbons were conducted. After 18 h only 5% of the original material (low concentration) was found in the water phase, while in the experiment with high initial hydrocarbon concentration only 6% of the hydrocarbons were left in the water column after 1 h.

By contrast, Gschwend *et al.* (1982) pointed out that adsorptive losses of alkylbenzenes to sediments (coastal sediments, Vineyard Sound, Massachusetts, USA) are small and that these sediments cannot be considered as an important sink for alkylbenzenes. They calculated that only a few percent of C2 alkylbenzenes, and less than 10% of C3 alkylbenzenes would be adsorbed. Wakeham *et al.* (1983) stated that for the chlorinated C2 hydrocarbons and some low chlorinated benzenes, sorption to particulate matter plays only a minor role ($< 0.1\%$). Moreover, it was clearly shown (Wakenham and Canuel, 1986) that sedimentation in mesocosmos experiments was negligible, and therefore removal of VOCs to the sediments seems not to be significant compared to the other removal processes (volatilization, biodegradation).

Adsorption/desorption onto sediments appears to be of minor importance in the cycling of VOCs in river waters (McDonald *et al.*, 1988).

Sorptive retardation of volatile hydrocarbons in ground water can be an important factor for the evaluation of a potential ground water contamination by VOCs. Generally, adsorption of dissolved organic compounds on solid surfaces increases as both the octanol/water partition coefficient of the compounds, and the organic carbon content of the solid surfaces increase.

4.6.4 Photo-induced degradation

Little is known about the photo-induced degradation of VOCs in waters. Photo-oxidation has been often considered as an important VOC elimination process in natural waters and as an important slick weathering process. However, these processes are not well quantified. Generally, the photo-assisted formation of very reactive oxygenated species might significantly affect the fate of volatile organic compounds in water.

It is well documented, that on irradiation with sunlight, and in the presence of humic material, nitrate and nitrite salts, oxygen species are formed in water, with concentrations ranging from 10^{-9} mol/l (RO_2) to 10^{-17} mol/l ($OH^{\bullet}$). These agents may successfully attack organic substances, initiating breakdown processes. Reaction with dissolved aromatic compounds, for example, leads to the formation of simple phenolic compounds, which are more polar and therefore more soluble in water (Russi *et al.*, 1982; Kotzias *et al.*, 1986; Moza and Parlar, 1987).

4.6.5 Oil in seawater

Considerable efforts have been made in documenting oil spill accidents and identifying and estimating the effects of these discharges. Several processes govern the behaviour and fate of oil discharged at sea, in particular spreading, evaporation, photolysis and biodegradation.

Spreading of oil on water is a very complex process, being influenced by various factors (evaporation, photolysis, etc.). Crude oil tends to spread to a mean thickness of approximately 0.1 mm (Mackay and McAuliff, 1988).

Evaporation is a well characterized process which is governed mostly by the oil vapour pressure. It has been calculated that about 30% of a crude oil spill may evaporate per day.

Photolysis can be seen as an important slick alteration process in regions with high insolation. Photo-induced generation of reactive oxygenated species might lead to the formation of water-soluble compounds, which results in the production of relatively toxic organic compounds into the water column.

Biodegradation of volatile organics has been regarded as one of the main elimination processes for these compounds. However, quantification of this process is very difficult and depends on many factors, e.g. hydrocarbon degradability, nutrient and oxygen status of water, etc.

4.7 Environmental aspects of selected organohalogen compounds

During the period 1984–1988 a Scientific Advisory Committee, assisting in the preparation of directives concerning the pollution caused by certain dangerous substances discharged into the aquatic environment, was requested by the Commission of European Communities (CEC) to advise on several compounds (CEC, 1990). Of these compounds, available information on the VOCs, including physicochemical properties and predicted environmental distribution, as well as other relevant data on toxicity and persistence, will now be given.

1,2-Dichloroethane ($Cl–CH_2–CH_2–Cl$) is moderately toxic for aquatic biota; longterm data, however, are not available. Persistence differs according to the environment. In the atmosphere the half-life can be very short due to evaporation, but in ground waters, where no degradation or evaporation occurs, the half-life is very long. 1,2-Dichloroethane has been shown to be mutagenic in *Salmonella* and *Drosophila*. It has also been shown to be carcinogenic in animals. According to CEC recommendations, concentration in fresh water should be as low as possible and should not exceed 10 μg/l. The same is valid for the concentration of $Cl{-}CH_2{-}CH_2{-}Cl$ in salt water.

Trichloroethylene ($Cl{-}CH{=}CCl_2$) must be considered as toxic for aquatic biota since it produces adverse effects at the mg/l level. Persistence differs according to the environment. In the atmosphere the half-life is expressed in days or weeks due to photo-oxidation. In water, the half-life can be very short due to evaporation but in ground water the half-life is very long. The concentration of trichloroethylene in fresh and salt water should be as low as possible and should not exceed 10 μg/l. Information and data for *tetrachlorethylene* ($Cl_2{-}C{=}C{-}Cl_2$) are the same as for trichloroethylene.

Carbon tetrachloride (CCl_4) is moderately toxic for juvenile and adult biota and is toxic to embryos and larvae of several species of aquatic organisms. CCl_4 does not biodegrade in the aquatic environment. It is known to be biodegradable only under anaerobic conditions and at a slow rate. CCl_4 evaporates rapidly from running water (half-life 0.3 to 3 days) but persists much longer in still water (30 up to 300 days). CCl_4 is not mutagenic but has been shown to be carcinogenic at high doses in one species of laboratory mammals. WHO issued a tentative guideline of 3 μg/l for drinking water. According to CEC recommendations the concentration of CCl_4 in fresh water and salt water should not exceed 10 μg/l.

Chloroform ($CHCl_3$) is moderately toxic for juvenile and adult biota and is toxic below 100 μg/l to embryos and larvae of several species of aquatic organisms. The potential for bioaccumulation of $CHCl_3$ is low due to its physicochemical characteristics. $CHCl_3$ is very resistant to biodegradation in the environment, although it can be metabolized in warm-blooded vertebrates. Chemical degradation in the aquatic environment is extremely slow. The compound evaporates rapidly from running waters (half-life 0.3 to 3 days) but persists much longer in still water (up to 300 days). $CHCl_3$ has been also found in ground water, where it is persistent. Re-entry of $CHCl_3$ into the aquatic environment by atmospheric fall-out after rainfall has often been observed.

In addition to industrial inputs, $CHCl_3$ is synthesized naturally, probably in the oceans, but also through chlorination of raw water in the preparation of drinking water, along with formation of other trihalomethanes. Chloroform is mutagenic but has been shown to be carcinogenic at high doses in laboratory animals. WHO issued a guideline value of 30 μg/l for drinking water to protect human health. The concentration of $CHCl_3$ in fresh water and salt water should however be as low as possible and should not exceed 10 μg/l.

Chloroform in drinking water. Addition of chlorine to water leads to the formation of hypochlorous acid (HOCl) and hypochlorite ions (OCl^-) according to the reactions:

$$Cl_2 + H_2O \rightarrow HOCl + H^+ + Cl^-$$
$$HOCl \rightarrow H^+ + OCl^-$$

The formation of each species, HOCl or OCl^-, depends on the pH. In strongly acidic soluion HOCl is dominant, whereas in alkaline solution formation of OCl^- dominates. However, most water supplies have a pH value between 6 and 8. In these cases hypochlorous acid is the most prevalent chlorine species. HOCl is the compound responsible for the formation of the chlorinated substances, in particular chloroform, during the chlorination of water.

As a possible formation mechanism for chloroform, it has been postulated that HOCl reacts with organic water contaminants containing carbonyl groups or alcohol groups:

$$R{-}\overset{\overset{\displaystyle O}{\|}}{C}{-}\underset{\underset{\displaystyle H}{|}}{\overset{\overset{\displaystyle H}{|}}{C}}{-}H + 3HOCl \rightarrow R{-}\overset{\overset{\displaystyle O}{\|}}{C}{-}OH + CHCl_3 + 2H_2O$$

forming an organic acid and chloroform

In the case of acetone:

$$CH_3{-}\overset{\overset{\displaystyle O}{\|}}{C}{-}CH_3$$

a common organic solvent, the reaction with HOCl leads specifically to acetic acid and chloroform.

Dichloromethane (CH_2Cl_2) is a highly volatile chemical, which is produced industrially in great volumes (EC production in the order of 2×10^5 tons per year) and is used as a solvent, as a propellant in aerosols, and as a chemical intermediate. Physicochemical properties of CH_2Cl_2 and field monitoring data indicate that the atmosphere is the major environmental compartment at risk. However, the concern for the aquatic environment is not negligible. Chemical and biological processes in the aquatic environment are few and slow (months, years) and are judged to be insignificant under environmental conditions. CH_2Cl_2 shows low acute toxicity for aquatic organisms, and the bioaccumulation potential is estimated to be low. Mammalian acute toxicity is low, while long-term (inhalation) exposure is reported to cause severe degenerative liver damages. According to the quality objectives, the concentration of CH_2Cl_2 in fresh water and salt water should not exceed 10 μg/l.

For most organohalogene compounds mentioned here the USEPA set the maximum contaminant level (MCL) at 5 μg/l, while for toluene and the xylenes the MCL was set at 2 mg/l.

4.8 Sampling of VOCs in water

4.8.1 Sampling procedure

Generally, concentrations of VOCs in water samples range from a few ng to μg/l. Therefore, in order to achieve reliable analytical results special attention should be given to the sampling procedure. This requires that the sampling equipment has to be extremely clean. To eliminate contamination in surface water sampling (seawater, river water, etc.) samples should be taken away from any kind of motorboat, or other facilities which might be a source of contamination. Sauer (1980) describes a procedure for surface water sampling and sampling of VOCs in the water column. In the meantime sampling equipment has been modified according to individual cases. The basic devices, however, remain the same. Samplers are usually glass bottles of 2 to 30 l volume, equipped with Teflon or glass stoppers. All the equipment should be acid-washed and rinsed with water free of VOCs prior to sampling.

The VOCs are collected using the 'grab sample' technique where the glass bottle is completely submerged in the water or filled with the water to be analysed, excluding all headspace. Sample conservation to avoid biological degradation is mostly done by using sodium azide or $HgCl_2$ (addition directly after sampling). Storage of the samples should be at 4°C. Usually, analysis of VOCs in water must be achieved within 3 to 14 days of collection.

4.8.2 Methods for sampling enrichment

Due to their volatility, extraction of VOCs from water samples by means of an organic solvent (the method used for most of the less volatile organic water pollutants) is not suitable. The best way to handle the VOCs present in the ppt–ppb range is to bring them via the vapour phase onto a solid sorbent, from which they are desorbed by rapid heating and transferred to the analytical system or extracted by a solvent. A comprehensive review on isolation and reconcentration techniques of VOCs from water has recently been published (Namiesnik *et al.*, 1990).

In recent years, two methods have been widely applied for the analysis of VOCs in water samples. One is closed-loop stripping analysis, first reported in 1973 (Grob). The other is the purge-and-trap method, which is based on work done by Bellar and Lichtenberg in 1974.

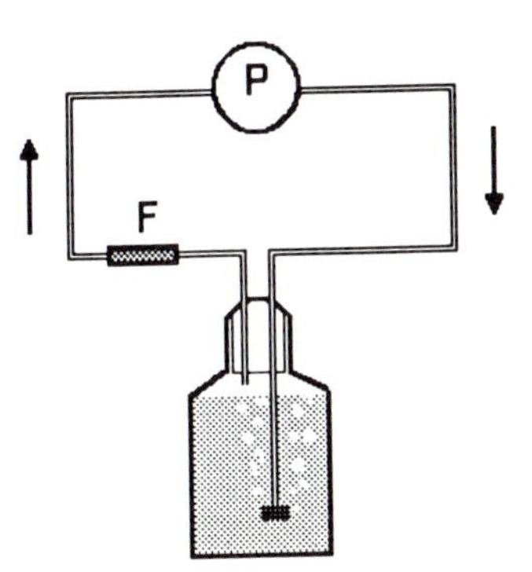

Figure 4.1 Closed circuit for stripping with inert gas (air). P = pump; F = holder for activated carbon filter trap. After Grob (1973).

4.8.2.1 Closed-loop stripping analysis. This method has been applied to various water samples for the measurement of purgeable organic compounds at the ng/l level. For stripping the organic contaminants from water a broad choice of conditions is available, ranging from evaporation by a stream of inert gas at room temperature up to the conditions of steam distillation (Grob, 1973). For VOCs, stripping with air at room temperature has been mostly employed, while steam distillation is more suitable for stripping polar compounds. By this technique the volatile compounds of the water are trapped in a sorbent (activated carbon filter trap) by pumping the stripping gas in a closed circuit (see Figure 4.1) via the trap and the water phase. The activated carbon trap retains the volatile organic compounds, while the purge gas passes through and is then returned to repurge the water sample via the pump.

The volume of the water sample varies from 0.5 to 2.0 l. Stripping temperature is approximately 30°C. The adsorbent filter is of pure wood charcoal, heat activated. The amount of activated charcoal varies from 1.5 mg for pure water samples to 5.0 mg for moderately contaminated samples. Disc geometry and diameter, and thickness of the filters vary according to the amount of charcoal used. The flow rate of stripping ranges from 1 to 2.5 l/min, and the stripping procedure takes 1 to 3 h.

After the collection of the volatile organics, the adsorbent filter is removed and fitted with a glass vial. The organic compounds are extracted with a small amount (5–100 μl) of a solvent, e.g. CS_2 or CH_2Cl_2. An aliquot of this solvent can be injected into a GC or GC-MS for further analysis. Recoveries were reported to be more than 80% (Grob and Zürcher, 1976). Advantages and disadvantages of the method can be summarized as follows:

Advantages

(i) The method is simple, rapid, very sensitive and relatively trouble free
(ii) The sample is extracted from the carbon filter trap with only few μl of the solvent (CS_2 or CH_2Cl_2)
(iii) The extraction requires no more than 10 min.

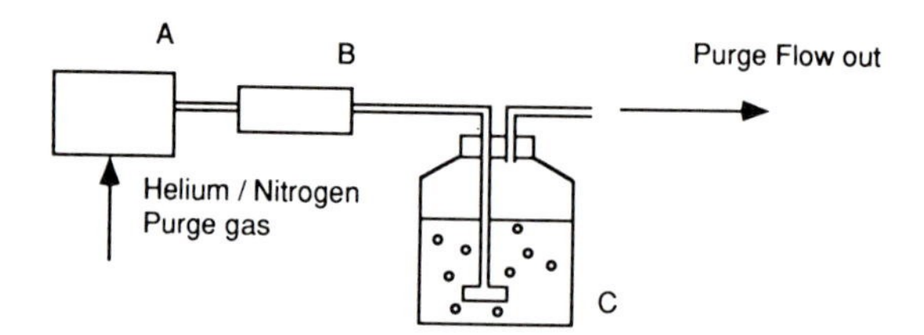

Figure 4.2 Stripping of volatile organic compounds from the water sample. A = flowmeter; B = purge valve; C = sample.

Disadvantages

(i) The range of compounds effectively determined is limited
(ii) Highly volatile compounds such as chloromethane, methylene chloride and chloroform are only poorly recovered by the CS_2 extraction
(iii) Highly contaminated samples (e.g. waste water) may overload the activated carbon filter trap.

4.8.2.2 Purge-and-trap method. In this method the VOCs are stripped directly from the water samples when helium is bubbled through the sample causing the VOCs to leave solution and dissolve in the gas phase (Figure 4.2).

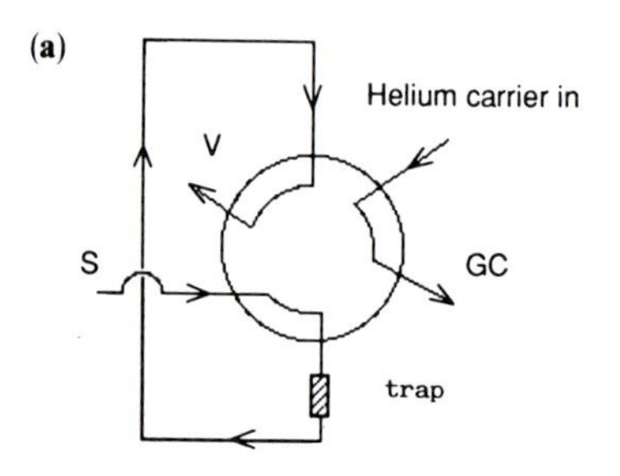

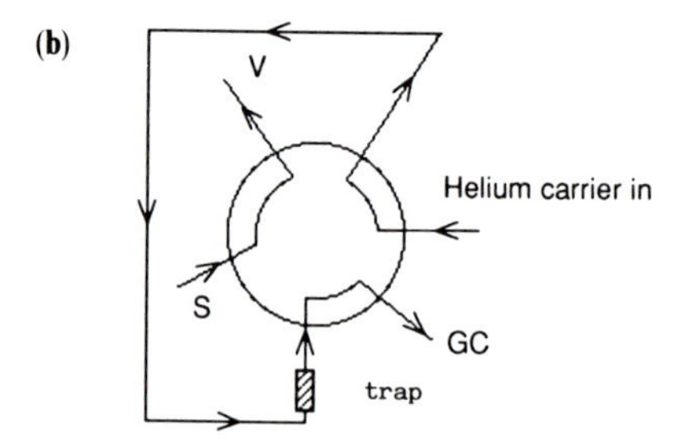

Figure 4.3 Purge-and-trap technique. (a) operation in purge mode. VOC adsorption onto the trap. (b) Desorption procedure. V = vent; S = from sample; GC = gas chromatography.

Table 4.5 Purgeable volatile organics.

Carbon tetrachloride	Methylchloride
1,2-Dichloroethane	Methylbromide
Chloroform	Dichloropropane
Dichloromethane	Dichloropropene
Tetrachloroethylene	Dichloroethylene
Trichloroethylene	Tetrachloroethane
Bromoform	Chlorobenzene
Benzene, toluene	Dichlorodibromomethane
Ethylbenzene	
Trimethylbenzene	

By means of a six-port valve the helium gas flow is directed onto a trap tube containing an organic polymer adsorbent (Tenax) or other materials (Carboxen, Carbotrap, activated charcoal) as adsorbent, where the VOCs from the purge gas are effectively retained. After the stripping/adsorbing procedure the trap is connected through the six-port valve with the analytical column and the trapped volatiles are thermally desorbed (at 250°C) and transferred directly onto the top of the column (Figure 4.3). In order to make the transfer of the trapped VOCs onto the GC column more efficient an additional step can be inserted in the above procedure; the trapped VOCs are thermally desorbed and transferred by helium as carrier gas to a liquid nitrogen cooled capillary trap (cryofocusing) for further sample enrichment. Finally, the adsorbed compounds are transferred by rapidly heating (200–250°C), as a plug, onto the GC column for the analysis. Generally, the stripping efficiency depends on the partition coefficients of the compounds, while the efficiency in adsorbing the stripped volatiles depends on the retention characteristics of the adsorbent, and the organic compounds being adsorbed (Table 4.5).

The volatile organics determined by the previously mentioned purge-and-trap method include the compounds with boiling points between those of *n*-hexane (69°C) and *n*-pentadecane (271°C). Sensitivity of the purge-and-trap method is mostly less than 1 ng/l. Recovery rates of more than 85% are reported for chlorinated and non-chlorinated hydrocarbons.

Several problems exist for the purge-and-trap method. One is due to cross-contamination in the stripping device when samples with low and high concentrations are analysed in succession. Another problem is sample foaming with samples obtained from highly contaminated waters.

4.9 Analysis of VOCs in water

The method of choice for VOC analysis is gas chromatography, with flame ionization detection (FID) or electron capture detection (ECD) for the analysis of halogenated volatile compounds. In addition, combined GC–MS

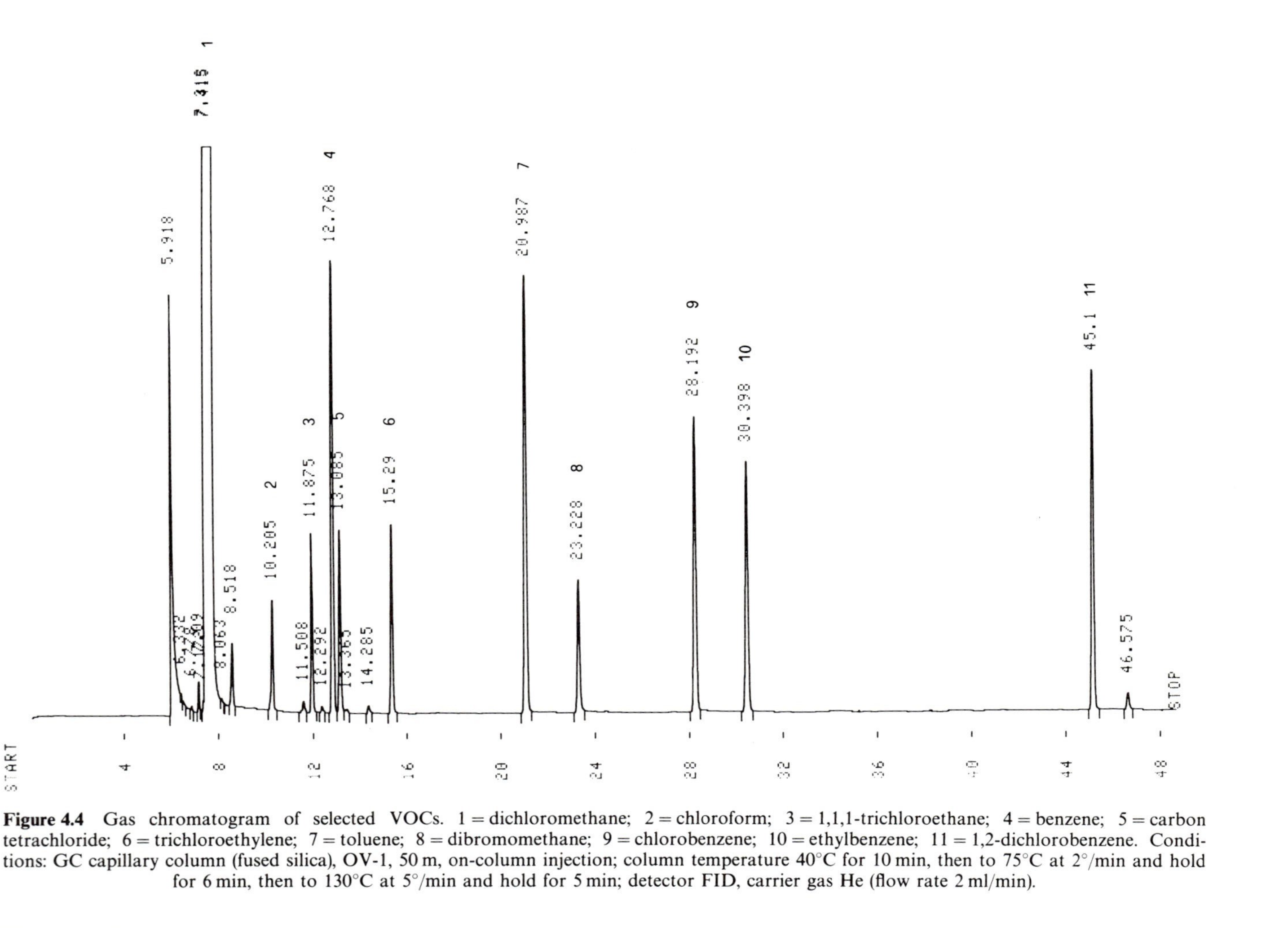

Figure 4.4 Gas chromatogram of selected VOCs. 1 = dichloromethane; 2 = chloroform; 3 = 1,1,1-trichloroethane; 4 = benzene; 5 = carbon tetrachloride; 6 = trichloroethylene; 7 = toluene; 8 = dibromomethane; 9 = chlorobenzene; 10 = ethylbenzene; 11 = 1,2-dichlorobenzene. Conditions: GC capillary column (fused silica), OV-1, 50 m, on-column injection; column temperature 40°C for 10 min, then to 75°C at 2°/min and hold for 6 min, then to 130°C at 5°/min and hold for 5 min; detector FID, carrier gas He (flow rate 2 ml/min).

Table 4.6 Columns used in gas chromatography for the analysis of VOCs

Type of column	Operating conditions
Packed columns	
SP-2100	0°C–180°C
OV-17	–10°C–150°C
SP-1000	40°C–220°C
Capillary columns	
SE-54	20°C–120°C
SE-52	50°C–220°C
5% Phenylmethylsilicone	30°C–300°C
Ucon HB 5100	25°C–175°C
VOCOL (diphenyldimethylpolysiloxane)	10°C–120°C
DB-624	5°C–115°C
OV-1	40°C–130°C

systems have often been used for the identification of VOCs. In Figure 4.4 a typical chromatogram of selected VOCs is shown. A selection of the GC columns most commonly used for VOC measurement is given in Table 4.6.

4.9.1 *Factors influencing VOC analysis*

In the last decade capillary columns (0.25 mm and 0.32 mm i.d.) have been used for the analysis of individual VOCs. However, recently the use of wide-bore (0.53 mm and 0.75 mm i.d.) capillary columns received considerable attention due to the possibility of operating them at flow rates more compatible with purge-and-trap systems (Mosesman *et al.*, 1987). Narrow-bore capillary columns are generally operated at flow rates of 1 ml to 3 ml/min. Higher flow rates substantially reduced the separation efficiency. Using purge-and-trap systems requires flow rates high enough to ensure the rapid transfer of the trapped compounds from the adsorbent to the column. Several techniques have previously been applied to overcome the flow incompatibility problem; for example splitting the purge-and-trap flow, or secondary trapping with a cold trap. Each technique introduces several drawbacks; for example, splitting the flow will reduce sensitivity, while secondary trapping, even with a specialized cryofocusing device, results in losses of highly volatile gases.

4.9.2 *Sum parameters (TOX, POX)*

Recently, ground water contamination by VOCs has received considerable attention. Of particular interest are the halogenated compounds, which may have a direct impact on drinking water quality. The analysis of individual compounds in water samples is often time-consuming and requires high analytical expertise. Therefore, recent work in analytical chemistry has led to the development of methods and instruments for the measurement of the sum of the halogenated compounds in water samples, avoiding any kind of separation of the individual VOCs. Total organic halide (TOX) analysis can be seen as a useful monitoring tool for water pollution, providing rapid information on the extent of potential water contamination.

TOX and POX (purgeable organic halide) measurements have been correlated with specific halogenated organic compound determinations to show the applicability of the sum parameter determination. TOX analysis is based on the adsorption of the volatile organics from the water onto activated charcoal followed by a combustion process and microcoulometric titration. POX analysis is similar and is accomplished by purging of the VOCs from the water followed by adsorption onto charcoal, combustion and microcoulometric titration. Amy *et al.* (1990) found that the organic halide parameter (TOX or POX) provides an accurate estimate of the sum of the chlorinated organic compounds present in samples of contaminated ground water containing complex mixtures. However, because of the different abundance, reactivity and toxicity of the VOCs in water, it seems reasonable to identify and quantify the individual compounds.

4.9.3 Description of a practical procedure for the analysis of VOCs in water samples

A 2 or 5 l sample bottle should be cleaned three times with 100 ml acetone and rinsed with distilled water. During filling of the bottle with the water to be analysed, mixing with ambient air should be minimal (Grob and Grob, 1974). After sampling, sodium azide or $HgCl_2$ should be added to the water to avoid biological degradation of the organic compounds (McDonald *et al.*, 1988).

Analysis of the sample should be done by stripping the VOCs from the water and transferring them onto an adsorbent placed in a glass or metal tube. Usually, Tenax-TA, Carboxen and Carbotrap, alone or in combination, are used. Stripping the VOC from the water sample is mostly done by purging with an inert gas (He, N_2). Duration of purging varies according to the sample. Usually the volume of the purge gas passed through the sample is five to ten times higher than the volume of the sample itself. Purging velocity should be adjusted to between 50 and 150 ml/min (Lopez-Avila *et al.*, 1987a,b; McDonald *et al.*, 1988).

Following the purge step, the adsorbent trap is desorbed at 200–250°C. The adsorbed compounds are derived either directly or are preconcentrated on a capillary trap with liquid nitrogen at −150°C (cryofocusing), and are then thermally desorbed and transferred onto the GC column for further analysis.

4.10 Conclusions

VOCs in natural waters are present in relatively low concentrations. After accidental events (e.g. the Exxon Valdez oil spill in Alaska and the IXTOC oil spill in the Gulf of Mexico), however, concentrations can reach extremely high values. Volatility, biodegradation and, to a lesser extent, photochemical degradation adsorption onto sediments are the processes which govern the behaviour and fate of VOCs in the aquatic environment.

Ground water contamination is a major concern because the above processes do not, or do not very effectively, eliminate VOCs. Ground water is widely used, after treatment, as a drinking water supply and therefore should be free from any kind of water micropollutants. Contaminated ground and tap water, even after costly cleaning procedures, can act as sources of pollutants, which are formed, for example, during chlorination and which might affect drinking water quality.

In the last decade the development of sophisticated analytical methods has enabled the determination of volatile organics, even in very low concentrations, in the different water compartments and has helped in recognizing pollution problems. Analysis of individual compounds should be in the foreground, even if to some extent measurements of sum parameters correlate well with the measurements of specific VOCs. Analysis of individual compounds helps in assessing in a more accurate way the hazardous potential of contaminated water.

Finally, the occurrence of VOCs in natural waters seems to be, from the environmental point of view, less significant than their presence in the atmosphere, where they are directly taking part in atmospheric reactions. Contamination of ground water and drinking water by VOCs, however, has a direct impact on human health and is therefore a serious problem which can be solved only by applying the best available technology.

References

Alexander, M. (1981) Biodegradation of chemicals of environmental concern. *Science* **211** 132–138.

Alexander, M. (1985) Biodegradation of organic chemicals. *Envir. Sci. Technol.* **18**(2) 106–111.

Alvarez, P.J.J. and Vogel, T.M. (1991) Substrate interaction of benzene, toluene, and *p*-xylene during microbial degradation by pure cultures and mixed culture aquifer slurries. *Appl. Environ. Microbiol.* **57**(10) 2981–2985.

Amy, G.L., Greenfield, J.M. and Cooper, W.J. (1990) Correlations between measurements of organic halide and specific halogenated volatile organic compounds. *Water Chlorination* **6** 691–701.

Bellar, T.A. and Lichtenberg, J.J. (1974) Determining volatile organics at microgram per liter levels by gas chromatography. *J. Am. Water Works Assoc.* **66** 739–744.

Bellar, T.A., Lichtenberg, J.J. and Kroner, R.C. (1974) The occurrence of organohalides in chlorinated drinking waters. *J. Am. Water Works Assoc.* **66** 703–706.

Bianchi, A.P., Bianchi, C.A. and Varney, M.S. (1989) Marina developments as sources of hydrocarbon inputs to estuaries. *Oil and Chemical Pollution* **5** 477–488.

Biziuk, M., Kozlowski, D. and Blasiak, A. (1991) Determination of volatile halogenated compounds sin tap and surface waters from the Gdánsk district. *J. Envir. Anal. Chem.* **44** 147–151.

Brooks, J.M., Wiesenburg, D.A., Burke, R.A. Jr. and Kennicutt, M.C. (1981) Gaseous and volatile hydrocarbon inputs from a subsurface oil spill in the Gulf of Mexico. *Envir. Sci. Technol.* **15**(8) 951–959.

Burns, K.A. and Saliot, A. (1986) Petroleum hydrocarbons in the Mediterranean Sea. *Marine Chemistry* **20** 141–157.

Commission of European Communities (CEC) (1990) Scientific Advisory Committee to examine the toxicity and ecotoxicity of chemical compounds. *Activity Report 1984–1988.*

Criddle, C.S., De Witt, J.T. and McCarty, P.L. (1990a) Reductive dehalogenation of carbon tetrachloride by *Escherichia coli* K-12. *Appl. Environ. Microbiol.* **56**(11) 3247–3254.

Criddle, C.S., De Witt, J.T., Grbić-Galić, D. and McCarty, P.L. (1990b) Transformation of carbon tetrachloride by *Pseudomonas* sp. strain KC under denitrification conditions. *Appl. Environ. Microbiol.* **456**(11) 3240–3246.

Derenbach, J.B. (1985) Investigation into a small fraction of volatile hydrocarbons. *Marine Chemistry* **15** 295–303.

Egli, C., Scholtz, R., Cook, A.M. and Leisenger, T. (1987) Anaerobic dechlorination of tetrachloromethane and 1,2-dichloroethane to degradable products by pure cultures of desulfobacterium sp. and methanobacterium sp. *Fems. Microbiol. Lett.* **43** 257–261.

Evans, P.J., Mangand, D.T. and Young, L.Y. (1991) Degradation of toluene and *m*-xylene and transformation of *o*-xylene by denitrifying enrichment cultures. *Appl. Environ. Microbiol.* **57**(2) 450–454.

Glaze, W.H. and Henderson, J.E. IV (175) Formation of organochlorine compounds from the chlorination of a municipal secondary effluent. *J. Water Poll. Control FED* **47**(10) 2511–2515.

Gomez-Belinchon, J.I., Grimalt, J.O. and Albaiges, J. (1991) Volatile organic compounds in two polluted rivers in Barcelona (Catalonia, Spain). *Wat. Res.* **25**(5) 577–589.

Grob, K. (1973) Organic substances in potable water and its precursors. *Journal of Chromatography* **84** 255–273.

Grob, K. and Grob, H. (1974) Organic substances in potable water and its precursor. Part II. Applications in the area of Zurich. *Journal of Chromatography* **90** 303–313.

Grob, K. and Zürcher, F. (1976) Stripping of trace organic substances from water equipment and procedure. *Journal of Chromatography* **117** 285–294.

Gschwend, P.M., Zafiriou, O.C. and Gagosian, R.B. (1980) Volatile organic compounds in seawater from the Peru upwelling region. *Limnol. Oceanogr.* **25**(6) 1044–1053.

Gschwend, P.M., Zafiriou, O.C., Mantoura, R.F., Schwarzenbach, R.P. and Gagosian, R.B. (1982) Volatile organic compounds at a coastal site. Seasonal variations. *Environ. Sci. Technol.* **16** 31–38.

Hurford, N., Law, R.J., Fileman, T.W., Payne, A.P. and Colcomb-Heiliger, K. (1990) Concentrations of chemicals in the North Sea due to operational discharges from chemical tankers—Results from the Second Survey. October 1988. *Oil Chemical Pollution* **7** 251–270.

Jensen, S. and Rosenberg, R. (1975) Degradability of some chlorinated aliphatic hydrocarbons in sea water and sterilized water. *Water Res.* **9**(7) 659–661.

Knap, A.H. and Williams, P.L. (1982) Experimental studies to determine the fate of petroleum hydrocarbons from refinery effluent on an estuarine system. *Environ. Sci. Technol.* **16**(1) 1–4.

Kotzias, D., Herrmann, M., Zsolnay, A., Russi, H. and Korte, F. (1986) Photochemical reactivity of humic materials. *Naturwissenschaften* **73** 35.

Lopez-Avila, V., Wood, R., Flanagan, M. and Scoot, R. (1987a) Determination of volatile priority pollutants in water by purge and trap and capillary column gas chromatography/mass spectrometry. *Journal of Chromatographic Science* **25** 286–291.

Lopez-Avila, V., Heath, N. and Hu, A. (1987b) Determination of purgeable halocarbons and aromatics by photoionization and Hall electrolytic conductivity detectors connected in series. *Journal of Chromatographic Science* **25** 356–363.

Mackay, D. and McAuliffe, C.D. (1988) Fate of hydrocarbons discharged at sea. *Oil and Chemical Pollution* **5** 1–20.

Mackay, D. and Shiu, W.X. (1981) *Phys. Chem. Ref. Data* **10**(4) 1175–1199.

McDonald, T.J., Kennicutt, II, M.C. and Brooks, J.M. (1988) Volatile organic compounds at a coastal Gulf of Mexico site. *Chemosphere* **17**(1) 123–136.
Morris, J.C. and McKay, G. (1975) Formation of halogenated organics by chorination of water supplies. *EPA-600/1-75-002*. US Environmental Protection Agency.
Mosesman, N.H., Sidisky, L.M. and Corman, S.D. (1987) Factors influencing capillary analyses of volatile pollutants. *Journal of Chromatographic Science* **25** 351–355.
Moza, P.N. and Parlar, H. (1987) Einfluß von Sauerstoff auf die Direktphotolyse einfacher Aromaten in wäßrigen Systemen. *Naturwissenschaften* **74** 137.
Namieśnik, J., Górecki, T., Biziuk, M. and Torres, L. (1990) Isolation and preconcentration of volatile organic compounds from water. *Analytica Chimica Acta* **237** 1–60.
Nirmalakhandan, N.N. and Speece, R.E. (1988) QSAR model for predicting Henry's law constant. *Environ. Sci. Technol.* **22** 1349–1357.
Oldenhuis, R., Oedzes, J.Y., Van der Waarde, J.J. and Janssen, D.B. (1989) Degradation of chlorinated aliphatic hydrocarbons by *Methylosinus trichosporium* OB3b expressing soluble methane monooxygenase. *Appl. Environ. Microbiol.* **55**(11) 2819–2826.
Pearson, C.R. and McConnell, F. (1975) Chlorinated C_1 and C_2 hydrocarbons in the marine environment. *Proc. R. Soc. London, Ser. B* **189** 305–332.
Pittigrew, C.A., Haigler, B.E. and Spain, J.C. (1991) Simultaneous biodegradation of chlorobenzene and toluene by a *Pseudomonas* strain. *Appl. Environ. Microbiol.* **57**(1) 157–162.
Russi, H., Kotzias, D. and Korte, F. (1982) Photoreduzierte Hydroxylierungsreaktionen organischer Chemikalien in natürlichen Gewassern. *Chemosphere* **11**(10) 1041–1048.
Saliot, A. *et al.* (1984) Biogéochimie de la Matière organique en milieu estuarien: stratégies d'échantillonnage et de recherche élaborées en Loire (France). *Oceanol. Acta* **7** 191–207.
Sauer, Jr., T.C. (1980) Volatile liquid hydrocarbons in waters of the Gulf of Mexico and Caribbean Sea. *Limnol. Oceanogr.* **25**(2) 338–351.
Sauer, Jr., T.C. (1981) Volatile organic compounds in open ocean and coastal surface waters. *Organic Geochemistry* **3** 91–101.
Scheele, B. (1980) Reference chemicals as aids in evaluating a research programme—selection aims and criteria. *Chemosphere* **9** 293–309.
Schnoor, J. (1991) The Alska oil spill: Its effects and lessons. *Environ. Sci. Technol.* **25**(1) 14.
Smith, J.G. (1975) Chloine in your water. *J. Chem. Educ.* **52**(10) 656–657.
Wackett, L.P., Brusseau, G.A., Householder, S.R. and Hanson, R.S. (1989) Survey of microbial oxygenases: Trichloroethylene degradation by propane-oxidizing bacteria. *Appl. Environ. Microbiol.* **55**(11) 2960–2964.
Wakeham, S.G. and Canuel, E.A. (1983) Geochemistry of volatile organic compounds in seawater: Mesocosm experiments with 14-C-model compounds. *Geochimica and Geochimica Acta* **50** 1163–1172.
Wakeham, S.G., Davis, A.C. and Karas, J.L. (1983) Mesocosm experiments to determine the fate and persistence of volatile organic compounds in coastal sea water. *Environ. Sci. Technol.* **17** 611–617.
Zoeteman, B..J., Harmsey, K. and Linders, J.B.H.J. (1980) Persistent organic pollutants in river water and ground water of the Netherlands. *Chemosphere* **9** 231–249.
Zsolnay, A. (1979) Hydrocarbons in the Mediterranean Sea, 1974–1975. *Marine Chemistry* **7** 343–352.

5 VOCs and soil pollution

J.J.G. KLIEST

5.1 Introduction

In this chapter the role of VOCs in soil pollution will be described. The soil is perhaps the most complex environmental compartment, while VOCs comprise an enormous number of compounds with strongly variable properties. Consequently, within the framework of this book it is impossible to highlight all relevant aspects of the topic. In this chapter therefore only those aspects in which soil pollution by VOCs has a direct relationship with the subjects of the other chapters of this book will be dealt with. In particular the relationship between the environmental compartments soil and air, and the possible effects of soil pollution on human health will be discussed.

After a short introduction on the subject, sections 5.3 and 5.4 will give some theoretical backgrounds with regard to the behaviour of VOCs in the soil environment. In particular the partitioning of the compounds over the different soil phases and the transport through the soil system will be discussed. Section 5.5 gives an overview of the different methods of sampling and analysis that can be of use during research into VOCs in soil. In section 5.6 the possible relationship between soil pollution with VOCs and human health will be discussed.

5.1.1 The soil as part of the environment

In environmental studies it is common to subdivide the environment into compartments. A frequently used subdivision for this purpose comprises six compartments: (i) air; (ii) soil; (iii) water; (iv) suspended solids; (v) sediment; and (vi) biota.

The soil system can subsequently be subdivided into: (i) a liquid phase; (ii) a solid phase; (iii) an air phase; and (iv) a biotic phase. The solid phase of the soil consists of inorganic material (sand, loam, clay) with a certain amount of organic matter (humic material, peat). In the upper part of the soil parts of the pores between the solid soil particles are filled with air and this zone is therefore called the unsaturated zone of the soil. In the saturated zone of the soil system the pores are filled completely with water. The water in the pores in the unsaturated zone of the soil is called the pore water. The water in the saturated zone is called ground water. The ground water can be

Table 5.1 Some basic data on different groups of volatile organic compounds

Group of compounds	Basic formula	Examples	Polarity
Alkanes and cycloalkanes	C_xH_y	*n*-Hexane Cyclohexane	Non-polar compounds
Unsaturated hydrocarbons	C_xH_y	Hexene 1,4-Butadiene	Polarity depends on level of saturation
Aromatic hydrocarbons	C_xH_y	Benzene Toluene	Non-polar compounds
Ketones and aldehydes	$C_xH_yO_z$	Formaldehyde	More polar compounds
Carboxylic acids	C_xH_yCOOH	Benzoic acid Phenylacetic acid	Polar compounds
Organic esters	$C_xH_yO_a$	Methyl acetate Acetic anhydride	Polarity depends on molecular composition
Halogenated hydrocarbons	$C_xH_yX_zR_a$	Trichloroethene Dichloromethane	Polarity depends on molecular composition
Phenols	$C_xH_y(OH)_z$	Phenol Cresol	Mostly more polar compounds
Chlorophenols	$C_xH_yX_z(OH)_a$	Chlorophenol 1,4-Dichlorophenol	Polarity depends on molecular composition
Nitrogen-containing compounds	$C_xH_yN_z$	Amines Amides Nitroso compounds	Polarity depends on molecular composition

divided in the phreatic ground water, which is influenced directly by rainfall and local water management, and deeper ground water, which is shielded by one or more impermeable layers.

5.1.2 VOCs and soil pollution

VOCs comprise a great number of compounds with strongly variable properties. In Table 5.1 an overview is given of the most important groups of VOCs relevant to soil pollution. With regard to the presence and behaviour of the compounds in the soil, several properties are of great importance. One of the most important is the polarity of a compound. This property determines for a large part both the fate of a compound when it is brought into the soil, and the equilibrium partitioning of a compound over the different soil phases. Non-polar compounds in general are hydrophobic and will show a relatively strong tendency to adsorb to the organic fraction of the solid phase of the soil or to volatilize. Polar compounds will have a relative preference for the liquid phase of the soil. Moreover, the sorption of polar compounds to the solid phase of the soil differs from the sorption of non-polar compounds.

5.2 Sources of soil contamination with VOCs: relationship with other environmental compartments

5.2.1 Sources of soil contamination

There are many sources of VOCs. Some of these sources are biogenic. Some VOCs can be produced by plants and animals and by bacterial degradation of organic matter. Soil contamination with VOCs, however, is mostly the result of pollutants orginating from anthropogenic sources. Huge quantities of VOCs are produced in the oil industry. Some of these VOCs are used as motor fuel or as a raw product for the chemical industry. The chemical industry itself is an important producer of VOCs, which are used in industrial and consumer products. Only a few of the VOCs which are brought into circulation are recycled. The rest are emitted into the environment and will distribute over the different environmental compartments.

Soil pollution with VOCs can be a direct result of an emission, for instance a local spill on an industrial site or near a fuel station. Very important, however, is secondary pollution of the soil by VOCs, as a result of the deposition of VOCs from polluted air. There are numerous activities in which VOCs are emitted in the air. Use of various household products, industrial emissions, traffic and painting are some examples. Once emitted in the air the compounds may be transported over variable distances. Some of the compounds will be degraded by UV light or by a reaction with OH radicals, or be oxidized by ozone or depleted in other ways. Some of these transformation processes will lead to others, in most cases more polar, VOCs.

Other VOCs will be deposited on the soil surface by wet or dry deposition, or will adsorb directly to soil components at the interface of soil and air. The process of deposition and adsorption leads to a relatively diffuse contamination of the soil with VOCs.

Once a compound has entered the soil system, it may adsorb to organic matter in the soil, solve in ground water or revolatize. Some of the components will be susceptible to bacterial degradation, others will be persistent.

The result of all these processes is a background concentration in soil. The level of this background concentration is influenced by regional differences, but also by soil-specific properties such as the organic carbon content of the soil.

5.2.2 Fate of VOCs in the environment: the fugacity concept

A useful concept to model the fate of VOCs in the environment is the fugacity concept, which was introduced by Lewis (1901) and applied to environmental systems by Mackay (1979). Fugacity can be defined as the tendency for a chemical to escape from one phase to another. At equilibrium all fugacities, which are expressed in units of pressure, are equal.

The advantage of the fugacity concept is that it makes it possible to bring together all transport, reaction and partitioning processes into one model by giving units that are comparable to these processes. In the fugacity concept the concentration of a compound in a phase is the ratio of its fugacity and the fugacity capacity:

$$C = \frac{f}{z} \tag{5.1}$$

where C = concentration (mol/m^3), f = fugacity (Pa) and z = fugacity capacity (mol/m^3.Pa). The value z is unique for a chemical in a phase at a defined temperature and can be calculated on the basis of properties such as solubility, vapour pressure and the octanol/water partition coefficient (K_{ow}).

Examples of the calculation of z values are given by Patterson and Mackay (1985) who define five levels of calculation. With the most simple calculation (level I) it is possible to calculate the equilibrium partitioning of a fixed amount of a chemical between well-mixed phases. The next step (level II) is to include first order reactions, transformations and advection. A level III calculation is applicable on a non-equilibrium but steady-state system. In a level IV calculation a non-equilibrium, non-steady-state system is accounted for and in a level V calculation it is possible to model systems with a heterogeneous concentration in space and time. An overview of the characteristics of each level of calculation is given in Patterson and Mackay (1985). The complexity increases with the level of calculation.

In recent years the fugacity models or closely related forms of modelling have been used in a number of studies for the modelling of a wide variety of systems (Mackay *et al.*, 1985; Hattermer-Frey, 1990; Rierder, 1990). The model is most frequently used to obtain a general understanding of the the fate of compounds in relatively large-scale environmental systems. A useful application is the determination of the expected environmental partitioning of new compounds.

Most of the necessary data for calculation of fugacities can presently be found in handbooks (Verschueren, 1983) or in databases, such as the database CESARS (published by the Ministry of the Environment, Ontario and distributed by CCINFO, Canada). For some of the compounds partitioning in the environment is calculated in CESARS using a level I fugacity model, which is the most generally applicable.

5.3 Partitioning of VOCs in the soil

5.3.1 Introduction

With regard to the behaviour of VOCs in the soil two aspects are of great importance: (i) partitioning of the VOCs over the different soil phases; and

(ii) transport through the soil system. When the soil is in equilibrium the mass of a contaminant will be distributed over the three soil phases, solids, water and air, in a constant ratio. The partitioning of a compound over the soil phases is determined by: (i) the extent to which the compound adsorbs to the solid phase of the soil; (ii) the solubility of the compound in the liquid phase; and (iii) the volatility of the compound.

5.3.2 *Adsorption and desorption to the solid phase*

The description of the adsorption/desorption process of VOCs to the solid phase of the soil is very complex. This is a result not only of the complexity of the soil system but also of the fact that the properties of VOCs that determine the behaviour can vary strongly.

Sorption of a compound can be the result of several processes. An overview of these processes is given by von Oepen *et al.* (1991). Van der Waals attraction is the most common. Other forms are hydrophobic bonding, hydrogen bonding, bonding by charge transfer, ligand exchange and ion bonding, ion–dipole and dipole–dipole interaction, and chemisorption.

Non-polar compounds will adsorb mainly as a result of van der Waals attraction and hydrophobic bonding. In the soil non-polar compounds will adsorb predominantly to the organic matter. The organic matter, however, is not a homogeneous matrix but consists of components of different polarities: humines, humic acid and fulvic acid. Of these the humin fraction is the strongest adsorbent for non-polar compounds. When the organic matter content of the soil is low the clay content of the soil can have a significant influence on the sorption of VOCs (Mingelgrin and Gerstl, 1983).

The adsorption of VOCs to soil particles can be described by:

$$\frac{X}{m} = K_d \times c_l \tag{5.2}$$

in which X/m = the amount of adsorbed material (kg per kg soil), K_d = the distribution coefficient (dm^3/kg) and c_l = the concentration of the compound in the liquid phase (kg/dm^3). This formula can be derived from: (i) the equation of Langmuir:

$$\frac{X}{m} = \frac{K_1 \times K_2 \times c_l}{1 + K_2 \times c_l} \tag{5.3}$$

in which K_1 and K_2 are constants; and (ii) the equation of Freundlich:

$$c_s = K \times c_e^n \tag{5.4}$$

in which c_s = the concentration in the solid phase, K = a constant, c_e = the equilibrium concentration in the solution and n = a constant depending on the solute and the solvent. This assumes a low concentration level of the compound in solution and $n = 1$, which implies a linear relationship between c_l and c_s.

For non-polar compounds the adsorption in the soil is determined mainly by the adsorption to the organic matter in the soil. For these compounds it is possible to rewrite equation (5.2) as:

$$\frac{X}{m} = K_{oc} \times f_{oc} \times c_l \tag{5.5}$$

in which K_{oc} = the coefficient of adsorption to organic carbon (dm^3/kg) and f_{oc} = the fraction of organic carbon in the soil.

For polar compounds the partitioning cannot be calculated using the K_{oc} because forms of sorption other than sorption to the organic matter in the soil predominate. Factors such as the clay content of the soil and pH are very significant for the adsorption of these compounds.

The K_{oc} value can be measured directly in batch experiments with different types of soil. In many cases however the K_{oc} is calculated from the octanol/water partitioning coefficient, K_{ow}. The K_{ow} value has been assessed for a great number of compounds. For the calculation of K_{oc} from K_{ow} several empirical equations exist:

$$\log K_{oc} = \log K_{ow} - 0.317 \text{ (Brown and Flagg, 1981)} \tag{5.6}$$

$$\log K_{oc} = \log K_{ow} - 0.21 \text{ (Kenaga, 1980)} \tag{5.7}$$

$$K_{oc} = 0.63 K_{ow} \text{ (Karickhoff \textit{et al.}, 1979)} \tag{5.8}$$

$$K_{oc} = 0.411 K_{ow} \text{ (Karickhoff, 1981)} \tag{5.9}$$

$$\log K_{oc} = 0.72 \log K_{ow} + 0.49 \text{ (Chiou \textit{et al.}, 1979)} \tag{5.10}$$

The results of the calculations using the equations can differ significantly, as is illustrated in Figures 5.1 and 5.2. Furthermore, the calculation itself is disputable, because the soil–water system is significantly different from the octanol– water system. Sterical differences between isomeric compounds will lead to a different sorption behaviour in the soil, although the K_{ow} values are essentially the same (Mingelgrin and Gerstl, 1983). The use of empirically assessed K_{oc} values is therefore preferable.

5.3.3 Partitioning between the liquid phase and the gas phase

A second important process is partitioning between the liquid and gas phases in the soil. The partitioning between these two phases can be described by the Henry coefficient:

$$H = \frac{V_p \times M}{760 \times S} \tag{5.11}$$

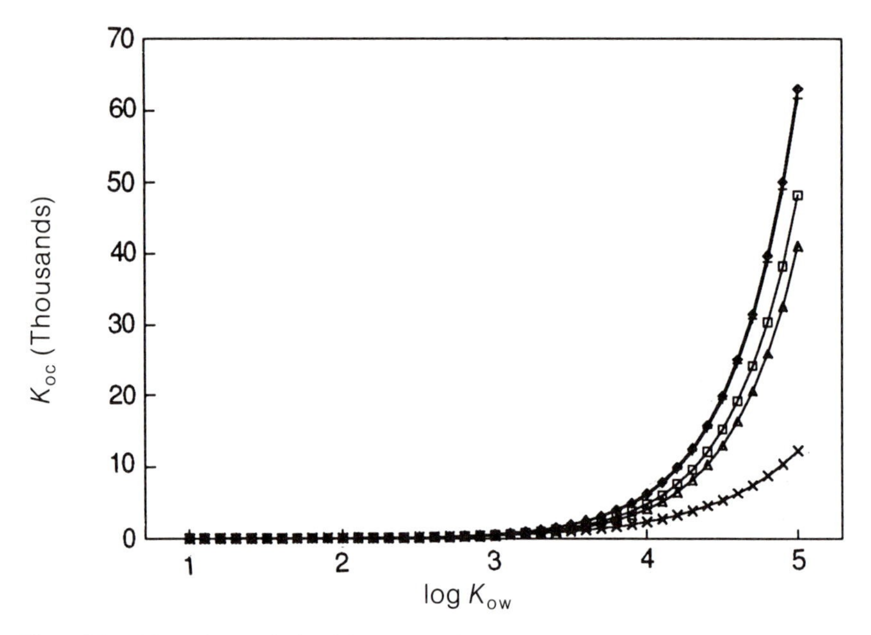

Figure 5.1 Values of K_{oc} calculated from K_{ow} values using equations (5.6) (□), (5.7) (+), (5.8) (◇), (5.9) (△) and (5.10) (×).

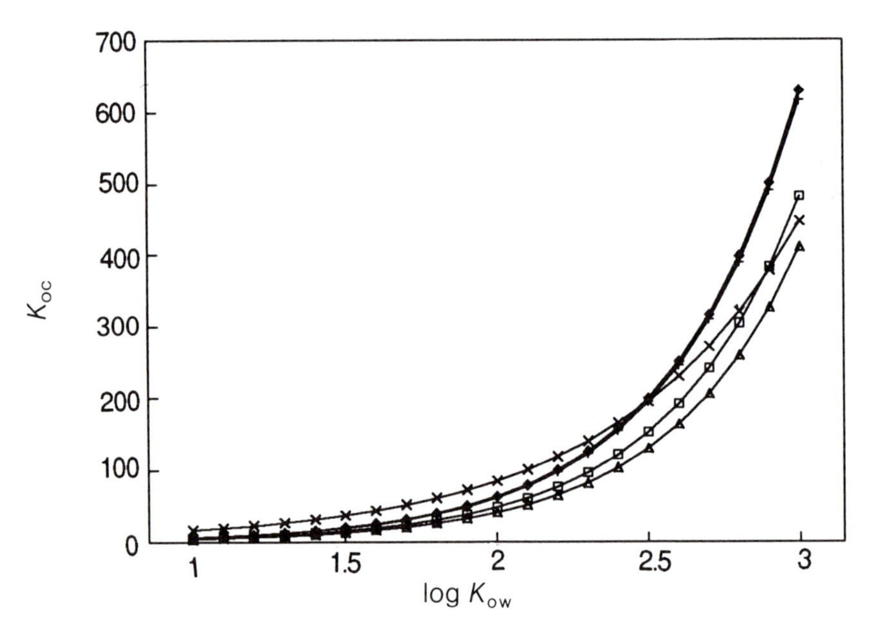

Figure 5.2 Values of K_{oc} calculated from K_{ow} values using equations (5.6) (□), (5.7) (+), (5.8) (◇), (5.9) (△) and (5.10) (×).

in which H = the Henry coefficient (atm. m^3/mol), V_p = the vapour pressure (mm Hg), M = the molecular weight (g/mol) and S = the solubility (mg/l).

More convenient in calculations is the use of the K_{lg} value, which is related to the Henry coefficient according to:

$$K_{lg} = \frac{R \times T}{H} = 0.0623 \times \frac{T \times S}{V_p \times M} \qquad (5.12)$$

in which R = the gas constant (atm.m^3/mol.K) and T = temperature (K).

5.3.4 *Calculation of partitioning for non-polar compounds*

Using the K_{oc} and K_{lg} values partitioning over the three soil phases can be described by the equations of Goring (1972):

$$\%\text{gas phase} = \frac{100 \times e_g}{e_g + K_{lg} \times e_w + \dfrac{(100 - (e_g + e_w)) \times d \times oc}{100}} \times K_{oc} \times K_{lg} \qquad (5.13)$$

$$\%\text{solid phase} = \frac{(100 - (e_g + e_w)) \times d \times oc \times K_{oc} \times K_{lg}}{e_g + K_{lg} \times e_w + \dfrac{(100 - (e_g + e_w)) \times d \times oc}{100}} \times K_{oc} \times K_{lg} \qquad (5.14)$$

$$\%\text{liquid phase} = \frac{100 \times K_{lg} \times e_w}{e_g + K_{lg} \times e_w + \dfrac{(100 - (e_g + e_w)) \times d \times oc}{100}} \times K_{oc} \times K_{lg} \qquad (5.15)$$

in which e_g = gas-filled pores (%), e_w = water-filled pores (%), d = the density of the solid phase (kg/dm^3) and oc = the organic carbon content (%). The equations are applicable for non-polar compounds under the conditions that each soil particle is surrounded by a stagnant water layer and that equilibrium exists between the soil phases.

5.3.5 *The kinetics of adsorption*

The kinetics of adsorption is a second important factor in the description of the adsorption process. The transport of compounds to the sorption sites can be divided into:

(1) transport of a molecule to the outer sorption sites (macrotransport):
(2) diffusion into the micropores or capillaries (microtransport); and
(3) diffusion through the solid phase itself to sites that are not accessible to the solvent (von Oepen *et al.*, 1991).

In a normal moist soil transport to the outer sorption sites in the soil will generally take place by diffusion through the stagnant water layers that surround most of the soil particles (Goring, 1972). After reaching the outer sorption sites the compound may diffuse into micropores. This second stage of the sorption process may take significant additional time.

Von Oepen *et al.* (1991) have carried out batch experiments with different soil types. Under fixed conditions with a number of organic acids, amines and amides, equilibrium was reached within 16 h for all compounds. It is not clear, however, if this equilibrium applies to all three processes of adsorption or just to the relatively rapid transport to the outer sorption sites.

Under optimal conditions the adsorption and desorption curves would be similar. In this case the adsorption is completely reversible. In most cases, however, the desorption curve will differ from the adsorption curve, partly by stronger or even non-reversible bonding to the solid phase, partly by inclusion in solid particles (Genuchten *et al.*, 1974; Bowman and Sans, 1985). As a result of this hysteresis effect desorption can be slower than adsorption and incomplete. The hysteresis effect has been found a significant factor in the rate of desorption of VOCs in several studies (Begeman, 1978; Duffy *et al.*, 1980; Rogers *et al.*, 1980).

5.4 Transport of VOCs in the soil

5.4.1 Introduction

The mobility of a contaminant in the soil is determined largely by the relative concentrations of the contaminant in the water and air phases of the soil. Transport of the contaminants can take place in both water and air by convection and diffusion. Convection is the process by which the contaminant moves with the phase in which it is present. In diffusion the contaminant moves relative to the phase in which it is present.

Convection in the water phase is a function of the concentration of the contaminant in the water phase and the movement of the water in the soil. The horizontal movement of water can be a result of the ground water current. A vertical flow in the upward direction can take place as a result of upward percolation, or capillary rise following evaporation of water at the soil surface. A vertical flow in the downward direction may be the result of leaching.

Convection of contaminants with the soil air phase takes place when the soil air moves as a result of a rising ground water level. Gas production in deeper soil layers, for example as a result of anaerobic digestion, may be a second driving force for the mobility of soil air.

In all cases of convection it must be remembered that new adsorption equilibria will be created along the route of transport. Depending on the local concentrations in the solid phase of the soil, contaminants could, for example, adsorb or desorb. This influences the total transport velocity of the contaminant.

Diffusion of a certain compound in the water and air phases of the soil is determined by the concentration gradient, the diffusion coefficient and a number of soil-specific properties. In comparison with diffusion in air, diffusion of VOCs in the water phase of the soil is a very slow process. For most compounds the velocity of diffusion in air is 10^5–10^6 times higher than in water.

5.4.2 Diffusion in the soil under steady-state conditions

The driving force for diffusion through the soil air is a difference in concentration between two locations in the soil. In its most simple form the diffusion flux can be described by Fick's first law (equation (5.16)).

$$F = -D\frac{\mathrm{d}C}{\mathrm{d}X} \tag{5.16}$$

in which F = the diffusion flux (g/m^2/s), D = the diffusion coefficient (m^2/s) and $\mathrm{d}C/\mathrm{d}X$ = the concentration gradient (g/m^3/m).

Fick's first law is applicable to a homogeneous soil layer, when there exists a linear concentration gradient. This implies that the diffusion flux is constant in time and space over the route under consideration (steady-state condition). In real situations the conditions necessary to apply Fick's first law will only be approximately met. For example, in most cases there will be no constancy of the diffusion flux with time, because of the depletion of material as a result of diffusion or other mechanisms of removal. The diffusion coefficient will vary with temperature. Moreover, most soils are far from homogeneous, which implies that the diffusion coefficient is not constant in space.

The coefficients of diffusion, which can be found in literature for a great number of compounds, are generally given in air. For application in soil air a correction is necessary for the tortuosity factor:

$$D_{sa} = D_a \times \frac{e_g^{10/3}}{(e_g + e_l)^2} \tag{5.17}$$

in which D_{sa} = the diffusion coefficient in soil air (m^2/s), D_a = the diffusion coefficient in air (m^2/s), e_g = the volume of air per volume of soil (fraction) and e_l = the volume of water per volume of soil (fraction)

Combination of equations (5.16) and (5.17) leads to:

$$F_{sa} = -D_{sa}\frac{\mathrm{d}C}{\mathrm{d}X}. \tag{5.18}$$

If the soil is not homogeneous with respect to the tortuosity or the air fraction,

D_{sa} will vary. Under steady-state conditions the flux will be constant along the route of diffusion. This implies that the concentration gradient will deviate from linearity.

The equations for the calculation of the flux in the liquid phase of the soil are analogous:

$$D_{sl} = D_l \times \frac{e_l^{10/3}}{(e_g + e_l)^2} \tag{5.19}$$

and

$$F_{sl} = -D_{sl} \times \frac{dC}{dX} \tag{5.20}$$

5.4.3 *Diffusion under non-steady-state conditions: Fick's second law*

Under non-steady-state conditions the process of diffusion in the soil air is much more complex. This is the case in most situations.

In a soil system which has been polluted with VOCs, continuous losses of VOCs may occur at the surface of the soil by volatilization. Although the loss of a compound in the soil air will party be compensated by desorption of material from the solid phase, the concentration gradient will change in time leading to a continuously decreasing flux.

Another example of non-steady-state diffusion is the situation in which a contaminant enters a soil layer which was initially not contaminated with the compound. In this case the establishment of an equilibrium between the soil phases implies that part of the contaminant will be removed from the soil air by sorption and solution. This affects the concentration gradient and decreases the flux of the contaminant.

For those cases in which the flux is not a constant in time and space, Fick's second law is valid:

$$\left(\frac{dC}{dt}\right)_x = D \times \frac{d^2C}{dX^2} \tag{5.21}$$

in which $(dC/dt)_x$ = the flux (g/m^2/s). Fick's second law is a second order differential equation, which can be solved using certain preconditions. Cranck (1975) gives a number of solutions under different preconditions for Fick's second law. Some of these solutions apply to situations which can be found in the soil. A few examples of these will be given in the following section.

5.4.4 *Applications of Fick's second law*

To define the preconditions that are necessary for application of Fick's second law a model should be set up in which assumptions are made for the stage of the diffusion process and the dimensions of the system. In the following paragraphs two models are given, each of which is significant for non-steady-state diffusion as a result of volatilization of compounds at the soil/air interface.

Model 1: Contamination initially uniformly distributed. At $t = 0$ the contaminant was uniformly distributed over a soil layer with a thickness L. As a result of volatilization significant depletion takes place, thus no steady-state conditions exist. Diffusion through the gas phase of the soil is the predominant route of transport in the unsaturated layer with thickness L. The diffusion is one-dimensional and upward. Horizontal transport and transport from deeper soil layers are neglected. From this model two submodels can be derived:

Model 1a Volatilization is in its initial phase. The concentration at the bottom of a soil layer with thickness L is equal to the initial concentration (semi-finite medium).

Model 1b Volatilization has proceeded significantly. As a result of this the concentration at the bottom of a soil layer with thickness L is lower than the initial concentration (finite medium).

Model 2: Upper soil layer with thickness L was initially not contaminated. Transport of pollutants takes place through the soil gas phase from deeper soil layers. The concentration at the bottom of a soil layer with thickness L is constant. Diffusion is one-dimensional and upward. The concentration at the soil surface equals zero. The soil layer with thickness L is homogeneous with regard to the tortuosity, e_g and the organic carbon content.

Solutions for Fick's second law for the above models can be found in Cranck (1975). In the case of a semi-finite medium (model 1a) integration of Fick's second law leads to the equations:

$$\frac{C_x}{C_0} = \mathrm{erf}\frac{C}{2\sqrt{D^*_{sa} \times t}} \tag{5.22}$$

and

$$M_t = 2 \times R_d \times C_0\sqrt{\frac{D^*_{sa} \times t}{\pi}} \tag{5.23}$$

in which M_t = the amount of compound which has left the soil at time t, C_x = the concentration in the soil air at depth x (g/m^3), C_0 = the uniform initial concentration in soil air (g/m^3), t = time (sec), D^*_{sa} = the corrected diffusion coefficient in the soil air (m^2/s) and R_d = the fraction of the total amount of compound that is present in the soil air.

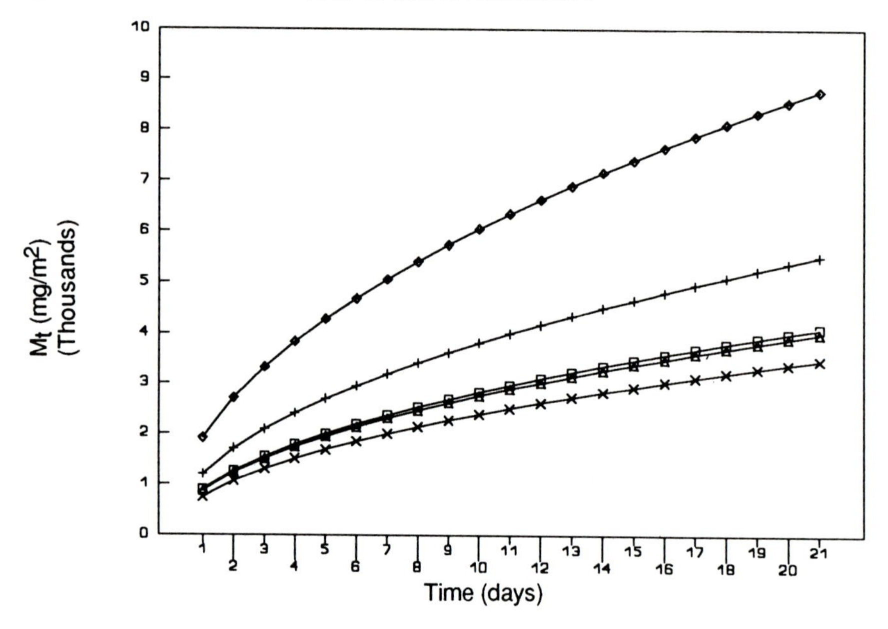

Figure 5.3 Cumulative amounts of VOCs that have left the soil at time t (days) with an initial uniform concentration $C_0 = 1000$ mg/m^3 in the soil air. Parameters used for calculations are given in Table 5.2. □ = Benzene; + = toluene; ◇ = m/p-xylene; △ = n-hexane; × = tetrachloroethylene.

D^*_{sa} can be calculated as:

$$D^*_{sa} = \frac{1}{R_d + 1} \times D_{sa} \tag{5.24}$$

and R_d can be derived from the equation of Goring (1972) as:

$$R_d = \frac{e_g + e_l \times K_{sa} + (100 - (e_g + e_l)) \times d \times oc \times K_{sa} \times K_{oc}}{e_g} \tag{5.25}$$

The results of a number of calculations are given in Figures 5.3 and 5.4. Figure 5.3 represents the cumulative amounts of pollutants that have left the soil system at time t. Calculations have been made for five different VOCs. In Figure 5.4 the flux from the soil system is given for the same five VOCs.

For model 1b this gives the solution:

$$\frac{C_{x,t}}{C_0} = \frac{4}{\pi} \times e^{-D^*_{sa} \times \pi^2 \times \frac{t}{4} \times l^2} \times \frac{\cos l - x}{2 \times l} \tag{5.26}$$

when

$$\frac{D^*_{sa} \times t}{l^2} > 0.15 \quad \text{and} \quad L \rightarrow \infty \tag{5.27}$$

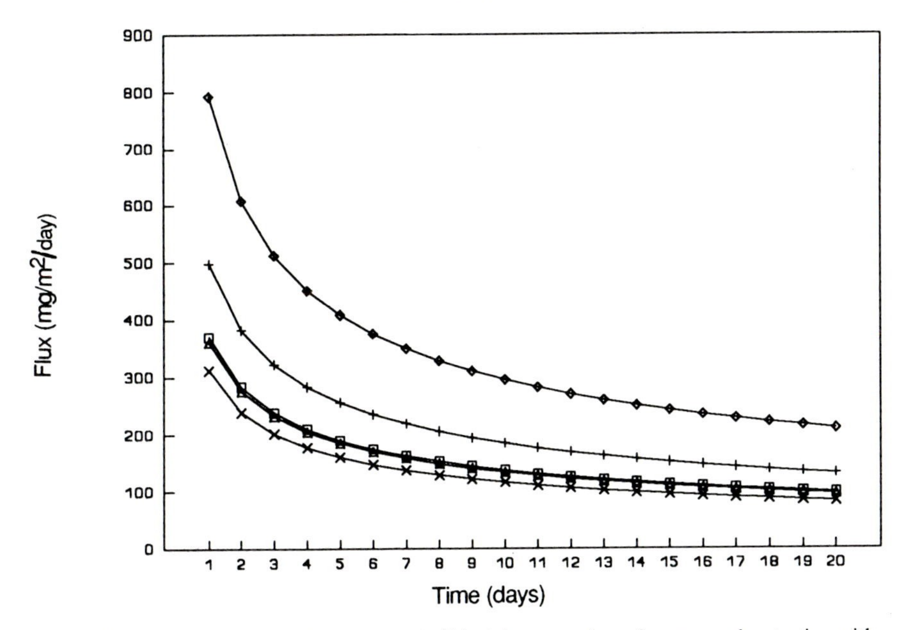

Figure 5.4 Flux from the soil system (mg/m^2/day) for a number of compounds, starting with an initial concentration of 1000 mg/m^3 in the soil air. Parameters used for calculations are shown in Table 5.2. □ = Benzene; + = toluene; ◇ = *m*/*p*-xylene; △ = *n*-hexane; × = tetrachloroethylene.

Table 5.2 Parameters used for calculations in Figures 5.3, 5.4, 5.5 and 5.6.

Parameter	Benzene	Toluene	*m*/*p*-Xylene	*n*-Hexane	Tetrachloroethylene
K_{oc} (dm^3/kg)	81[1,2,3]	267[1,4,5]	814[1,4,5]	6270[6]	488[7]
K_{lg} (g/m^3/g/m^3)	7[8]	6.1[8]	6.3[8]	0.16[8]	1.4[8]
D_a (m^2/day)	0.708[9]	0.636[9]	0.583[9]	0.645[9]	0.602[9]
d (soil) (kg/dm^3)	2.6				
Organic carbon (%)	0.2 (Figures 5.3, 5.4 and 5.6)	0.1 (Figure 5.5)			
C_0 (soil air) (mg/m^3)	1000				
e_g (%)	35				
e_l (%)	15				

[1]Platford (1983); [2]Verschueren (1983); [3]Karickhoff *et al.* (1979); [4]Thus and Kraak (1985); [5]Bruggeman *et al.* (1982); [6]Miller *et al.* (1985); [7]Trouwborst (1982); [8]calculated according to equation (5.12); [9]calculated according to Reid *et al.* (1977).

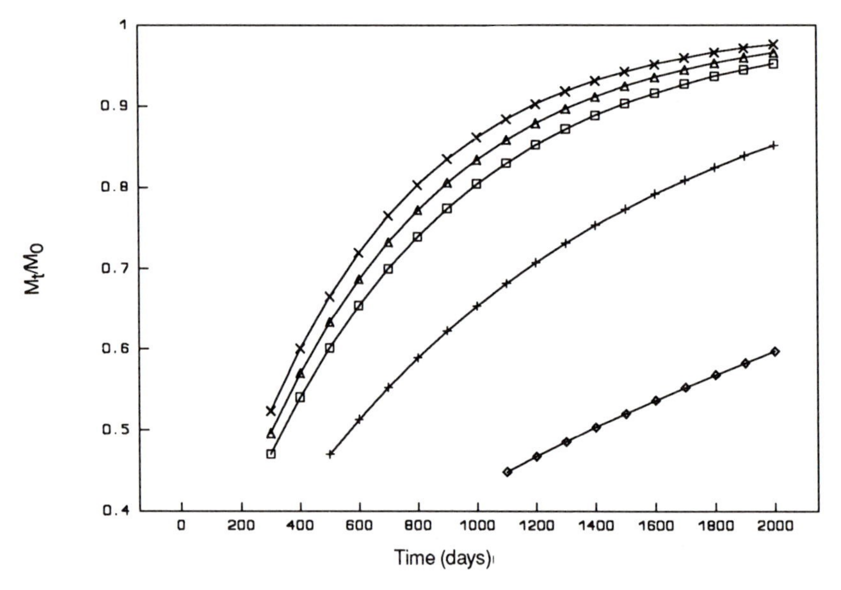

Figure 5.5 Fractions of the total amounts of VOCs that have left the soil at time t (days). Parameters used for the calculations are shown in Table 5.2. □ = Benzene; + = toluene; ◇ = m/p-xylene; △ = n-hexane; × = tetrachloroethane.

and

$$\frac{M_t}{M_0} = 1 - \frac{8}{\pi^2} \times e^{\frac{-D^*_{sa} \times \pi^2 \times t}{4 \times l^2}} \tag{5.28}$$

in which l = the soil layer under consideration (m) and M_t/M_0 = the fraction of the total amount of the pollutant that has left the soil at time t.

L can be calculated as

$$L = \frac{D_a/\sigma}{D^*_{sa}/l} \tag{5.29}$$

in which σ is the thickness of the laminar layer of ambient air at the soil surface. L thus represents the ratio of the resistance for diffusion in the laminar air layer and the soil air layer consideration.

Calculation of the concentration in relation to the depth according to equation (5.28) leads to Figure 5.5.

Model 2 (surface layer initially clean) can be described by:

$$\frac{Q_t}{lC_2} = \frac{D_{sa}t}{l^2} - \frac{1}{6} - \frac{2}{\pi^2} \times \sum_{n=1}^{\infty} \frac{(-1)^n}{n^2} \times e^{-D^*_{sa} \times n^2 \times \pi^2 \times t/l^2} \tag{5.30}$$

in which Q_t = the amount of pollutant that has passed soil layer l at time t, and C_2 = the concentration of the compound at depth $x = l$. The graphical presentation of equation (5.30) is given in Figure 5.6.

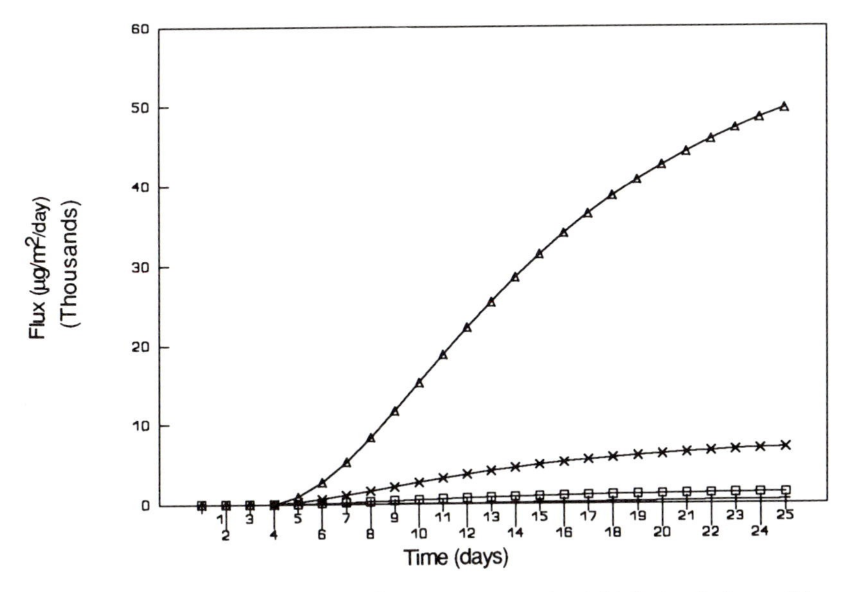

Figure 5.6 Flux at the soil surface ($\mu g/m^2/day$) in the case of an initially ($t = 0$) clean soil layer of 1 m thickness on a polluted soil layer. Parameters used for the calculations are shown in Table 5.2. □ = Benzene; + = toluene; △ = *n*-hexane; × = tetrachloroethane.

5.4.5 Other methods for calculation of diffusional transport

The calculation of diffusional transport in the soil, using the mathematical formulae that have been deduced from Fick's laws, is theoretically correct. For use in practical situations, however, this method of calculation has serious limitations. As stated in section 5.4.2, Fick's first law is only applicable in situations where there is steady state. The solutions for Fick's second law, given in section 5.4.4, presume that the soil layer under consideration is homogeneous with regard to porosity and organic carbon content, while there is constancy in time with regard to temperature and moisture content. In practice the soil is far from homogeneous and there is no constancy in time, as a result of temperature change and weather conditions. The results of the calculations will therefore only be approximations and are especially useful to compare the behaviour of compounds.

In theory it is possible to include more preconditions in the analytical solutions for Fick's second law (Cranck, 1975). The resulting formulae, however, quickly become too complex to be practical. If it is necessary to imply inhomogeneity and changing conditions it is much more convenient to use simulation languages or other forms of numerical methods. An example of the use of a simulation language is the description of the evaporation of 1,3-dichloropropene (Leistra, 1972).

5.4.6 *Convection with the liquid phase*

Transport by convection with the liquid phase plays an essential part in the horizontal transport of most VOCs. Pollutants that reach the ground water will partly solve or, if the solubility in water is not high enough, form floating layers. In both cases the compounds will be transported with the local ground water current. The velocity of the transport is mainly a function of local hydrology. Non-solvable compounds which are heavier than water may leach through the ground water to deeper soil layers.

When the saturated zone of the soil reaches the soil surface, convective transport is the only significant upward transport process. In this case the driving force is evaporation of water at the soil surface. Mackay (1980) has worked out the theoretical basis of the phase transition of pollutants from the liquid to the gas phase. For compounds with a Henry coefficient $H > 10^{-3}$ diffusion through the liquid film is the rate-limiting factor. Most non-polar VOCs belong to this group. If $H < 2 \times 10^{-5}$, which applies to most of the polar VOCs, resistance in the gas film is limiting. It can be deduced, however, that in both cases upward transport of water is the actual limiting factor. Diffusion through the water phase is generally much too slow to make up for the losses by volatilization of non-polar VOCs at the soil surface. The result is that exhaustion will occur in the upper ground water layer. Volatilization of these VOCs at the soil surface will therefore equal the upward convective transport as a result of evaporation of ground water. For compounds with a low Henry coefficient voltalization may be slower than the supply from deeper ground water layers. In this case accumulation may occur at the water/gas interface. As a result of this accumulation the partial vapour pressure at the interface will rise until voltalization equals supply, or until the maximum solubility of the compound is reached.

Upward transport is relevant at locations where evaporation of ground water is higher than precipitation of water, for example soil under buildings. Evaporation in crawl spaces from houses in The Netherlands was empirically assessed by Fast *et al.* (1987). Under winter conditions the mean net evaporation of water appeared to be 104 ml/m^2 per day.

5.5 The assessment of concentrations of VOCs in the soil

5.5.1 *Introduction*

The assessment of soil concentrations of VOCs consists of several steps. The first choice that should be made is in which phase(s) of the soil the concentration of a VOC should be assessed: the solid, liquid or gas phase of the soil. The choice depends on the aim of the investigation. If the aim is to get information on the extent of a pollution, measuring concentrations in the solid phase will in most cases be the obvious choice. (As has been shown in section 5.3.1 the greatest mass fraction of most volatile compounds will, in a situation of equilibrium, be adsorbed to the solid phase.) When the aim is to gain a better understanding of the transport of VOCs through the soil system, it is useful also to assess the concentrations in the liquid and gas phase. The concentrations in these phases could theoretically be calculated from the solid phase under the assumption of equilibrium. Measuring the concentrations of the compounds in the liquid and gas phase, however, will often give more accurate results.

5.5.2 Sampling and analysis of the solid phase of the soil

5.5.2.1 Sampling strategy. Sampling the solid phase of the soil with the intention of assessing the concentration of certain VOCs starts with making a sampling plan. Important questions that should be dealt with in this sampling plan are:

- The number of samples
- The size of the samples
- The selection of the location of the samples
- The sampling technique
- The extent to which samples can be mixed before analysis

The number of samples that should be taken depends on the aim of the investigation, the size of the investigated area and the homogeneity of the soil system and the pollution. To prevent the necessity of taking and analysing a large number of samples it is useful to plan the investigation in phases. First, all historical and site-specific information regarding the area under investigation should be gathered. The objectives of this preliminary investigation are to get answers to the questions:

- Is there a significant pollution?
- How is the pollutant distributed over the polluted site?
- Are there any hot spots of pollutant present?

Using the information from the preliminary investigation, and keeping in mind the aim of the investigation, the number of samples which should be taken in the indicative phase of the investigation can be determined.

There are many possible reasons for investigating the VOC content of a soil system. First, the aim may be to get an indication of the extent of pollution in a certain area. In the case of a pollutant that has a homogeneous or unknown distribution a good strategy is to take a number of subsamples in a grid over the polluted site and to mix these subsamples to form one or more analytical samples.

The size of each subsample depends on the homogeneity of the soil system. If the assumed local variation of the concentration of a VOC is large, for example as a result of considerable local variation in composition of the soil system, relatively large subsamples are necessary. More information about sampling strategy and theory can be found in Gilbert and Doctor (1985), Gilbert (1987) and Lame and Bosman (1991a, b).

When the presence of relatively serious pollution has been indicated, the next step is to investigate more thoroughly the extent of the pollution, the dispersion of the pollutants, the extent to which ground water layers have been or could be affected and the actual or potential danger for human health. Assessment of the dispersion of a pollutant and the actual or potential ground water pollution will often make it necessary to take a relatively large number of soil samples and to take samples from different soil layers. The strategy that should be followed in this phase of the investigation depends strongly on the information from the first two phases of the investigation. A description of a possible strategy is given in Lame and Bosman (1991b).

The study of human exposure and possible related health effects requires a different strategy. In this case the possible routes of exposure should be screened and the sampling strategy should be based on this exposure analysis. These possible exposure routes are discussed in detail in chapter 1. As for the solid phase of the soil, the possible exposure routes consist mainly of ingestion and dermal uptake. The samples that are taken should represent those parts of the soil with which people can come into contact. In most cases this will be the upper layers of the soil.

5.5.2.2 Sampling methods. During the collection, storage and handling of soil samples losses of VOCs can occur, as a result of volatilization and degradation of compounds. To achieve accurate results it is therefore necessary to act with care.

With regard to the collection of soil samples for analysis of VOCs, no standardized sampling procedures exists. Samples can be taken with a shovel (upper parts of the soil) or an auger. There are many types of auger, the most commonly used being the Gouge auger, the Edelman auger and the Riverside auger. Gouge augers can be used to take samples with minimal disturbance. With this kind of sampling, soil profiles of a length of 50 cm and more can be obtained. Edelman augers can be used to take relatively undisturbed samples at greater depths in all kind of soils (sand, clay, loam, marsh). Riverside augers are especially useful for augering in hard, stiff soils, both above and below the ground water level.

In many cases the sampling can be performed manually. In some cases however (deeper subsurface borings), it will be necessary to use, for example, truck-mounted drilling rigs.

Sampling with shovels or augers always leads to some disturbance of the soil matrix and to losses of VOCs which are not easy to quantify. The losses will be greater for compounds with a relatively high soil gas concentration.

For compounds with a high soil sorption affinity the losses will be low to insignificant.

A method of sampling that will reduce losses due to volatilization is to take undisturbed soil samples by means of a soil sample ring. This ring is pushed into the soil manually or by using a hammer. Afterwards the ring, containing the undisturbed soil sample, is taken out, closed with Teflon caps and transported to the laboratory. For very accurate results or when investigating compounds with a very low soil sorption affinity, it is necessary to immerse the undisturbed sample in-field in methanol, before transporting it to the laboratory (Siegrist and Jenssen, 1990).

Transport and temporary storage of samples should take place in containers of glass or stainless steel. These containers should be airtight with Teflon or Teflon-coated caps. Storage in plastic bags or open containers will lead to unacceptable losses. Losses of VOCs from soil samples during transport can be reduced by minimizing the headspace in the sample containers and by refrigerating the samples at 4°C during transport and storage. In a study by Siegrist and Jensen (1990) it was shown that the method of sampling and transport greatly influenced the 'negative bias', i.e. the extent of losses of compounds such as methylene chloride, dichloroethane, 1,1,1-trichloroethane, trichloroethylene and toluene. The influence on the recovery of chlorobenzene was much less. The most accurate method in all cases appeared to be undisturbed sampling, transport in a glass container without headspace and in-field immersion in methanol.

5.5.2.3 Sample pretreatment. In most cases the size of the sample taken to the laboratory will be greater than the amount of material which can be analysed. It is then necessary to take a representative subsample. In many cases the material will need homogenization, for instance by grinding. During the process of homogenization, however, serious losses may occur as a result of volatilization. A method often applied to reduce this problem is grinding of the soil material while adding liquid nitrogen (cryogenic grinding). Although this method will reduce volatilization, there is little quantitative information about the losses that may still occur.

5.5.2.4 Extraction. When a representative subsample is obtained, the next step is to detach the compounds from the soil matrix. In most cases this is achieved by extraction with a proper solvent. The choice of solvent depends on the chemical characteristics of the compounds of interest. Non-polar compounds are most efficiently desorbed with non-polar solvents such as pentane, hexane, toluene or dichloromethane. Polar compounds, on the other hand, can be desorbed with more polar solvents, such as methanol and acetone. In both cases, however, the presence of water in the soil sample can be a problem. The desorption of non-polar compounds which are sorbed to the soil material can be greatly reduced by the fact that thin water layers can close up the adsorbed compounds, thus making them inaccessible to the solvent. This inhibits recovery of the compounds during extraction. The problem can be solved by adding a more polar solvent to the soil sample before or during the extraction.

Polar compounds often show an affinity for water, which reduces the efficiency of extraction with a polar solvent. In the case of phenols and aldehydes, this problem can be solved by derivatizing the compounds to substances which are non-polar and thus can be extracted by a non-polar solvent such as toluene.

Other techniques often employed are headspace analysis and the so-called purge-and-trap technique. For both techniques no solvents are added to the soil samples.

The headspace technique is a semi-quantitative method of measurement. The concentration of a compound is measured in a well-defined headspace of a subsample at a defined temperature. The concentration in the solid phase of the soil is then calculated as an equilibrium concentration, using the theory of partitioning.

In the purge-and-trap technique the compounds are purged from the solid sample by a gas, for instance nitrogen or clean air. The stream of gas is then passed over an adsorbent, on which the compounds are trapped and concentrated. An advantage of the purge-and-trap technique is that a low limit of detection can be reached as a result of the concentration step. Moreover, the method can be automated relatively easily. A disadvantage is that extraction methods usually yield better recoveries.

5.5.2.5 Analysis. The actual analysis of VOCs usually takes place using chromatography. Many choices can be made with respect to the type of chromatography, the column type and the method of detection. Many of the choices depend on the compounds that should be measured. A number of different techniques will be mentioned briefly in the following section.

5.5.2.6 Chromatography. Although there are many types of chromatography, those most frequently used for the analysis of VOCs are high performance liquid chromatography (HPLC) and gas chromatography (GC). HPLC is used most often for the analysis of molecules with a relatively high molecular weight, such as the derivatized phenols and aldehydes. GC can be used to analyse most non-polar VOCs. The polarity of the column and the gas chromatographic conditions, such as temperature program, depend on the type of VOC. Non-polar compounds are separated best on a non-polar silicon phase column, such as CPsil-5. For good separation of more polar compounds a polar column is necessary, for example CPsil 19.

There are many types of detection. Flame ionization detection (FID) is a universal method of detection for VOCs. Electron capture detection (ECD) is much more sensitive for halogenated compounds and gives much lower limits of detection than does FID. Photoionization detection (PID) is highly sensitive for aromatic compounds. With all these types of detection, however, combination with gas chromatography does not enable fully positive identification of compounds due to the possibility of coelution of compounds.

Detection using mass spectrometry has the advantage that an unequivocal identification of compounds is possible, especially when the technique is used in combination with GC (GC–MS). This technique, however, is more difficult and prone to errors when used by less experienced analysts. Modern instruments have low detection limits, which allows this technique to be used for trace analysis in soil samples.

5.5.3 *Sampling and analysis of the liquid phase of the soil*

5.5.3.1 Introduction. With regard to the liquid phase of the soil it is useful to distinguish the liquid phase in the unsaturated zone and the ground water in the saturated zone of the soil. Sampling and analysis of the water in the unsaturated zone of the soil (pore water) is most frequently used to measure equilibrium concentrations as part of studies to fit models for transport and partitioning, and for assessing the availability of pollutants for plants. For a number of compounds the capillary rise of water is an important means of upward transport of the solutes.

Ground water can act as a medium for horizontal transport of VOCs. Moreover, ground water is used for drinking water and irrigation. Ground water samples can thus give information about the dispersion of a contaminant and the quality of the ground water to be used as a raw material for drinking water.

Sampling of pore water is a tedious activity, especially when the aim is to measure the VOC content. The first problem is to extract a volume of water from a well-defined depth. For this purpose ceramic cups are often used, and are brought into the soil at the desired depth. The actual sampling takes place with a pump. This creates a further problem. Most pumps operate by creating a vacuum above the water column. In the case of VOCs this vacuum could lead to losses of compounds by stripping and volatilization. Depending on the type of compound, these losses can be very significant. The method of sampling ground water depends on the type and behaviour of the pollutant. When a compound reaches the ground water it can solve completely. When the solubility coefficient is exceeded however, a layer can be formed, which floats on the ground water and can be transported in a horizontal direction by the local ground water current. When the compound is heavier than water, it can submerge to deeper ground water layers, while in the meantime it is transported in a horizontal direction.

The strategy of sampling should suit the expected behaviour of the contaminant. When a floating layer is expected special effort should be made to find this layer and sample it, because often the bulk of the pollution will be found in this floating layer. A submerged plume of a high-specific-weight contaminant will make it necessary to sample much deeper ground water layers to assess the dispersion of the contaminant. In all cases it will be necessary to gather information about the hydrology of the location, to make it possible to select the right sampling locations.

Sampling takes place using a measuring well, which is brought into the ground at the desired depth. The installation of a filter can take place hydraulically or manually, using a split barrel or drill. After installation the ground water should be allowed to stabilize for several hours. Following this, several litres of ground water should be pumped up and discarded, before the sample can be collected.

Sampling floating layers demands a special technique. The sample should be taken at the interface of the ground water and the unsaturated zone. The exact depth of this interface should thus be known and the filter of the measuring well should be brought into the soil at exactly this depth. Sampling the floating layer has primarily an indicative purpose. The existence of a floating layer can be proven, the compounds in the floating layer can be analysed and a raw estimate can be made of the amount of pollutant involved.

5.5.3.2 Sample collection. As stated previously, the collection of water samples for analysis of VOCs demands special care to prevent losses as a result of volatilization. The use of pumps that function by creating a vacuum above the water column can for this reason hold a risk, especially when air is sucked. This is likely to happen when the sampling is carried out with a flow that is too high in less permeable soil layers. When air sucking is avoided, losses of most VOCs are within acceptable limits (Oosterom *et al.*, 1988).

An alternative is to use special immersion pumps, which are brought into the measuring well. A disadvantage is that these pumps are often made of synthetic material. This could cause losses by sorption of compounds to the pump material. Another possibility is the use of diaphragm pumps. These pumps, which are designed specifically to sample low levels of VOCs are gas operated. During compression of the diaphragm, there is no contact between the gas and the sample. Disadvantages of this kind of sampling are the long cycle times while sampling deep wells, and the relatively high cost (Hatayama, 1986).

The hose by which the sample is brought to the surface should ideally be made of Teflon, although PFTE will often suffice. The use of other synthetic materials will in many cases lead to unacceptable losses by sorption.

5.5.3.3 Sample storage and transport. The best way to store samples is to fill glass bottles completely with the sample and close them with a Teflon or Teflon-lined cap. There should be no headspace above the sample, as this induces volatilization. The samples should be cooled at 4°C and analysed as soon as possible.

Another possibility for non-polar compounds is to mix the sample on-site with a non-polar organic solvent of specific weight lower than that of water. In this way the VOCs are, to some extent, already extracted from the water phase during transport, therefore losses are minimized. Shaking or ultrasonic extraction in the laboratory can complete the extraction procedure.

A number of compounds are easily degraded by microorganisms. In this case sample conservation should take place. In the case of phenols and chlorophenols this conservation is realized by adding 1 g copper sulphate to one litre of sample.

Sample pretreatment consists of extraction with an appropriate solvent and shaking (by hand or machine) or ultrasonic wave treatment. The headspace technique and the purge-and-trap method are also applied, especially for non-polar compounds. The final analysis is similar to the analysis of VOCs extracted from the solid phase of the soil.

5.5.4 Sampling of soil air

5.5.4.1 Introduction. There has been growing interest in the sampling and analysis of soil air during the last decade. One of the first publications with regard to soil air sampling was from Neumayr (1971). In The Netherlands the method has been used with varying success in a study in which the relation between soil contamination and indoor air pollution was assessed (Fast *et al.*, 1987). Lately the method has been use on a larger scale. The main reason for the growing use of the soil air sampling method is the fact that it is convenient. In addition, especially for VOCs, the results are sometimes more relevant than the results of measurements of the solid phase and the ground water. For most VOCs in the unsaturated zone of the soil, diffusion through the air phase is by far the most important route of upward transport, and thus of volatilization of compounds from the soil system. As shown in section 5.4.2 this form of diffusional transport is fully dependent on the gradient of the concentration in the unsaturated zone. Measuring this concentration gradient is in theory a much more direct way of assessment than calculating it from solid phase data. The uncertainties with regard to the application of the partition theory support this opinion. For the same reason the soil air concentrations should not be used to calculate solid or liquid phase concentrations. As an indication or as a tool for mapping the dispersion of a pollutant the method can be very useful.

5.5.4.2 Sample collection. An important factor during collection of a soil air sample for quantitative analysis is that the sample is coming from a well-defined spot in the soil. The maximum sample size is therefore limited. This problem can be partly overcome by using special sampling and analysis techniques, such as thermal desorption. With this technique relatively low detection limits are possible with only a very small amount of sample. The lowest limits of detection will, however, be significantly higher than with regular air sampling.

In the method used by Fast *et al.* (1987) for the sampling of non-polar volatile organics, a funnel is brought into the soil at the desired depth. To achieve this, a hole is made using an auger with a diameter equal to the diameter of the funnel. A septum is put in the neck of the funnel and through this septum is fed a supply tube of stainless steel or Teflon with a small diameter. After installation in the soil, the hole is filled tightly with soil and the soil system is allowed to stabilize for some days. Next the sample is taken by drawing air through the supply tube using a pump or syringe. The air is led over an adsorption tube filled with Tenax. This tube is cooled and transported to the laboratory. Analysis takes place after thermal desorption of the compounds.

When using this technique, losses of compounds may occur as a result of condensation or sorption by the tube material. These losses are difficult to quantify. A solution for this problem could be to omit the use of the supply tube and to bring the adsorption tube itself into the soil. This implies much more sophisticated technical facilities, however.

If the method is used for screening purposes only, the application of a portable gas chromatograph can be very helpful, and makes it possible to measure compounds directly in the field and to trace spots with high concentrations of pollutants. In this way the pollution on a site can be mapped indicatively. Instead of a funnel, a simple hollow probe might be used, which is driven into the soil using a hammer (Hatayama, 1986). However, this enlarges the possibility for short-cut air streams along the wall of the tube.

5.5.4.3 Analysis. The methods of analysis are identical to the methods of analysis of solid and liquid phase sample (see section 5.5.2.5).

5.6 Soil pollution with VOCs: human health risks

5.6.1 Introduction

In theory soil pollution with VOCs could pose a risk for human health. Whether a risk actually exists depends on the possibility of exposure of humans to the pollutants and the health effects that may be the result of this exposure.

Exposure of humans to a soil pollutant can take place by means of several routes. An overview of these routes of exposure is given in Figure 5.7. The level of exposure depends on the distribution of the compounds over the different soil phases but also, for example, on the depth at which the pollutant is present in the soil. Exposure to VOCs depends to a large extent on the mobility of the compounds in the soil system. The mobility in soil is

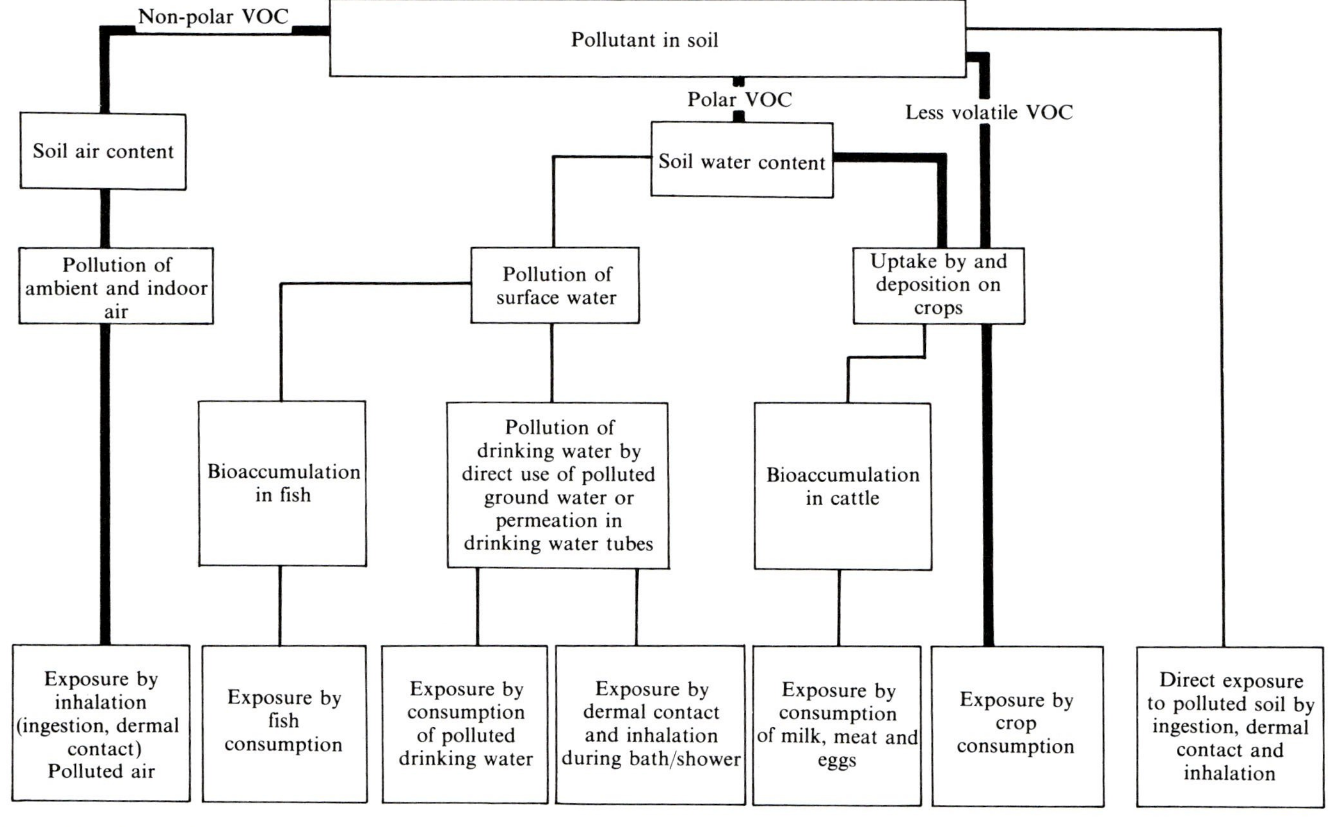

Figure 5.7 Routes of exposure of humans to soil pollutants (see sections 5.6.1 and 5.6.3). For highly volatile, non-polar VOCs inhalation of polluted air is the dominant route. For less volatile and more polar VOCs crop consumption is most important.

determined almost completely by the mobility in the liquid and air phases.

As stated in section 5.4.1 compounds can move upward as a result of diffusion in the soil air or convection with upward moving water or air. After reaching the surface of the soil, such compounds may volatilize and pollute the ambient or indoor air.

Downward transport of a compound in the soil system may be the result of convection with downward moving (leaching) water or transport as pure solute. Ultimately the pollutant will reach the ground water, where it can solve or form a floating layer. In both cases the movement of the pollutant depends on the local hydrology. Use of ground water as a raw material for potable water or even as a direct source of drinking water is a second important route of exposure.

Vertical transport with ground water may spread the pollutant over a wider area. This may lead to a secondary pollution of spots that were initially clean.

Consumption of crops grown on polluted soil or irrigated with polluted ground water, or consumption of meat or milk from cattle grazed on a polluted spot or drenched with polluted water is a third exposure route. Finally, direct contact with polluted soil or ground water can lead to ingestion of contaminated soil particles or dermal uptake of pollutants.

The dominant route of exposure is determined by the properties of the compound and the local situation. For non-polar VOCs the most likely dominant route is volatilization/inhalation. Polar VOCs will volatilize less easily as a result of their affinity for the liquid phase. For these compounds the ingestion of polluted crops and/or drinking water could be more important. It should also be noted that exposure can take place by several routes simultaneously, each route adding to the total exposure.

Van den Berg (1991) has given a comprehensive model for the calculation of total human exposure to a great number of pollutants, including VOCs. The aim of this model, called 'Csoil', is to calculate human toxicologically based standard values for soil pollutants. These so called C-standard values have a legal significance in The Netherlands when reclaiming polluted areas. The starting point of the C-standard values is a human *tolerable daily intake* (TDI) value, which has been assessed by Vermeire *et al.* (1991) for a number of priority pollutants on the basis of toxicological data.

The TDI value is based on a defined *maximal tolerable risk* (MTR). This is defined as the amount of a compound, expressed as mg/kg body weight at oral exposure, to which people can be exposed without adverse health effects. In the case of genotoxic carcinogens the MTR level is defined as the amount of a compound, expressed as mg/kg body weight for oral exposure or as an air concentration for exposure via the respiratory tract, which causes one additional death in 10^4 people with life-long exposure (Vermeire *et al.*, 1991). When the C-standard value is related to an indoor air concentration

Table 5.3 Tolerable daily intake (TDI) values and concept C-standard values for a number of VOCs. Sources: Vermeire *et al.* (1991), van den Berg (1991) and van den Berg *et al.* (1993).

Compound	TDI (mg/kg body weight)	Concept human toxicological C-standard value (mg/kg) in standard soil with $f_{oc} = 0.02$
Benzene	0.0043	1.1
Toluene	0.43	340
Ethylbenzene	0.136	50
m-Xylene	0.01	26
Styrene	0.077	153
Heptane	3.1	3.5
Phenol	0.06	46
p-Cresol	0.05	4.7
1,2-Dichloroethane	0.014	3.9
Dichloromethane	0.06	19
Trichloromethane	0.03	8.9
Tetrachloromethane	0.004	0.92
Trichloroethene	0.54	300
Tetrachloroethene	0.016	3.9
Vinylchloride	0.0035	0.078
Chlorobenzene	0.3	530
p-Dichlorobenzene	0.19	1100
Monochlorophenol	0.003	14
Tetrahydrofurane	0.01	0.4

which is higher than an available tolerable concentration in air or an odour threshold for a specific compound, it is corrected downward.

The theoretical basis of the soil transport part of the model is formed by the fugacity theory of Mackay (1979) and the soil transport modelling of Jury *et al.* (1983, 1984a, b). Moreover, equations are given for the calculation of ingestion, dermal exposure, inhalation of both gaseous compounds and soil particles, and consumption of secondary polluted crops and drinking water.

5.6.2 Calculation of human toxicologically based C-standard values

In Table 5.3 TDI-values, assessed by Vermeire *et al.* (1991), and corresponding concept C-standard values, calculated by van den Berg (1991), for the solid phase of the soil are given for a number of VOCs. The table indicates that low C-standard values are found for the chlorinated aliphatic and chlorinated aromatic compounds. The values for the aromatic compounds benzene, toluene, ethylbenzene and *m*-xylene are remarkably high, considering the attention often paid to these compounds in environmental policy.

In van den Berg and Roels (1991) the human toxicologically based C-standard values were combined with ecotoxicological MTR values. This resulted in integrated C-standard values. With regard to VOCs, these integrated C-standard values were lower than the C-standard values, based solely on human toxicology, in only a few cases.

It should be emphasized that the C-standard values are meant as a guideline for detecting potential risk for human health under standard conditions. The existence of any actual risk depends on the use that is made of the polluted location, for example if people are living on the polluted spot or if there are crop-growing activities. If the use of a polluted spot is limited, the actual exposure and health risk will be lower. On the other hand, it is possible that conditions are worse than the standard conditions. The presence of pure solute in the form of a floating layer, for example, will lead to much higher volatilization than is calculated under equilibrium conditions. Furthermore, calculation of the potential risk according to Csoil depends heavily on the reliability of the assessed solid phase concentrations. When the pollutant is distributed homogeneously over a polluted spot a reliable assessment of the mean soil concentration is possible. A heterogeneously distributed pollutant will be much harder to map accurately, and it is often the case that much higher local concentrations are present.

5.6.3 *Potential and actual risk*

If, on a polluted location, concentrations of VOCs are found that imply a potential risk for human health, the actual risk should be evaluated on the basis of both the use that is made of the polluted spot, and the local conditions, If the pollution involves a residential area or a crop-growing area, measuring the actual exposure by means of air measurements and crop analysis will often give the most reliable data to evaluate the actual exposure.

Table 5.4 shows that using the Csoil model of van den Berg (1991), inhalation and crop consumption are by far the most dominant routes of human exposure to VOCs from soil. For the aromatic, aliphatic and chlorinated aliphatic compounds volatilization/inhalation is dominant, while for the more polar compounds and the chlorinated aromatics the consumption of crops is relatively important.

5.6.4 *Validation of exposure modelling and discussion*

Just like the basic theories of Mackay and Jury, the Csoil model is a theoretical model. Most of the parameters that are used in it have a semi-empirical basis. The uncertainty of the results of calculations, applied to specific cases of pollution, are unknown. Validation of the model on a pilot scale and in practical situations is therefore necessary.

With regard to volatilization/inhalation an extensive study has been carried out by Fast *et al.* (1987). In this study the relationship between soil pollution by VOCs and indoor air quality was examined by means of measurements

Table 5.4 Relative contributions of different routes of human exposure to VOCs. Source: van den Berg (1991).

Compound	Inhalation	Crop consumption	Dermal uptake	Consumption of drinking water
Benzene	84	13	2	
Toluene	76	22	1	1
Ethylbenzene	65	30	3	2
m-Xylene	56	38	3	2
Styrene	43	49	5	3
Heptane	98	2	1	
Phenol	15	85		
p-Cresol	9	81	1	9
1,2-Dichloroethane	67	31		2
Dichloromethane	76	22		1
Trichloromethane	79	18		2
Tetrachloromethane	92	7		
Trichloroethene	82	16	1	1
Tetrachloroethene	92	8		
Vinylchloride	99	1		
Chlorobenzene	52	37	4	5
p-Dichlorobenzene	31	62	4	2
Monochlorophenol	7	92		
Tetrahydrofurane	99	1		

at various locations with known soil contamination. Measurements were carried out in 77 houses at 11 locations with different types of soil pollution. In each house the VOC content of the indoor air and the air in the crawl space was assessed. Samples were also taken of the solid phase of the solid and of the soil air under the basement of the house. Air and soil samples were analysed quantitatively for the presence of 45 non-polar VOCs. A number of soil-specific parameters, such as the organic carbon content and the porosity of the soil were also assessed, to enable fitting of models for transport and exposure.

In a limited number of the investigated houses there appeared to be a significant influence of the polluted soil on the indoor air quality. The highest contribution to the indoor air concentrations was found at two locations, where a spill from a fuel station had occurred. Typical concentration patterns in the soil near the gasoline station, and in the crawl space and living room of houses in the neighbourhood of this gasoline station, are shown in Figures 5.8 and 5.9, respectively.

However, in many houses, built on clearly polluted soil, the indoor air quality had not been affected. It appears that when there is no significant contribution from the soil, the concentrations in the crawl space are comparable to those in the ambient air. Figure 5.10 shows the results of the measurements in the air of the crawl spaces and living rooms of 20 reference houses. In Figure 5.11 the mean concentrations in the upper parts of the soil near these houses are shown.

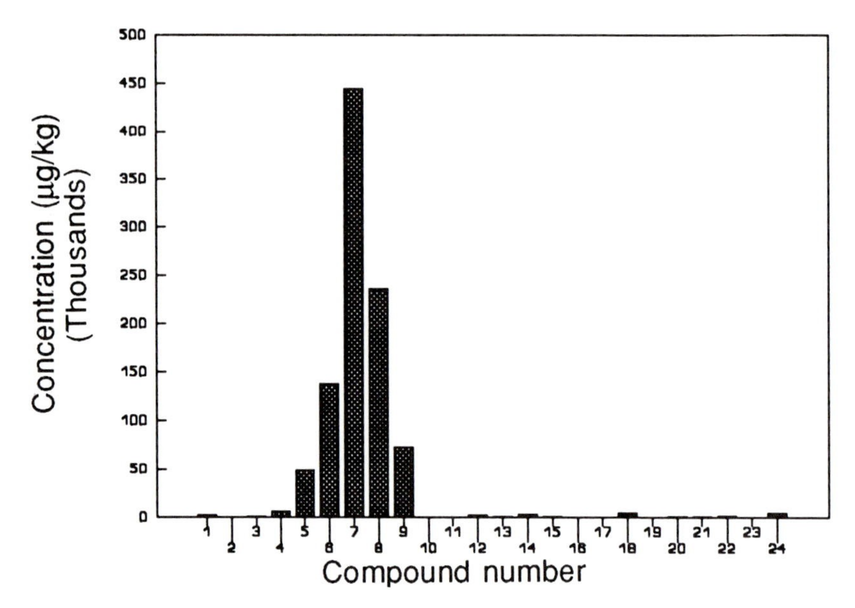

Figure 5.8 Soil concentrations of selected VOCs near a gasoline spill. 1 = *n*-hexane; 2 = *n*-heptane; 3 = *n*-octane; 4 = *n*-nonane; 5 = *n*-decane; 6 = *n*-undecane; 7 = *n*-dodecane; 8 = *n*-tridecane; 9 = *n*-tetradecane; 10 = *n*-hexadecane; 11 = benzene; 12 = toluene; 13 = ethylbenzene; 14 = *m*/*p*-xylene; 15 = *o*-xylene; 16 = isopropylbenzene; 17 = *n*-propylbenzene; 18 = *m*-methylethylbenzene; 19 = *p*-methylethylbenzene; 20 = *o*-methylethylbenzene; 21 = 1,3,5-trimethylbenzene; 22 = 1,2,4-trimethylbenzene; 23 = 1,2,3-trimethylbenzene; 24 = naphthalene. Source: Fast *et al.* (1987).

When the results of the measurements were fitted to the partitioning and transport model discussed in chapter 4, it appeared that no quantitative predictions of indoor air concentrations could be made on the basis of the established solid and air phase concentrations in the soil. There were large differences between the concentrations calculated on the basis of the soil phase concentrations and the concentrations actually measured in the crawl spaces and living rooms. To explain these differences the concentration patterns of the compounds in the solid phase of the soil and the crawl space air were statistically compared. The comparison was made with and without correction for compound characteristics (D, K_{oc} and K_{ow}), soil characteristics such as the *oc* content and porosity, and case-specific factors such ventilation rate of the crawl space and air temperature. The most remarkable result of this comparison was that in most cases correction of the pattern with the K_{oc} and K_{ow} values did not improve, but actually worsened the comparability of the two patterns.

Evaluation of both the quantitative and statistical analyses leads to the conclusion that, at least for non-polar VOCs, transport in the soil cannot adequately be described by the frequently used partitioning and diffusion theory. The discrepancy between the theory and reality is partly a result of

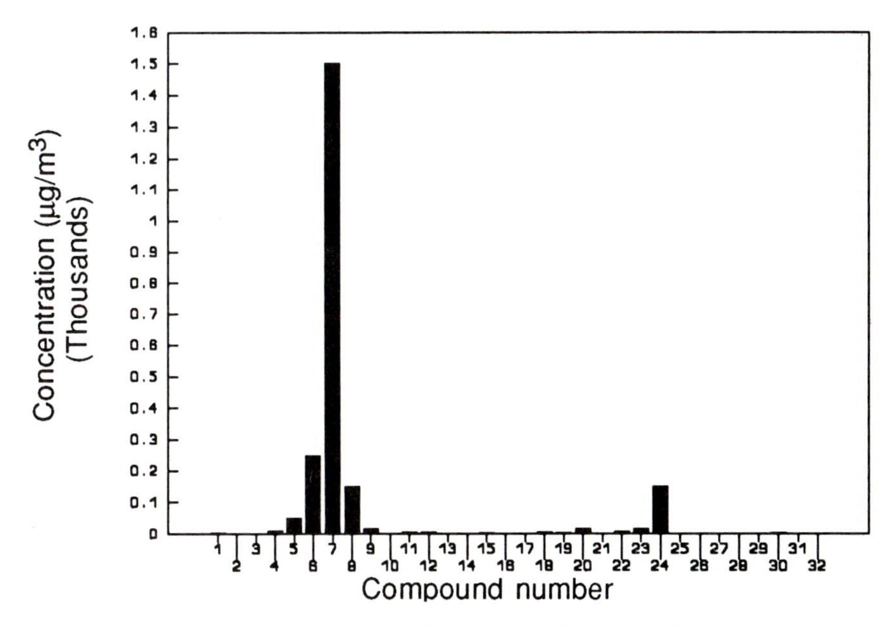

Figure 5.9 Concentrations of selected VOCs in the crawl space of a house near a gasoline spill. Numbering of compounds 1 to 24 as in Figure 5.8. 25 = 1,2-dichloroethane; 26 = chloroform; 27 = 1,1,1-trichloroethane; 28 = 1,1,2-trichloroethane; 29 = trichloroethene; 31 = tetrachloroethane; 31 = *p*-dichlorobenzene; 32 = limonene. Source: Fast *et al.* (1987).

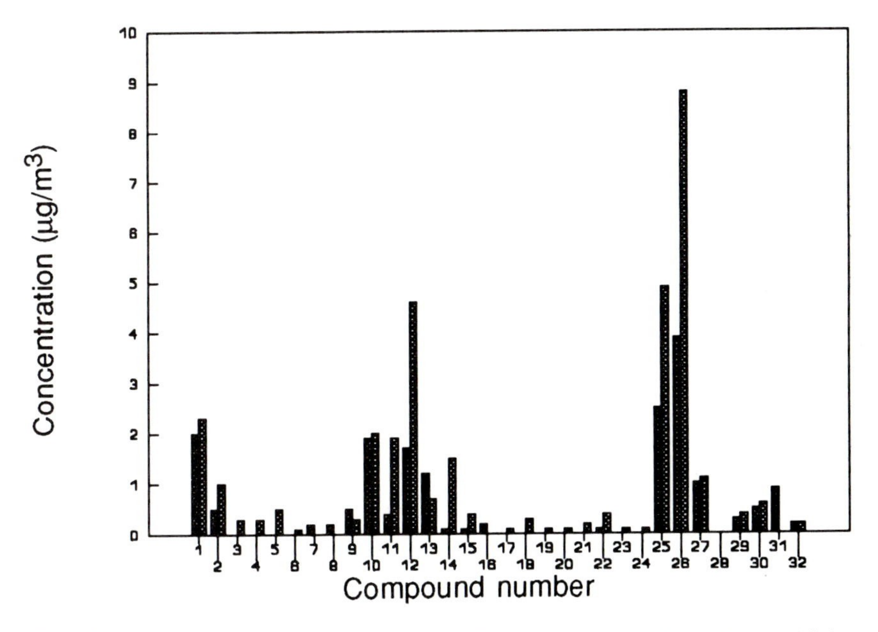

Figure 5.10 Concentrations of selected VOCs in the crawl spaces (left-hand bars) and living rooms (right-hand bars) of 20 reference houses. Numbering of compounds as in Figures 5.8 and 5.9. Source: Fast *et al.* (1987).

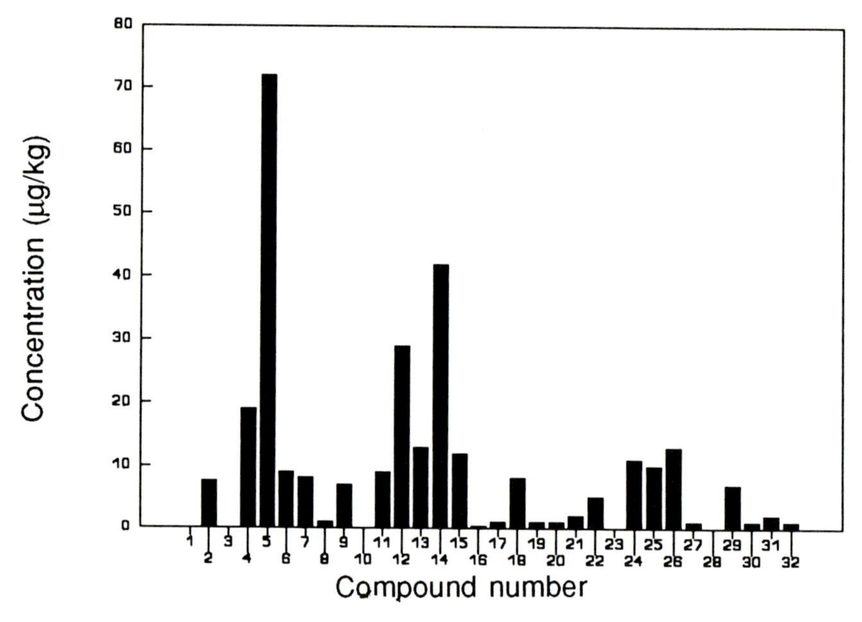

Figure 5.11 Mean soil concentrations of selected VOCs near 20 reference houses. Numbering of compounds as in Figures 5.8 and 5.9. Source: Fast *et al.* (1987).

inconsistencies in the theoretical concepts, as described in section 5.3.2. In addition, practical cases of soil pollution are much more complex than can be described by comparatively simple models. These considerations have serious implications for the applicability of models for soil transport of VOCs. Indeed, the reliability of the modelling is too low to justify its use for calculation of the actual exposure.

In existing houses, measuring the concentrations in the crawl spaces, cellars or living rooms is a much more reliable means of assessing the actual exposure. When the concentrations in the cellars or crawl spaces are higher than the ambient air concentrations, this is a strong indication of an abnormal situation. Soil pollution might then be one of the possible causes.

The assessed concentrations in living rooms can be interpreted by comparison with a data set of well-defined reference measurements, for example those given by Lebret (1985) or Fast *et al.* (1987).

When there are no houses, but only plans to build houses on polluted soil, modelling the potential transport of compounds from the soil to the indoor air is the only possibility. Decision making on this basis, however, will always hold a risk, which is difficult to predict.

References

Begeman, E. (1978) Transport en afbraak van oile in bodem- en grondwater. *Nota 1056*, ICW, Wageningen, The Netherlands (in Dutch).

Berg, R. van den (1991) Blootstelling van de mens aan bodemverontreiniging. Een kwalitatieve en kwantitatieve analyse, leidend tot voorstellen voor humaan toxicologische C-waarden. National Institute of Public Health and Environmental Protection, Bilthoven, The Netherlands (RIVM). *Rapport nr. 725201006* (in Dutch, English summary).

Berg, R. van den and Roels, J.M. (1991) Beoordeling van risico's voor mens en milieu bij blootstelling aan bodemverontreininging. Integratie van deelaspecten National Institute of Public Health and Environmental Protection, Bilthoven, The Netherlands (RIVM). *Rapport nr. 725201007* (in Dutch, English summary).

Berg, R. van den, Dennman, C.A.J. and Roels, J.M. (1993) Risk assessment of contaminated soil: Proposals for adjusted, toxicologically based Dutch soil clean-up criteria. In *Contaminated Soil '93*, pp. 349–364, Kluwer Academic Publishers, Dordrecht.

Bowman, M.C. and Sans, W.W (1985) Partitioning behaviour of insecticides in soil water systems. II. Desorption hysteresis effects. *J. Environ. Qual.* **14** 360–365.

Brown, D. and Flagg, E. (1981) Empirical prediction of organic pollutant sorption in natural sediments. *J. Environ. Qual.* **10** 382–386.

Bruggeman, W.A., Steen, J. van der, and Hutzinger, O. (1982) Reversed phase thin-layer chromatography of polynuclear aromatic hydrocarbons and chlorinated biphenyls. Relationship with hydrophobicity as measured by aqueous solubility and octanol-water partition coefficient. *J. Chromatogr.* **238** 335–346.

Chiou, G.T., Peters, L.J. and Freed, V.H. (1979) A physical concept of soil–water equilibria for nonionic organic compounds. *Science* **206** 831–832.

Cranck, J. (1975) *The Mathematics of Diffusion*, 2nd edn. Clarendon, Oxford.

Duffy, J.J., Peake, E. and Mothadi, M.F. (1980) Oil spills on land as potential sources of groundwater contamination. *Environ. Int.* **3** 107–122.

Fast, T., Kliest, J.J.G. and Wiel, H.J. van de (1987) De bijdrage van bodemverontreiniging aan de verontreiniging van de lucht in woningen. Reeks Milieubeheer **6**. Ministerie van Volkshuisvesting, Ruimtelijke Ordening en Milieubeheer, Leidschendam (in Dutch).

Genuchten, M.Th. van, Davidson, J.M. and Wierenga, P.J. (1974) An evaluation of kinetic and equilibrium equations for the prediction of pesticide movement through porous media. *Soil Sci. Soc. Am. J.* **43** 839.

Gilbert, R.O. (1987) *Statistical Methods for Environmental Pollution Montitoring.* Van Nostrand Reinhold Company, New York, ISBN 0 442 23050 8.

Gilbert, R.O. and Doctor, P.G. (1985) Determining the number and size of soil aliquots for assessing contaminant concentrations, *J. Environ. Qual.* **14-2** 286–292.

Goring, C.A.I. (1972) Agricultural chemicals and the environment. A qualitative viewpoint. In Goring, C.A.I. and Hamaker, J.W. (eds) *Organic Chemicals in the Soil Environment.* Vol. II, pp. 793–863. Marcel Dekker Inc., New York.

Hatayama, H.K. (1986) Site investigation: A review of current methods and techniques. In *Contaminated Soil*, Martinus Nijhoff Publishers, Dordrecht.

Hattemer-Frey H.A., Travis, C.C and Land, M.L. (1990) Benzene: Environmental partitioning and human exposure. *Environ. Res.* **53** 221–232.

Jury, W.A., Spencer, W.F. and Farmer, W.J. (1983) Behaviour assessment model for trace organics in soil: I. Model description. *J. Environ. Qual.* **12** 558–564.

Jury W.A., Spencer, W.F. and Farmer, W.J. (1984a) Behaviour assessment model for trace organics in soil: II. Chemical classification and parameter sensitivity. *J. Environ. Qual.* **13** 567–572.

Jury, W.A., Spencer, W.F. and Farmer, W.J. (1984b) Behaviour assessment model for trace organics in soil: IV. Review of experimental evidence. *J. Environ. Qual.* **13** 580–585.

Karickhoff, S.W. (1981) Semi empirical estimation of sorption of hydrophobic pollutants on natural sediments and soils. *Chemosphere* **8** 833–846.

Karickhoff, S.W., Brown, D.S. and Scott, T.A. (1979) Sorption of hydrophobic pollutants on natural sediments. *Water Res.* **13** 241–248.

Kenaga, E. (1980) Predicted bioconcentration coefficients of pesticides and other chemicals. *Ecotox. and Environ. Safety* **4** 26–38.

Lame, F.J.P. and Bosman, R. (1991a) Leidraad Bodemsanering: Oriënterend onderzoek naar de aard concentratie en omvang van bodemverontreiniging. *Rapport nr R 89/170 MT/TNO*, Delft, The Netherlands (in Dutch).

Lame, F.J.P. and Bosman, R. (1991b) Leidraad Bodemsanering: Oriënterend onderzoek naar de aard concentratie en omvang van bodemverontreiniging. *Rapport nr R 89/171 MT/TNO*, Delft, The Netherlands (in Dutch).

Lebret, E. (1985) *Air Pollution in Dutch Homes. An Exploratory Study in Environmental Epidemiology.* Agricultural University, Wageningen, Department of Air Pollution and Department of Environmental Health.

Leistra, M. (1972) *Diffusion and Adsorption of the Nematicide 1,3-Dichloropropene in Soil.* Dissertation, Agricultural University, Wageningen.

Lewis, G.N. (1901) The law of physico-chemical change. Daedalus. *Proc. Am. Acad.* **37**, 49.

Mackay, D. (1979) Finding fugacity feasible. *Environ. Sci. Technol.* **13** 1218.

Mackay, D. (1980) Solubility, partition coefficients, volatility and evaporation rates. In Hutzinger, O. (ed.) *The Handbook of Environmental Chemistry*, Vol. 2, Part A. Springer-Verlag, Berlin.

Mackay, D. *et al.* (1985) Evaluating the environmental behaviour of chemicals with a level III fugacity model. *Chemosphere* **14** 335–374.

Miller, M.M. *et al.* (1985) Relationships between octanol–water partition coefficients and aqueous solubility. *Environ. Sci. Technol.* **19** 522–529.

Mingelgrin, U. and Gerstl, Z. (1983) Reevaluation of partitioning as a mechanism of nonionic chemicals adsorption in soils. *J. Environ. Qual.* **12** 1–11.

Neumayr, V. (1971) Verteilungs- und transportmechanismen von chlorierten Kohlenwasserstoffen in der Umwelt. *WaBoLu Berichte* **3** 24–40.

Oepen, B. von, Kordel, W. and Klein, W. (1991) Sorption of nonpolar and polar compounds to soils: Processes, measurements and experience with the applicability of the modified OECD-Guideline 106. *Chemosphere* **22** (3–4) 285–304.

Oosterom, W.P., van *et al.* (1988) Sampling and conservation of volatile organic micropollutants in groundwater. *Contaminated Soil '88*, pp. 227–229. Kluwer Academic Publishers, Dordrecht.

Patterson, S. and Mackay, D. (1985) The fugacity concept in environmental modelling. In Hutzinger, O. (ed.) *The Handbook of Environmental Chemistry*, Vol. 2, Part C, pp. 121–140. Springer Verlag, Heidelberg.

Platford, R.F. (1983) The octanol–water partitioning of some hydrofobic and hydrofilic compounds. *Chemosphere* **12** 1107–1111.

Reid, R.C., Prausnitz, J.M. and Sherwood, T.K. (1977) *The Properties of Gases and Liquids*, 3rd end. McGraw-Hill, New York.

Rierder, M. (1990) Estimating partitioning and transport of organic chemicals in the foliage/atmosphere system: Discussion of a fugacity-based model, *Environ. Sci. Technol.* **24** 829–837.

Rogers, R.D., Macfarlane, J.C. and Cross, A.J. (1980) Adsorption and desorption of benzene in two soils and montmorrilonite clay. *Environ. Sci. Techn.* **14** 457.

Siegrist, R.L. and Jenssen, P.D. (1990) Evaluation of sampling method effects on volatile organic compound measurements in contaminated soil. *Environ. Sci. Techn.* **24** 1387–1392.

Thus, J.L.G. and Kraak, J.C. (1985) Comparison of phenyl- and octadecyl-modified silicagel as stationary phase for the prediction of *n*-octanol–water partition coefficients by HPLC. *J. Chromatogr.* **320** 271–279.

Trouwborst, T. (1982) Een vergelijkende analyse van de opname van voc door de means als gevolg van lucht rep. drinkwaterverontreiniging en de factoren die de lichaamsbelasting bepalen. *H_2O* **15** 208.

Vermeire, T.G. *et al.* (1991) Voorstel voor de humaan-toxicologische onderbouwing van C-(toetsings)waarden. National Institute of Public Health and Environmental Protection, Bilthoven, The Netherlands (RIVM). *Rapport nr. 725201005* (in Dutch).

Verschueren, K. (1983) *Handbook of Environmental Data on Organic Chemicals.* 2nd. edn. Van Nostrand Reinhold Company, New York.

6 Future monitoring techniques for VOCs

W.A. McCLENNY

6.1 Automated gas chromatographs (autoGCs)

6.1.1 What are autoGCs and why are they being developed?

In the present context, the term 'automated gas chromatographs' (autoGCs) refers to systems that are beginning to be used in air monitoring network stations to provide frequent, usually hourly, analysis of ambient air for VOCs, over monitoring periods of days to months. Typically, the hardware for an autoGC consists of a benchtop gas chromatograph equipped with: (1) a sample enrichment unit (preconcentrator) for VOCs; (2) non-specific detectors such as the flame ionization detector (FID), electron capture detector (ECD), or photoionization detector (PID); and (3) a data acquisition system for formatting, storage, retrieval and transmission of monitoring data. The preconcentrator can be single- or multi-stage. The most common configuration is the two-stage preconcentrator, as shown in Figure 6.1(a). This consists of: (i) a primary trap (usually a multi-adsorbent trap, sometimes cooled to assist in trapping) for VOCs, through which a known volume of sample air is passed; and (ii) a secondary trap on which VOCs are 'focused' before release into the GC. The primary trap is backflushed with an inert carrier gas to clear oxygen and other non-adsorbed gases and then heated to cause thermal desorption of VOCs to the secondary trap. The secondary trap may consist of a cooled inert surface such as silanized glass wool in a short section of deactivated fused silica capillary column, or a small volume of solid adsorbent in a narrow diameter tube. The objective for using the secondary trap is to deliver VOCs to the column in a small volume so that chromatographic resolution is limited only by the inherent resolving power of the GC column. In a single-stage preconcentrator (Figure 6.1(b)) the VOCs can be condensed onto a very cold trap such as a 5 cm length of standard wall 1/8 inch tubing filled with silanized glass beads, or onto a solid adsorbent trap. For this configuration, cooling the GC column essentially duplicates the function of a focusing trap in a two-stage system, but extravagant use of liquid cryogen is required to cool the entire gas chromatographic oven.

The autoGC system is operated under the control of a microprocessor, which is programmed by the system operator with instruction sets that automatically control the various temperature and pressure zones within the

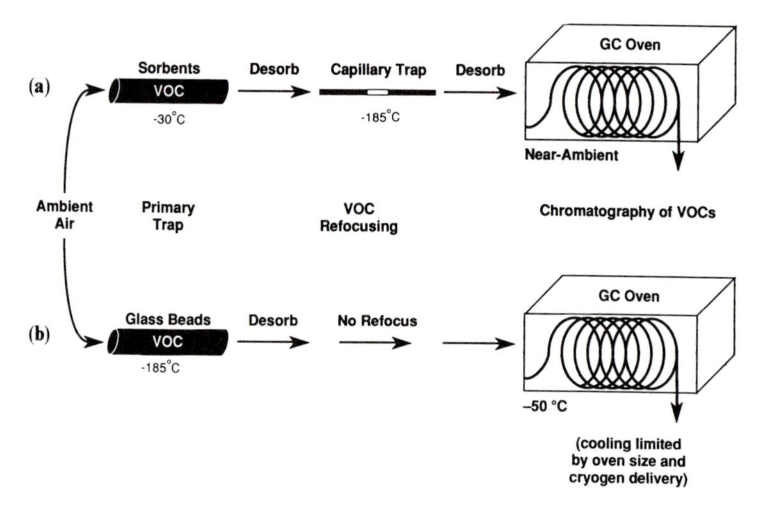

Figure 6.1 Preconcentrator/GC configurations for autoGCs: (a) a two-stage preconcentrator; (b) a single-stage preconcentrator.

GC as well as switches and valves so that the sample is properly routed through the system. Calibration of the system is established by analysis of a known mixture of targeted VOCs in humidified, scientific grade zero air. Retention times of target VOCs are used for identification, and response factors (parts per billion by volume per unit of peak area) are used for quantitation. Internal standards may be added to each sample so that relative retention times and relative response factors can be used. The use of internal standards essentially recalibrates the GC for every run and is the basis for continual quality control of the data. To ensure correct identifications, a sample of air in a container, e.g. a canister, is sometimes returned to the laboratory for analysis by GC–MS. The ultimate result of autoGC monitoring is a data base of target compounds plus unknowns, usually 25 to 100 in number, with a complete listing every hour.

The autoGCs are used to establish the prevalence, mix, and temporal and spatial variability of VOCs. This information is in turn related through mathematical modeling to the presence and location of VOC sources, including local sources such as nearby industrial plants (microscale sources), major urban sources such as the exhausts and evaporative emissions from internal combustion engines (mesoscale sources), and regional sources such as biogenic emissions (macroscale sources). The VOCs can be apportioned among different types of sources through the use of some type of source apportionment modeling. The relative importance of anthropogenic and biogenic hydrocarbons to the ozone producing potential of urban atmospheres is one question being addressed. Since health risk from exposure to individual toxic VOCs has been estimated in many cases, data bases developed with autoGCs can also be used in combination with census data on population distributions to obtain the cumulative risk from exposure to cancer-causing VOCs through the air pathway.

In the USA, requirements for 'enhanced ozone monitoring' as specified in Title I of the Clean Air Act Amendments (CAAA) of 1990 include the monitoring of ozone precursors. Based on the implementation plan as provided by EPA's Office of Air Quality Planning and Standards (OAQPS, 1993), monitoring frequency will be every three hours. AutoGCs are a likely choice for this task, although the use of the canister-based sequential sampling and the associated analytical methodology (see section 6.1.2) is a viable alternative. Guidance on the use of both these approaches is given in a document provided by the USEPA (Purdue *et al.*, 1991).

In the USEPA's 1990 Atlanta Ozone Precursor Study (Purdue, 1991) autoGCs were used for the first time at multiple sites, in this case six sites placed strategically at urban, upwind and downwind sites. The data resulting from this network are being applied to the study of ozone production and to the source apportionment of ozone precursors. The Atlanta Study experience has formed the basis for strategies to support the 'enhanced ozone monitoring' requirements stated in Title I of the CAAA. In particular, the same criteria for site selection are being suggested for urban areas that currently exceed the ozone standard.

Title III of the CAAA requires the monitoring of certain hazardous air pollutants in order to document the residual human health risk to maximally exposed individuals from fugitive emissions and from source-related emissions after emission controls have been placed in operation. A total of 189 entries (compounds or categories of compounds) are listed in Title III as hazardous. Many of the VOCs in this list have been monitored routinely by the USEPA in one of their monitoring networks, either in the Toxic Air Monitoring System (TAMS) (Evans, 1991), which was a network for gaining field experience with monitoring techniques, or in the Urban Air Toxics Monitoring Program (UATMP) (McCallister *et al.*, 1989), which was designed to develop a data base for use in photochemical modeling of ozone production in urban settings. Some of the same compounds are proposed for monitoring in the country-wide Superfund Contract Laboratory Program (CLP) for Air, a program that will likely, although unintentionally, provide a measure of certification for commercial laboratories in the USA. This previous experience has established the concentration levels expected in various scenarios and the precision with which the concentrations of individual compounds must be monitored to distinguish significant departures from background levels. To the extent that the target compound sets for the Title I and Title III requirements overlap, some of the data from the enhanced ozone monitoring will be useful for studies of toxic VOCs.

6.1.2 *Applications for autoGCs*

AutoGCs exist in a number of interesting application-specific system configurations. As mentioned in section 6.1.1, one of the applications is the *monitoring of ozone precursors*, in which the autoGCs are used to collect and separate ozone precursors, that is, a set of C2–C10 hydrocarbons at typical urban levels. Since the light hydrocarbons are quite volatile, one concern has been how to reliably trap and retain these compounds over all or a major portion of an hour.

Preconcentration of C2 compounds by condensation requires temperatures at −185 °C or below. Chrompack[1] offers a system, the Model 9000 gas chromatograph with AutoTCT, that uses reduced temperature for refocusing volatile organics after collection of VOCs from the ambient air onto a multi-adsorbent preconcentrator cooled to −30 °C. Controlled release of liquid nitrogen is used to cool a 10 cm length of tubing, which is then heated rapidly to release organics into a small volume plug of carrier gas to be carried onto the gas chromatographic column. This system has been used as an autoGC in the USEPA's Atlanta Ozone Precursor Study. One improvement needed in this type of system (which uses liquid nitrogen to cool the gas chromatographic column also) is the elimination or reduction in usage of liquid nitrogen. Not only is the liquid nitrogen a continuing cost item, but its delivery and handling is inconvenient.

The use of a prototype closed cycle cooler has recently been reported (Smith, 1991). The cooler is based on a standard refrigeration cycle with a helium working gas. A section of deactivated fused silica capillary column is threaded through a narrow metal tube and placed in close proximity to the cold head of the cooler. Temperature cycling is achieved with the continual operation of the cooler and the intermittent resistive heating of the tube. After an initial cooldown of the interface between the closed cycle cooler head and the column, the system can be cycled in temperature between 135°C and −185°C within a 15 min period. A conceptual drawing of a proposed system using this approach is shown in Figure 6.2.

The use of liquid nitrogen has been eliminated entirely in a system designed by Perkin-Elmer Ltd[2] of the United Kingdom. The design is an adaptation of their Model 400 automated thermal desorber (for adsorbent-filled tubes) to direct sampling of the ambient air. The system is a single-stage system with condensation of VOCs occurring on the adsorbent-filled trap held at −30°C (by the use of a multi-stage thermoelectric cooler). The VOCs are rapidly desorbed (40°C/sec) directly from the trap onto a fused silica capillary tube and thereby onto the GC column. A preliminary separation of the ozone precursors occurs on a single column (BP-5 column) with the early eluting compounds through benzene being diverted to a second column (Al_2O_3

[1]For details of equipment manufacturers, see p. 264.

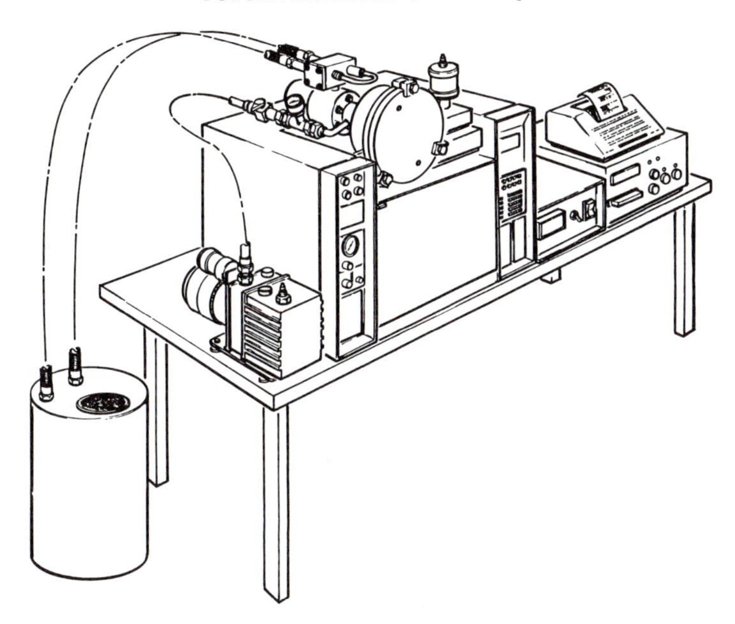

Figure 6.2 Conceptual drawing of one type of autoGC equipped with a closed cycle cooler.

PLOT column); later eluting compounds are diverted to a deactivated fused silica column. Twin flame column ionization detectors are used. Chromatographic separation of the ozone precursors (hydrocarbons) with this arrangement is shown in Figure 6.3.

A system offered by Dynatherm[3] avoids the use of any cold surface in a two-stage system design using solid adsorbent primary and secondary (focusing) traps. Depending on the composition of the traps, target gas sets can be adsorbed at ambient or above ambient temperatures, although an adsorbent combination for 100% retention and release of C2 hydrocarbons has not yet been found. By avoiding procedures typically used for water vapor removal, the Dynatherm approach ensures the survival of polar VOCs.

A second application for autoGCs is the *monitoring of toxic VOCs*, the lightest target VOC being methyl chloride. This application is to identify concentration hotspots and to thoroughly characterize the VOCs in terms of the occurrence, duration and intensity of pollution events. The data base is used to respond to odor complaints from concerned citizens, to approximate the concentrations of toxic compounds across an urban area (Buxton and Pate, 1991), or to identify likely local (< 5 km radius) sources of the VOCs (McClenny *et al.*, 1989). For this last application, identification of a specific source is best done with two monitoring locations. This type of scenario is shown schematically in Figure 6.4 using data from the 1989 Delaware SITE (Superfund Innovative Technology Evaluation) Study (McClenny *et al.*, 1992). In this case the trajectories of air masses carrying two source-related

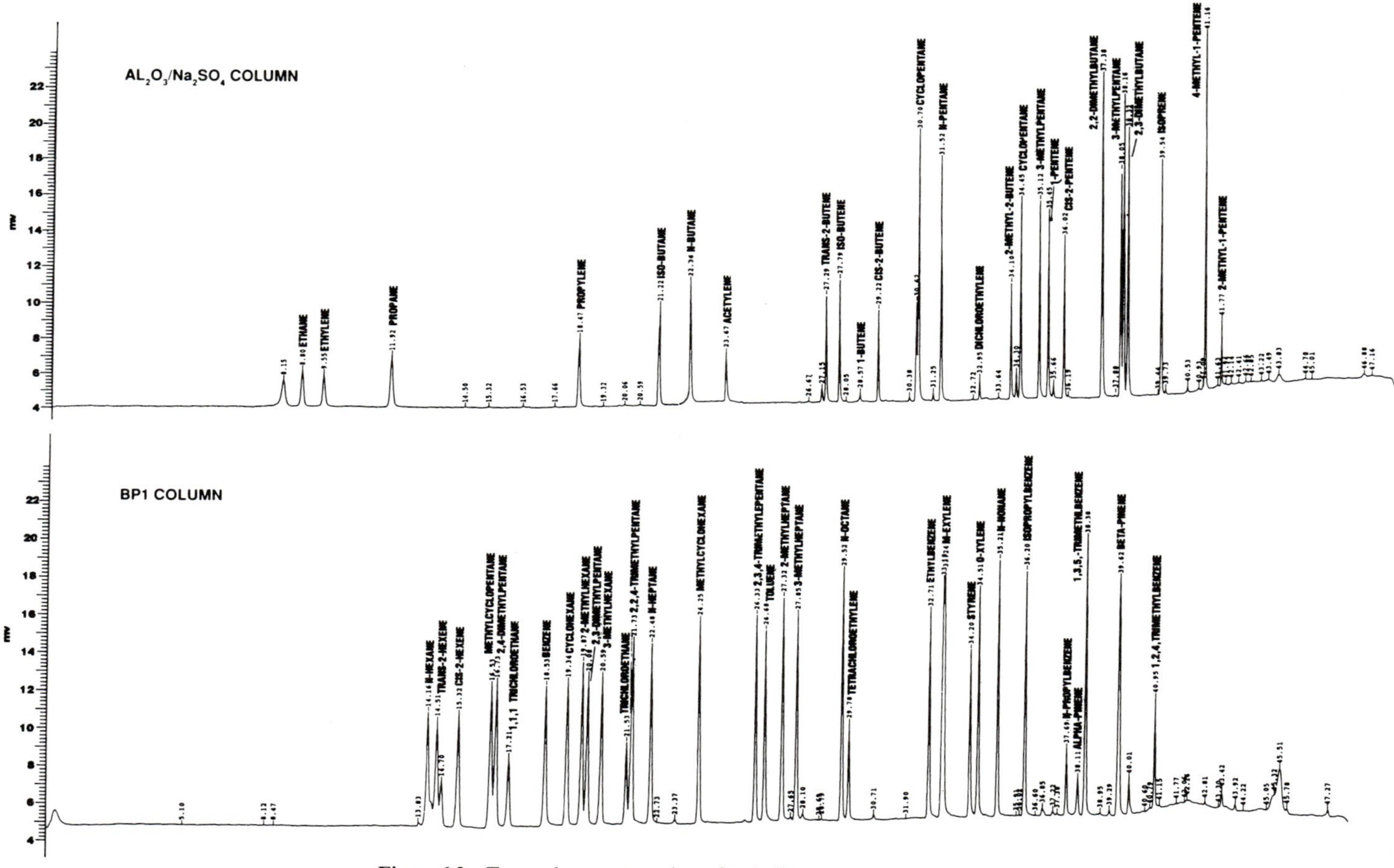

Figure 6.3 Two-column separation of volatile ozone precursors.

Figure 6.4 Trajectories of air masses carrying source-related compound groups.

compound groups, as identified by the interpretation of autoGC data from the site marked P4 and auto-sampler data from the site marked R1, were calculated based on a simple two-dimensional wind field inferred from a wind speed and direction sensor placed at the P4 site. Both trajectories appear to cross the same large industrial complex.

The first autoGC used by the USEPA for toxic VOCs is described in the open literature (McClenny *et al.*, 1984). This system is a modification of a Nutech Inc.[4] scheme for reduced temperature trapping and uses controlled release of liquid nitrogen directly to cool the trap. The system is often used to determine the target set of VOCs referred to as the TO-14 list. This list is a selected list of compounds, mostly non-polar, as described in the USEPA's *Compendium of Methods for the Detection of Toxic Organic Compounds in Air* (Winberry *et al.*, 1988). Other systems have been described that use solid sorbents to preconcentrate VOCs, followed by thermal desorption directly onto the gas chromatographic column. Examples include the combination of the XonTech[5] Model 930 solid adsorbent preconcentrator and the Hewlett Packard Model[6] 5890 GC (Hazlett and Bailey, 1991).

A third application is to *monitor at a plant boundary* in order to identify fugitive emissions and to anticipate possible health and ecological risk in the nearest neighborhood. The comparison of the emission inventories from specific sources, as provided by industry under the requirements of the Superfund Amendments and Reauthorization Act of 1986 (SARA, 1986) Title III, to actual measurements is also a possibility with autoGCs.

6.1.3 *Treatment of water vapor*

Condensation of water vapor from the sample may cause blockage of capillary columns in autoGCs. In two-stage systems such as shown in Figure 6.1(a), this can be eliminated by leaving the primary trap at or near room temperature and forward purging it with a volume of inert gas after sample collection. The hydrophobic nature of some adsorbents is such that water vapor can be preferentially removed from the adsorbent tube by including a carrier gas purge of the adsorbent prior to thermal desorption. Figure 6.5 shows the results of purging a triple-adsorbent tube (the primary tube in a Dynatherm Model ACEM 900) with different volumes of carrier gas, at different tube temperatures. Most of the water vapor can be removed in this

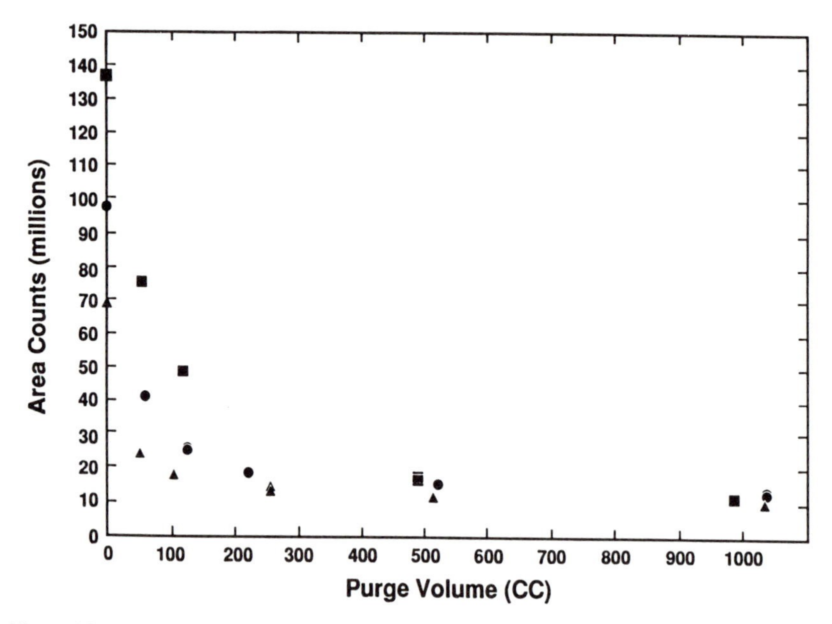

Figure 6.5 Example of water vapor reduction by purging multi-adsorbent after preconcentration of VOCs from ambient air sample: hydrogen response vs. purge volume. (■) 40°C; (●) 55°C; (▲) 65°C.

case without any appreciable loss of toxic VOCs. Considerable additional drying can occur during transfer of the sample to a focusing tube. In the system shown in Figure 6.1(b), a sample dryer of some type is often used to reduce the water vapor; unfortunately this process can also eliminate polar VOCs. There are several other methods for removing water from the sample. Entech[7] uses a short packed column which elutes water vapor first, after which the remaining sample can be backflushed onto a cold trap. Hsu *et al.* (1991) have described a system that allows reduced temperature trapping of VOCs from a large sample volume and eliminates excess co-collected water by purging the sample through water at room temperature prior to retrapping on a solid adsorbent.

Water vapor can also be reduced by trapping excess water on a slightly cold surface placed downstream of a primary trap (Ogle *et al.*, 1992). The 'water control module' offered by Tekmar[8] also uses such a technique.

6.1.4 Data processing to improve the retention time variability and quantitation of VOCs

The autoGCs identify compounds by retention time and use response factors for quantitation. One common scheme to ensure better retention time identifications is to use relative retention times and relative response factors based on marker compounds (or internal standards). Recently an exploratory data processing software package called MetaChrom[9] (Bloemen, 1991) has been developed for improving identification and quantitation of compounds. This procedure is an iterative, interactive process that is carried out on an existing data base of GC run results. A 'design' set of compounds (so-called time-calibrating peaks, e.g. the most prominent members of diluted automobile exhaust) is selected as retention time markers. The retention times of the time-calibrating peaks for each chromatogram in the data set are adjusted to coincide with reference retention times. An algorithm is then used to adjust the retention time of other compounds using the information on adjustments needed for the nearest two time-calibrating peaks. The process of adjustment of retention times can be graphically examined in the form of frequentograms, i.e. a histogram plot of retention time vs. the frequency with which this retention time occurs in all the chromatograms. Application of the algorithm to correct retention times has been demonstrated to improve the comparability of individual chromatograms with the overall data base and to improve resolution of neighboring chromatographic peaks. Adjustment of calibration factors to improve quantitation is approached in a similar manner.

6.1.5 Future directions for autoGCs

The best combination of components for general use of autoGCs has yet to be determined. The units must provide identification and quantitation of individual compounds from a target list while at the same time being cost effective and easy to establish and operate on a continuous basis. Reliable operation over periods exceeding several days has not yet been demonstrated. However, reliability of operation has often depended on the availability of liquid nitrogen for the cooling of the VOC preconcentrators. Elimination of this need by redesign of the systems while maintaining column-limiting resolution of compounds is a main objective in the continuing methods development process. Benchtop mass spectrometers are expected to be used as detectors for autoGCs since typical GC–MS configurations are now no more than a factor of two more expensive than the GC–special-detector systems referred to here. Of particular interest is the Varian[10] Saturn II ion trap detector. In one application, this system is used to preconcentrate VOCs from only 60 ml of ambient air. Because the amount of water vapor in this volume is so small, no dryer is needed and polar VOCs survive the sampling process. The sensitivity of the ion trap is such that detection limits well below 1 ppbv can be obtained.

6.2 Open path monitoring of VOCs

6.2.1 The fundamentals of open path monitoring

Open path monitors (OPMs) systems usually contain two stationary elements: (i) a source/receiver of radiation at one end of the path; and (ii) a retro-reflector at the other end, like the FTIR-based OPM represented in Figure 6.6. In this case the retro-reflector is an array of corner cubes, placed at distances up to 500 m. The beam is treated as a line of radiation attenuated as a function of distance by the atmosphere between the source and receiver. Under favorable conditions, the OPM measures the total number of molecules of specific target gases contained within the open path, this number being, over a typical measurement period of several minutes, a time and spatial average. The measurement of a particular gas is expressed as a gas burden in units of concentration times distance, such as ppmv-m or equivalently ppbv-km. The average concentration is then obtained by dividing the gas burden by distance.

Open path monitoring provides many potential advantages for monitoring of trace gases. Measurements are made *in situ* without the need for air sampling and the attendant considerations of sample integrity. Spectral data for a number of gases are available simultaneously and their concentrations can be inferred in near real-time by comparison with reference absorption spectra. Complete space–time coverage of large air volumes from a central location can indicate concentration trends that apply across large urban or suburban areas. In some cases monitoring paths bridge areas that would be impossible to monitor otherwise. Taken together these advantages make a strong case in favor of the use of open path monitors.

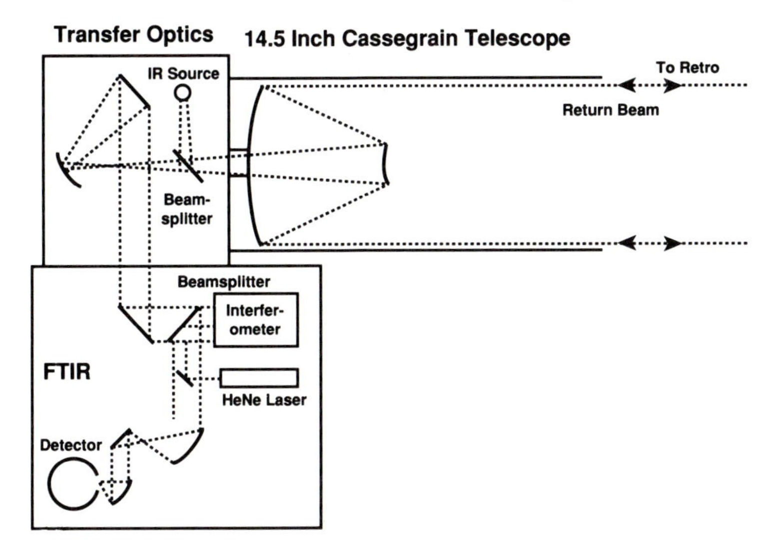

Figure 6.6 Diagram of an FTIR-based open path monitor: an array of corner cube retro-reflectors is used to return the radiation beam.

On the other hand, the transmission spectra taken by OPMs contain the superimposed spectra from several gases and interference problems may result. In the middle infrared, the water vapor and carbon dioxide spectra preclude the use of certain broad-spectral regions for typical pathlengths under a range of expected atmospheric conditions. To observe changes in the transmission spectrum, I, of the radiation source due to absorption by trace gases, the transmission spectrum without the trace gases of interest, I_o, must be known. The relationship between I and I_o at any given wavenumber is then:

$$I = I_o \exp(-\alpha c L) \text{ for transmission; and} \tag{6.1}$$

$$-\ln(I/I_o) = \alpha c L \text{ for absorbance} \tag{6.2}$$

where α denotes the absorption coefficient, c denotes the concentration of the trace gas and L is the pathlength the light travels between the radiation source and the detector. I_o is assumed to contain any contributions from permanent atmospheric gases as well as contributions from trace gases other than the target gas. The target gas concentration is simply obtained by solving equation (6.2) for c using known values of α and a value of L determined by the experimental situation.

One primary difficulty arises related to the determination of I_o. How does one obtain a spectrum without the target gas which represents correctly spectral absorption features in the field spectrum? In the simplest case, a prominent absorption feature of the target gas is not coincident with significant absorption features of any other gas. One way to obtain I_o, then, is to artificially supply the baseline transmission values across the target gas absorption band being considered. The resulting absorbance can then be compared to a library absorbance spectrum of the pure trace gas taken under controlled, known conditions. To improve the precision of the concentration determination, the comparison is made at a number of wavelengths across the absorption band. The gas burden cL of the target gas is then equal to the gas burden used for the library reference times the ratio of the absorbance of the atmospheric burden for the target gas to the absorbance obtained for the reference burden of target gas, i.e.

$$cL = (cL)_r[\ln(I/I_o)/\ln(I/I_o)_r] \quad (6.3)$$

where the subscript 'r' denotes reference spectra. By plotting one absorbance against the other, the ratio of the two as shown in equation (6.3) can be determined by a least squares fitting routine. Slight modifications in the artificial baseline can even be tried in order to find the best possible fit.

The situation can become much more complicated if absorption features of other gases overlap with those of the target gas. In the case of known interferences, a multiple least squares fitting routine has been used, resulting in the determination of all gases involved, although with detection limits that can be significantly less than determined for the individual gases only (Li-Shi and Levine, 1989). If some interferences are from unknown gases, significant residuals in the least squares fitting routine will result and additional data processing is necessary.

Recent articles (Spartz *et al.*, 1989; Spellicy, 1991; Grant *et al.*, 1992) in the open literature relate the characteristics of long path monitors to potential applications in addressing the requirements of the US Clean Air Act Amendments of 1990. Interest in Europe is evident by the widespread use of the OPSIS OPM (see below) and the convening of speciality conferences devoted to remote sensing (*International Symposium on Environmental Sensing*, 1992). In the USA, an international speciality conference on *Optical Remote Sensing and Applications to Environmental and Industrial Safety Problems* was convened in April 1992 to discuss the issues related to remote sensing.

Comparisons between the new generation OPMs and either conventional point monitors such as those in the USA for the 'criteria' pollutant gases, or routine GC–MS identifications of volatile organics have been carried out (Stevens *et al.*, 1990; Russwurm *et al.*, 1991), and comparisons between different OPMs are being discussed (Hudson *et al.*, 1992).

6.2.2 *Systems based on differential optical absorption spectroscopy (DOAS)*

DOAS systems have been reported in the open literature by Platt and Perner (1980) and converted into commercial reality by OPSIS of Sweden[11]. DOAS systems operate with a source of ultraviolet radiation to measure gases such as SO_2, O_3, C_6H_6 and HONO. Many interesting applications have already occurred both in Europe and, to a certain extent, in the USA. The small size of the ultraviolet light source facilitates the collimation of radiation and its transmission over distances up to 10 km. The source/receiver can be centrally positioned and programmed to find different retro-flectors that are located within range along any clear line of sight. Plane and Nien (1992) have described a recent version of a DOAS system that uses a folded light path, a diode array detector (1024 diodes), and computer-controlled operation of the entire instrument package. The remarkable detection limits for NO_3, NO_2 and CH_2O are given as 1 pptv, 0.6 ppbv and 0.8 ppbv, respectively, for a 5 km path length and 4 min integration time. To achieve these detection limits the relative quantum efficiencies of diodes in the array detector must be calibrated individually.

6.2.3 *Systems based on infrared absorption*

6.2.3.1 Gas filter correlation (GFC) techniques. Gas filter correlation techniques are now being used to monitor on-roadway vehicles (Chaney, 1983) and vehicles in inspection and maintenance stations (Lawson *et al.*, 1990). With GFC systems no spectra are obtained, but instead the target gas itself is used as a spectral filter. Usually one, but sometimes two or more, target pollutant gases is/are monitored. The GFC technique is best used for small molecules such as HCl, HF, CO, CH_4, etc., which have well-resolved fine structure in the infrared. Because of the simplicity of their design, GFC instruments are the least costly of OPMs.

6.2.3.2 Fourier transform infrared (FTIR) spectroscopy. FTIR open path monitoring systems are now entering a commercial phase. The versatility of these systems is essentially the same as those using DOAS except that the nature of the radiation source limits the range to 1 km and the transmission of radiation is not affected as much by haze and fog. In the current systems, the radiation beam is modulated before transmission in order to discriminate against background signals, and the transmission and receiving telescopes are separated in order to increase radiation throughput. The absorption coefficients for typical trace gases are generally lower in the infrared than in the ultraviolet, although the number of gases that can be monitored is greater. Water vapor and carbon dioxide absorption restrict the effective use of FTIR systems to certain portions of the infrared spectrum (740–1300 cm^{-1} and 2500–3000 cm^{-1}), although in some cases interference from these two gases can be reduced significantly by scaling and subtracting their reference spectra. This is demonstrated in Figure 6.7 for methane absorption. Figure 6.7(c) shows the result of subtracting a reference spectrum of water vapor from a spectrum taken in the field (Figure 6.7(b)). The resulting spectrum is clearly

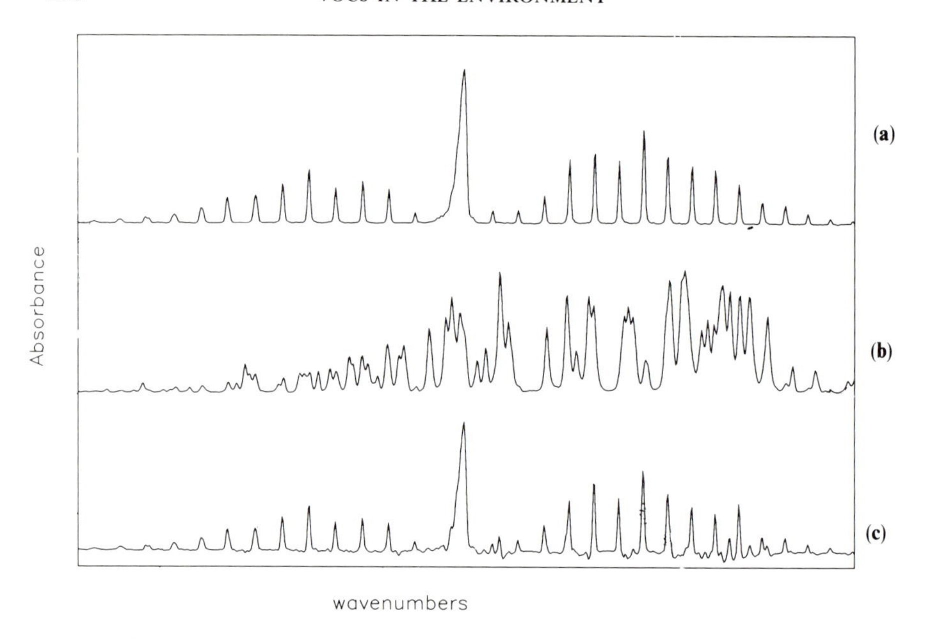

Figure 6.7 Example of recovery of methane spectrum from field spectrum: (a) methane reference spectrum; (b) field spectrum showing water vapor dominance; (c) field minus water vapor spectrum.

that of methane (Figure 6.7(a)). The creation of the reference spectrum for water vapor involved an iterative scaling of the methane spectrum from a reference library to a field spectrum, using two minor methane absorption lines, which were relatively free from water vapor interference. Refinement of subtraction schemes is needed to make this possibility generally useful.

A recent presentation at the Air and Waste Management Association speciality conference on *Optical Remote Sensing and Applications to Environmental and Industrial Safety Problems* (Hunt, 1992) included the design of a scanning FTIR system that is centrally located on top of a plant building to provide line-of-sight monitoring along paths defined by broadband light sources located within and along the boundary of the plant. The system was designed as a stand-alone system to monitor multiple gases, to scan multiple fence lines or areas of a production unit, and to operate continuously with low maintenance. This system, shown schematically in Figure 6.8, is uniquely designed to include a turret-mounted receiving mirror (2) with $\pm 30°$ vertical and 355° horizontal scan ranges. The positioning of the turret by the turret motors (4) is software controlled with automated pointing of the mirror to optimize signal return. The interferometer (5–7) is mounted directly below the receiving mirror, at the end of a hollow, vertical column through which the return radiation passes. A closed cycle cooler is used to cool the MCT

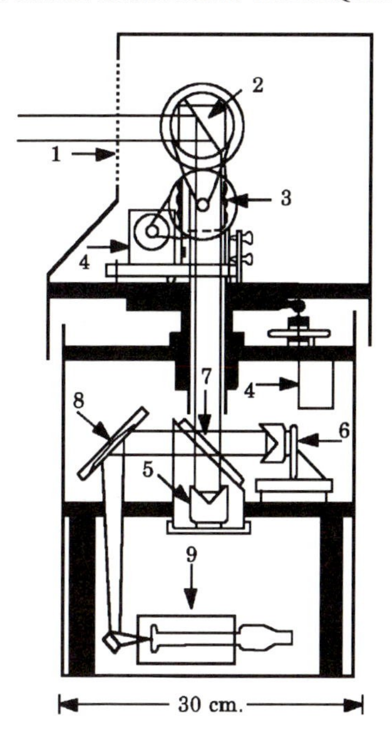

Figure 6.8 Multiple path FTIR-based open path monitor for detection of fugitive emissions. 1, IR window; 2, turret mirror; 3, video camera; 4, turret motors; 5, fixed mirror; 6, moving mirror; 7, beam splitter; 8, parabolic mirror; 9, IR detector.

detector (9). The system is designed along the vertical axis, specifically for the automated detection of fugitive emissions. Its successful demonstration suggests that monitoring from within a plant by plant personnel, using guidelines established by the responsible regulatory agency, could be an effective way to warn of fugitive emissions.

At the same A&WMA conference, a number of simulations were used to predict system results. These were obtained by co-adding reference spectra for target gases, water vapor and carbon dioxide and then simulating complications that could occur in field monitoring. These complications included spectral noise and relative wavenumber displacements of reference and simulated field spectra. Russwurm (1992) presented results obtained with a software program for 'spectral test and evaluation of absorption by least squares' (STEALS) which can simulate the effects of spectral shifts and interfering species in FTIR analysis. Results indicated that measurement of target gas concentrations was especially sensitive to spectral shifts in field spectra with respect to reference spectra.

Hanst and Hanst (1992) have discussed the availability of a spectral reference library of approximately 120 gaseous compounds taken with a sufficiently high spectral resolution to give true spectra shapes for atmospheric pressure, i.e. unaltered by the limited spectral resolution of the FTIR system. Spectra taken with lower resolution FTIR instruments, typically used in field measurements, do not resolve the fine structure of light gases; because of this they show apparent deviations from the logarithmic absorption law (equation (6.1)). However, a new spectral library can be obtained by degrading the high spectral resolution spectra, thereby tailoring the new reference library to the field FTIR. The degrading of a given high resolution reference spectrum is strictly a mathematical operation, obtained by truncating its interferogram. Practically, subtle differences in the optical design of individual FTIR-based systems may still result in slight differences between spectra obtained from field FTIR systems and those obtained mathematically. The importance of this remains to be determined.

6.2.3.3 Differential absorption lidar (DIAL) systems. DIAL systems are pulsed multi-wavelength laser systems. The laser light backscatters from the particulate matter in the ambient air as the short-duration pulse of radiation recedes from the laser source. At any one time the backscattered radiation is being received from a region of length ct, where c denotes the speed of light and t denotes the duration of the pulse. This is like having a retro-reflector moving away from the source at the speed of light. By processing the return signal from wavelengths on and off strong absorption features, the variations of concentration with distance can be obtained, a distinguishing feature of DIAL systems. Nielsen *et al.* (1991) have recently demonstrated the use of a DIAL system for mapping of ozone distributions in the lower atmosphere.

Several papers on DIAL technology were presented at the Air and Waste Management Association speciality conference on *Optical Remote Sensing and Applications to Environmental and Industrial Safety Problems.* These included the use of DIAL techniques by Optech Inc. of Canada[12] to monitor vertical distribution of ozone to altitudes of 50 km using a ground-based, excimer laser system (Smith, 1992). In another paper, the experimental results of flux measurements of atomic Hg performed at Lund Institute of Technology were presented (Edner *et al.*, 1992). For this system the source was a modified, commercially available dye laser. Concentration isopleths were shown and flux determinations were made for Hg emissions from a geothermal energy source.

6.3 Real-time and near real-time monitoring of VOCs

6.3.1 Analyzers equipped with diffusion scrubbers, gas denuders or nebulization/reflux concentrators

A number of improved, selective, gas scrubbing techniques for processing samples of ambient air are being used to obtain highly efficient removal and concentration of trace gases in a continuous collection process with periodic on-line analysis. These techniques replace bubblers and impingers and lead to near real-time monitoring capabilities. Large volumes of air must be processed and concentrated in a small volume of trapping solution to obtain sufficiently high concentrations for analysis. The techniques can be divided into three categories: (i) diffusion scrubbers; (ii) gas denuders; and (iii) nebulization/reflux concentrators.

Dasgupta (1988) has pioneered air monitoring methods development for diffusion scrubbers (membrane-based diffusion denuders) and most recently has disclosed an instrument design which couples a diffusion scrubber with an ion chromatograph equipped with an ultraviolet detector to give analytical results for ambient nitrous acid every six minutes (Vecera and Dasgupta, 1991). Other development of the diffusion scrubber for continuous monitoring includes a hydrogen peroxide monitor (Tanner *et al.*, 1986).

Vecera and Janak (1987) and Cofer and Edahl (1986) have developed nebulization/reflux concentrators for combinations of atmospheric trace gases and liquids in which the gases are soluble, and have demonstrated their use for determination of tropospheric sulfur dioxide and formaldehyde, respectively. In this type of system, sample air is pulled past one end of a capillary tube leading to a reservoir of liquid. Reduced pressure pulls the liquid from the reservoir, through the capillary and into the air stream as a fog of small aerosols. The aerosols efficiently scavenge trace gases that will dissolve in them. The aerosols impact upon a filter or solid surface and collect to eventually drop back into the reservoir of liquid. This process recycles the liquid, which becomes enriched with the target pollutants and is periodically measured.

The use of tubular denuders for collection of atmospheric gases is based on the Gormley–Kennedy equation (Gormley and Kennedy, 1949). Tubular denuders were first used in the 1970s (Durham *et al.*, 1978) to collect gases for air quality monitoring. The most recent efforts use improved designs with annular denuders (Possanzini *et al.*, 1983; Vossler *et al.*, 1988) and compound annular denuders (Lane *et al.*, 1988). Allegrini and De Santis (1989) have reviewed the status of denuder work in the measurement of atmospheric pollutants, including their use for formaldehyde and organic acids. Kelly and Barnes (1991) have demonstrated a simple and efficient glass coil scrubber for collection of gaseous formaldehyde. The detection scheme for this monitor is based on derivatization of formaldehyde in aqueous solution via the Hantzsch reaction, the cyclization of a β-diketone, an amine and an aldehyde to form a highly fluorescent product. Cobb *et al.* (1986) have examined the use of tubular denuders coated with carbon soot as collectors in thermal desorption–gas chromatographic analysis of organic compounds. Coutant *et al.* (1989) demonstrated the removal of trace organics from an air stream by the use of a wax-like coating applied to the walls of a compound annular denuder and have examined mathematically the case in which the efficiency of collection of gases on the wall coating is less than unity.

The extension of these techniques to a number of VOCs is likely to provide, at least for special studies, real-time or near real-time monitoring options. Commercially available instrumentation in this area includes the CEA[13] Model 555 monitor for formaldehyde, which uses a wet-surface glass coil for collection, followed by an analytical scheme based on the reverse pararosaline method and colorimetry detection. This method has been improved with respect to stability and modified to eliminate the use of mercurate in the analytical scheme (Walters, 1983).

6.3.2 *Fast gas chromatography and portable GCs*

Ke *et al.* (1992) have shown the results of analysis of complex mixtures of vapors in ambient air by fast gas chromatography, using isothermal chromatographic conditions. Mixtures of up to 34 components were separated in a total analysis time ranging from 8 to 100 seconds. This represents a 50- to 100-fold reduction in analysis time compared to conventional systems. Carrier gas flow rate was determined by computer modeling (Mouradian, 1989; Puig, 1990) subject to the constraints that chromatographic resolution be maintained while analysis time was reduced. Experimental test results agreed well with the model predictions. A similar experimental system, designed for portability, is shown in Figure 6.9. The system was configured around an HNU Systems, Inc.[14] Model 301 gas chromatograph modified to use a standard HNU flame ionization detector (FID) and an HNU Systems Nordien election capture detector (ECD) with a cell volume of 90 μl.

The results of analysis for a 32 component mixture of organic vapors in air is shown in Figure 6.10. The specific analysis conditions were: 0.25 mm i.d., 10 m column; 0.1 μm methyl silicone stationary phase with hydrogen carrier gas at 85 cm/sec. The oven temperature was 60°C. A second analysis at an oven temperature of 40°C was necessary to resolve the light compounds. Peak identities are given in Table 6.1. Compounds 33 and 34 (1,2,4-trichlorobenzene and hexachlorobutadiene) were indistinguishable from the baseline at both 40°C and 60°C, a fact which illustrates the importance of understanding the limitations of an isothermal system. Under other conditions of temperature and column flow rate, these compounds are resolved. These results for fast GCs indicate the possibilities for rapid detection of trace gases in ambient air monitoring situations where acute health effects may occur. Such systems would alert authorities to toxic releases of gases in the workplace, at the fenceline and/or in the community. In future some of the limitations of the systems, such as the relatively slow response time of detector electronics (100 ms for the ECD in the example given above vs. 5 ms for the FID), will be removed while others such as the restrictions imposed by isothermal operation on chromatographic resolution seem less certain. Limitations in field use will be the need for liquid nitrogen for cold trap operation, and the need for special electrical power supplies for trap heating.

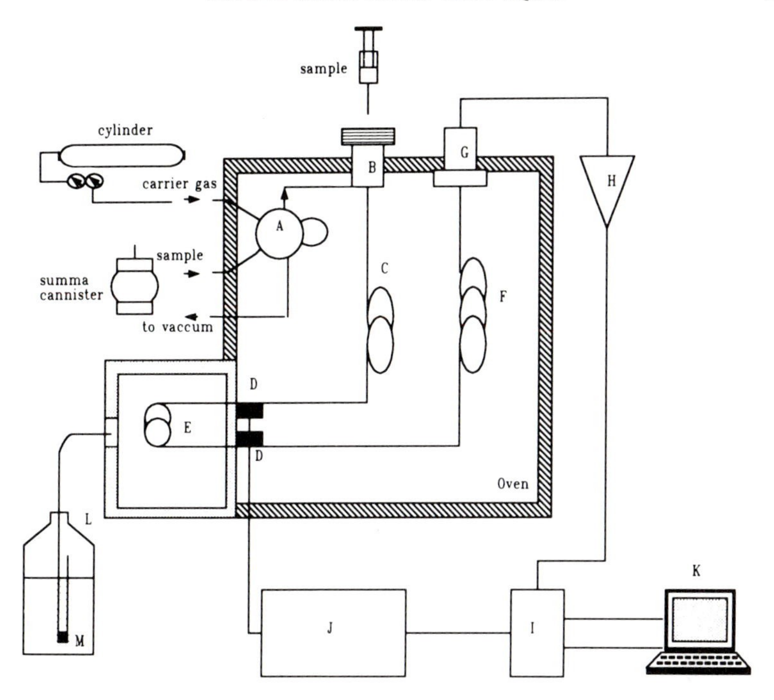

Figure 6.9 Schematic of a portable fast GC system. (A) six-port rotary valve; (B) injector; (C) buffer column; (D) heated copper electrodes; (E) cold trap and chamber; (F) capillary column; (G) detector; (H) fast electrometer; (I) A/D and D/A converter; (J) capacitor discharge power supply; (K) laptop computer; (L) liquid nitrogen Dewar; (M) heater.

Figure 6.9 includes the schematic of a liquid nitrogen system (L) with a heater (M) that can deliver a controlled release of liquid nitrogen in the field.

6.4 Passive sampling devices (PSDs) for VOCs

6.4.1 New developments in PSDs

The advantages of PSDs are compelling. These simple devices are low cost, trivial to deploy and use, and can be developed for any trace gas for which a suitable adsorbent material exists. The characteristics of PSDs have been established in articles by Lewis *et al.* (1985) and Coutant *et al.* (1985). PSDs depend on the diffusion of target gases from the ambient air through a diffusion barrier and onto a collection surface as given mathematically by Fick's first law of diffusion:

$$\text{Rate of sampling} = (AD/L)\,(C - C_o) \qquad (6.4)$$

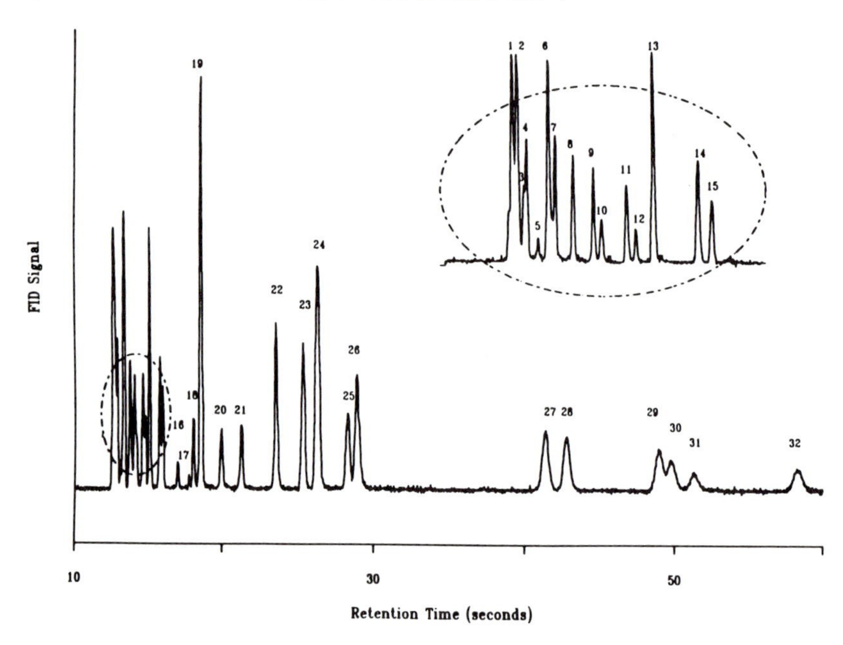

Figure 6.10 Fast gas chromatogram with 60 s total scan time. Peak identities are given in Table 6.1.

where A is the area of the diffusion barrier opening, D is the diffusion coefficient for the target gas, L is the effective length of the diffusion barrier, and C and C_o are the concentrations of the target gas in the ambient air and at the collection surface, respectively. PSDs are widely used in workplace environments and a protocol for evaluation has been developed by NIOSH (Cassinelli *et al.*, 1987).

Surprisingly, the effective sampling rate of the badges can be predicted accurately using a model developed by Coutant *et al.* (1985), a fact that has been established by the comparison of model predictions with experimental results from controlled experiments. Generally, the device works well if the boundary conditions for a constant air sampling rate are established, and if the adsorbing material remains an infinite sink for the target gas, i.e. $C_o = 0$. PSDs for many individual VOCs, SO_2, O_3, NO_2, NO_x, and HCHO have been demonstrated. Extension of this list of compounds is expected.

PSDs can be used either as dosimeters, much like radiation dosimeters, or as 'fencepost' monitors for deployment near suspected sources for extended monitoring periods. Several commercial concerns offer PSD analyzers. One of the most recent additions to this market is the Perkin-Elmer Model ATD 400 (replacing the ATD 50). This unit is designed to automatically analyze up to 50 of the Perkin-Elmer sampling tubes. These tubes can be adapted to use either as a PSD or for pumped sampling. Thermal desorption is used to

Table 6.1 Peak identities for Figure 6.10.

1	Methyl chloride
2	Vinyl chloride
3	Methyl bromide
4	Ethyl chloride
5	Dichloromethane
6	1,1-Dichloroethene
7	1,1,2-Trichloro-1,2,2-trifluoroethane
8	1,1-Dichloroethane
9	*cis*-1,2-Dichloroethene
10	Trichloromethane
11	1,2-Dichloroethane
12	1,1,1-Trichloroethane
13	Benzene
14	1,2-Dichloropropene
15	Trichloroethene
16	*cis*-1,3-Dichloropropene
17	*trans*-1,3-Dichloropropene
18	1,1,2-Trichloroethane
19	Toluene
20	1,2-Dibromoethane
21	Tetrachloroethene
22	Chlorobenzene
23	Ethyl benzene
24	*m,p*-Xylene
25	Styrene
26	*o*-Xylene
27	4-Ethyltoluene
28	1,3,5-Trimethylbenzene
29	1,2,4-Trimethylbenzene
30	Benzyl chloride
31	*m,p*-Dichlorobenzene
32	*o*-Dichlorobenzene
33	1,2,4-Trichlorobenzene
34	Hexachlorobutadiene

release VOCs from the tubes. The VOCs are then trapped onto an adsorbent-filled trap, thermoelectrically cooled to a temperature of $-30°C$, and thermally released again for separation and detection on a gas chromatograph.

6.4.2 Typical applications for PSDs

Successful comparisons of PSDs with conventional monitors are desirable to establish the credibility of this approach. Such comparisons have been carried out for selected VOCs (Mulik *et al.*, 1991). The PSDs for O_3, SO_2 and NO_x have been used recently to establish comparability with real-time monitoring instruments at field station network monitoring sites as part of an effort by the EPA to identify cost-effective monitoring strategies for use at rural monitoring sites. As an example of the use of PSDs, Figure 6.11 shows the results of the comparability testing for O_3 using a coated filter as described by Koutrakis *et al.* (1993) as the trace gas collector. Sample collection periods were one week; the real-time monitoring results were averaged and compared to the average result from three PSDs. Standard deviations for the PSD results are shown. The experiments are being extended to include monitoring sites at other locations where different atmospheric chemistry and weather conditions might change the comparability testing results.

6.5 Sample containment and transport

6.5.1 *Whole air sample containers and solid adsorbents*

In many of the monitoring programs in the USA and in several other countries the use of specially prepared, electropolished, stainless steel canisters has become the method of choice for sampling non-polar VOCs in ambient air (McClenny *et al.*, 1991). In the USA, these containers have, to a large degree, replaced tubes containing Tenax solid adsorbent for ambient air sampling, although Tenax is still used quite extensively in industrial hygiene applications and forms the basis for a significant and well-based effort in the United Kingdom (Brown and Purnell, 1979).

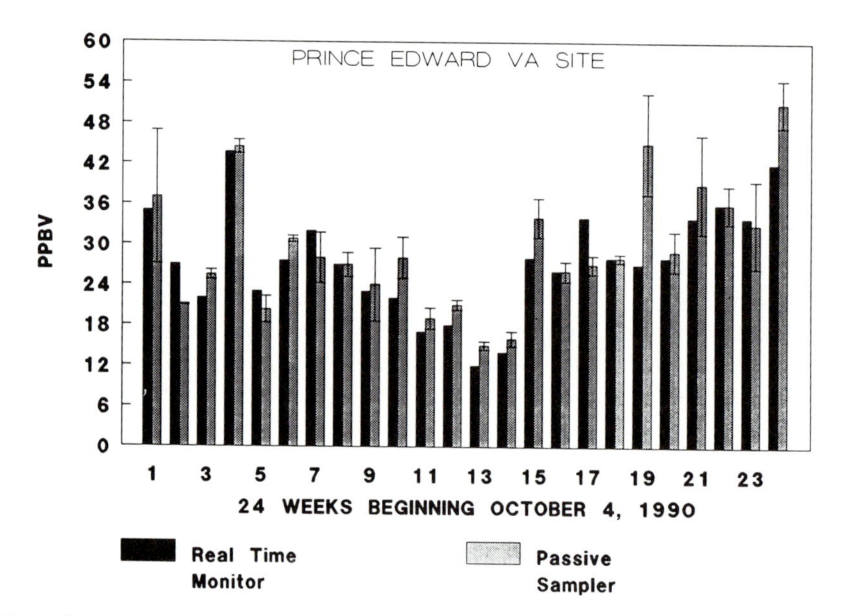

Figure 6.11 Comparability testing of PSDs with ozone photometer real-time monitor; average weekly results.

The popularity of the canisters has spawned a commercial market for canister-based samplers. Samplers are usually designed to include a weatherized housing containing the canisters, a manifold to channel ambient air through the housing at a high flow rate, a sample inlet line attached to the manifold, a flow rate controller in the sample inlet line, an optional in-line pump for pressurizing the canister, and a bi-stable valve to open/close the inlet to the canister. The sampler electronics can be set to initiate and terminate sampling over preselected intervals as long as 24 h. Simple, battery-operated systems with no pump (ambient air is pulled into the canister by vacuum), or canisters equipped with mechanical flow controllers and requiring no power, are also frequently used. Since the canisters are the last element in the sampling train, sample integrity is a potential problem, and quality control procedures to insure against contamination are a must.

A 'sector' sampling strategy for identifying the source of elevated ambient air concentrations of VOCs has evolved with the use of canisters (Pleil *et al.* 1988). The 360 degrees around the sampling location are divided into two sectors: (i) the 'in' sector bracketing the suspected source of emissions; and (ii) the 'out' sector defined as the remainder. The wind direction is used to trigger the routing of ambient air in one of two paths in the sector sampler. One route leads to the 'in' sector canister and the other to the 'out' sector canister. Sector samplers placed around the suspected source identify the common 'in' sector compounds as originating from the suspected source. The wind speed sensor is used to identify still winds and to cause sampling to pause until a persistent wind occurs.

While sector sampling is related to directional variations in concentrations, sequential sampling using a 1 h sampling interval can be used to identify temporal variations in VOC concentrations. This type of system is available commercially (SIS)[15] for fully automated sequential sampling using up to 18 canisters.

6.5.2 *The question of storage stability*

6.5.2.1 Canisters. An experimental effort to establish the stability of all possible combinations of important VOCs at all concentration levels of interest and under all realistic sampling conditions is not practical. However, theoretical work based on the implications of fundamental physicochemical processes is providing guidance to predict the stability of VOCs in canisters (Coutant, 1991). In this work, the potential for physical adsorption as a mechanism for loss was assessed using the principles embodied in the Dubinin–Radushkevich isotherm. Compound/sample specific properties such as polarizability, vapor concentration, temperature, and equilibrium vapor pressure are considered, as well as distinguishing properties of different surfaces. A menu-driven computer model has been developed to predict the adsorption behavior and vapor phase losses in multicomponent systems. The data base for the model contains relevant physicochemical data for 42 VOCs, 19 polar VOCs, and water, and has provisions for inclusion of additional compounds. Figure 6.12 shows the results of calculated recoveries of *o*-xylene from a canister as a function of relative humidity, at sample pressures of 1 atm and 5 atm. A certain level of humidity is required to keep the target gases in the gas phase; the amount depends on the specific target gas and the pressure of the sample. This effect is well established experimentally. Continuation of this work to calibrate the model and to compare model results with the results of controlled experimental work will establish the ultimate utility of this development.

6.5.2.2 Solid adsorbents. Tubes filled with solid adsorbents are usually the first elements in a sampling train, therefore contamination by the in-line elements during sampling is not a problem. As mentioned earlier, most two-stage autoGCs use a solid adsorbent primary trap. The selection of the type and amount of adsorbents, and of sample volume is crucial in order to prevent sample loss due to breakthrough of the target compounds and sample retention on the adsorbent during thermal desorption. In multi-adsorbent tubes diffusion of VOCs from one adsorbent to another during storage is a potential problem. The development of new adsorbents for multi-adsorbent tubes is expected to provide better options for sampling of VOCs in the future.

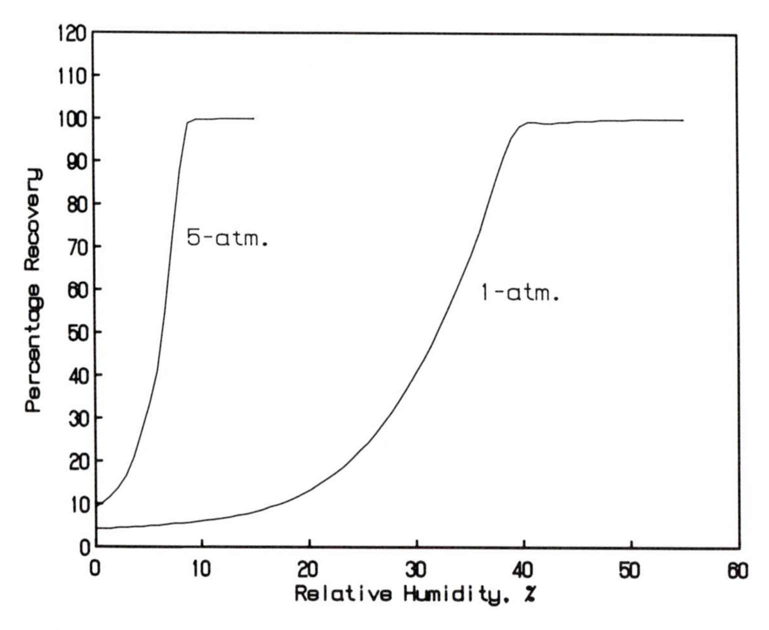

Figure 6.12 Recovery predictions for *o*-xylene (5 ppbv) using model of canister storage stability.

6.6 Other important VOC monitoring method developments

6.6.1 *Direct sampling inlets for GS–MS systems*

Arnold *et al.* (1991) have developed an inlet system for VOC analytical instrumentation that eliminates the need for lines and valves prior to introduction of an air sample onto a GC column. In a technique referred to as transfer line chromatography, the inlet uses the equivalent of a 'Deans' switch to momentarily introduce ambient air onto a short section of column and hence into an ion trap detector. Pleil (1991) has improved upon this approach by cryogenically preconcentrating VOCs from the sample directly onto the analytical column. The use of direct sampling or 'valveless' inlets can be particularly important for polar VOCs, as well as for those with higher molecular weights, to ensure their delivery to the gas chromatographic column.

6.6.2 *Multi-dimensional gas chromatography*

Multi-dimensional gas chromatography is used for separation of VOCs that coelute, e.g. polar compounds that are not resolved by separation on a non-polar column. The simplest form of multi-dimensional gas chromatography is the two-dimensional gas chromatography or 'heart-cutting' technique in which a portion of a chromatogram from a first column is introduced onto a second column with a different type of stationary phase for further separation (Van Hese and Grant, 1988). To make the approach more comprehensive, all of the sample should be subjected to two independent separation mechanisms. One such approach is to use a very high speed second column that successively processes all the sample as it eludes from the first column. The use of on-column thermal modulation as a means to introduce sample onto the second column has been demonstrated (Liu and Phillips, 1989). Such a system has two different columns which are connected in series. Peaks from the first column are chopped into a series of concentration pulses as they are introduced onto the second column. The second column is operated under fast GC conditions in which a chromatogram is generated in a few seconds.

6.6.3 *Time-of-flight mass spectrometry (TOFMS)*

New generation, rapid scanning time-of-flight mass spectrometry (TOFMS), equipped with time-array detection (TAD), may prove to be the optimum type of detection for fast gas chromatographs, making possible high sample throughput rates for VOC analysis while identifying unresolved compounds based on deconvolution of their mass spectra. For example, high speed analysis of complex mixtures by GC–TOFMS–TAD has been demonstrated by McLane *et al.* (1991) to reduce the run time for a mixture of 60 VOCs from 30 min, using standard capillary column gas chromatography, to 80 s, while identifying individual compounds by their mass spectra. This is accomplished by using the rapid mass spectral scan rates possible with TOFMS (Grix, 1991) (up to 5000 scans per second) to provide complete mass spectra in a period much shorter than the elution bandwidths of individual compounds from typical capillary columns. This means that the mass spectrum of each compound always has the same ratio of mass fragments. With conventional mass spectrometers, the mass spectrum is skewed since, for an individual compound, the scan time is comparable to the elution bandwidth so that the concentration of the compound in the ion source will change during the scan time. For scans during the leading edge of an elution peak, the mass spectrum is skewed towards higher masses; for scans during the trailing edge, the skew is towards the lighter masses.

TOFMS instruments have been designed to compensate for the small amount of material available on each scan by: (1) specially designed storage electron ionization sources to store ions between mass spectral scans (Grix *et al.*, 1989); (2) special flight tube designs incorporating an inhomogeneous field ion mirror for mass-independent lateral and longitudinal ion focusing (Grix *et al.*, 1988). Individual mass scans can be co-added to improve signal-to-noise ratios.

Given the scan-time advantage, a set of deconvolution algorithms have been developed to determine the number of compounds present, their actual retention times, and the pure mass spectra of individual compounds (Yefchak, 1991). That is, the chromatographic resolution sacrificed by shortened analysis time is recovered mathematically.

The TOFMS being operated at the Michigan State University, Ann Arbor, MI, USA, where most of the referenced work originated, has the following characteristics: (1) a mass range of 1–1000 units; (2) unit mass resolving power throughout the entire mass range in each transient with a 1 m flight tube and reflectron; (3) acquisition of at least 30 spectra per GC peak; (4) efficient ion storage between extractions; and (5) source conductance adequate to maintain the integrity of narrow elution profiles. The commercial possibilities of TOFMS would seem to be outstanding.

6.6.4 Direct sampling MS–MS

Wise *et al.* (1991) have developed two different types of direct sampling mass spectrometers for use in screening air samples for VOCs. Direct sampling inlets have been demonstrated both with a commercially available ion trap mass spectrometer (ITMS) and with a tandem source glow discharge quadrupole mass spectrometer, operating for detection at ppbv concentration levels. These instruments do not use chromatographic separation, hence the sample analysis time is shortened considerably (< 2 min). The system based on the

ITMS is tandem in time, in that the ITMS is capable of selective ion storage followed in time by multiple stages of collision-induced dissociation. The tandem source glow discharge quadrupole MS is tandem in space, using glow discharge as the ion source (McLuckey *et al.*, 1988) followed by a triple quadrupole MS. Atmospheric pressure ionization (API) sources for triple quadrupole MS have been successfully demonstrated and commercialized in the past. Interesting applications include the study of human breath analysis (Kelly *et al.*, 1989) and the atmospheric chemistry of industrial plumes.

6.6.5 *Microsensor arrays and gas chromatographs*

Several microsensors are currently being used for the detection of VOCs. Included in this list are surface acoustic wave arrays (Jarvis *et al.*, 1991), electrochemical arrays (Penrose *et al.*, 1991), and miniature gas chromatographs (Overton *et al.*, 1989). Sensor arrays are used to distinguish between different gases by the pattern of response of different individual sensors. Arrays are preselected for compound class-specific response. The use of microchip GCs with miniaturized injector and detector sub-assemblies can produce rapid and reliable analysis of VOCs. One example of this is the Michromonitor 200 microchip gas chromatograph, produced by Microsensor Systems, Inc.[16], which can be used with a preconcentrator to achieve ppbv detection levels.

6.7 Conclusions

Sampling and analysis techniques for monitoring VOCs are subject to a rapid evolution spurred by the requirements of new legislation such as the Clean Air Act Amendments of 1990 in the USA. Monitor design is evolving along lines that will permit multi-pollutant characterization (autoGCs), high sample throughput (fast GCs, TOFMS), reliability (canister-based sampling and analysis), ease of use (PSDs), cost-effectiveness (PSDs), and special characterizations such as fugitive emissions monitoring (remote sensing). Emphasis on the use of GC–MS in mobile, field units for definitive monitoring of individual VOCs is expected in the near future. Such systems would determine the prevalence of hazardous VOCs, allow more accurate estimates of human and ecosystem exposures, establish data bases for VOC apportionment to source types, reconcile ambient air concentrations with concentrations predicted from source emission inventories, and identify and characterize individual sources using monitoring strategies such as sector sampling and temporal profile analysis. For personal exposure monitoring, personal sampling devices for VOCs, in some cases coupled with type-of-activity sensors, are now available and are being developed further. Remote sensing monitors are finding useful applications in, for example: (1) the monitoring of on-roadway vehicles; (2) the use of DIAL techniques along with Doppler

LIDAR (for wind vector determination) to estimate industrial emission rates of various chemicals; and (3) the use of DOAS and FTIR-based systems to encompass urban communities and industrial areas with spatial and temporal scanning capabilities.

Equipment manufacturers

1. Chrompack International B.V., Middelburg, The Netherlands.
2. Perkin-Elmer Ltd., Maxwell Road, Beaconsfield, Buckinghamshire HP9 1QA, UK.
3. Dynatherm Analytical Instruments, Inc, PO Box 159, Kelton, PA 19346, USA.
4. NuTech Corporation, 2806 Cheek Road, Durham, NC 27704, USA.
5. XonTech, Inc., 6862 Hayvenhurst Avenue, Van Nuys, California 1406, USA.
6. Hewlett Packard, Avondate, PA, USA.
7. EnTech Laboratory Automation, 950 Enchanted Way #101, Simi Valley, CA 93065, USA.
8. Tekmar Company, 7143 East Kemper Road, PO Box 429576, Cincinnati, OH 45242-9576, USA.
9. Meta Four Software, Meekrap oord 6, 3991 VE Houten, The Netherlands.
10. Varian Instrument Group, Walnut Creek Division, 2700 Mitchell Dr., Walnut Creek, CA, USA.
11. OPSIS AB, Ideon, S-223 70 Lund, Sweden.
12. Optech Incorporated, 701 Petrolia Road, Downsview (Toronto) Ontario, Canada M3J2N6.
13. CEA Instruments, Inc., 16 Chestnut Street, Emerson, NJ 07630, USA.
14. HNU Systems Inc., 30 Ossipee Road, Newton Upper Falls, Massachusetts 02184, USA.
15. Scientific Instrumentation Specialists, PO Box 8941, 815 Courtney St., Moscow, Idaho 834843, USA.
16. Microsensor Systems Inc., Freemont, CA, USA.

References

Allegrini, I and De Santis, F. (1989) Measurement of atmospheric pollutants relevant to dry acid deposition. *Crit. Rev. Anal. Chem.* **21**(3) 237–255.

Arnold, N.S., McClennen, W.H. and Meuzelaar, Henk L.C. (1991) Vapor sampling device for direct short column gas chromatography/mass spectrometry analysis of atmospheric vapors. *Anal. Chem.* **63** 299–304.

Bloemen, H. (1991) Exploratory chromatographic data processing. Paper No. 91-68.4, Air and Waste Management Association, 84th Annual Meeting, Vancouver, BC, June 16–21, 1991.

Brown, R.H. and Purnell, C.J. (1979) Collection and analysis of trace organic vapor pollutants in ambient atmospheres. The performance of a Tenax GC adsorbent tube. *J. Chrom.* **178** 79–90.

Buxton, B.E. and Pate, A.D. (1991) Statistical modeling of spatial and temporal variations for the Atlanta ozone precursor study. Paper 91-68.10, Air and Waste Management Association, 84th Annual Meeting, Vancouver, BC, June 1991.

Cassinelli, M.E., Hull, R.D., Crable, J.V. and Teass, A.W. (1987) NIOSH protocol for the evaluation of passive monitors. In Berlin, A., Brown, R.H. and Saunders, K.J. (eds) *Diffusion Sampling: An Alternative to Workplace Air Monitoring*, pp. 190–202. Royal Society of Chemistry, London.

Chaney, L.W. (1983) The remote measurement of traffic generated carbon monoxide. *JAPCA* **33** 220–222.

Cobb, G.P., Hua, K.M. and Braman, R.S. (1986) Carbon hollow tubes as collectors in thermal desorption/gas chromatography analysis of atmospheric organic compounds. *Anal. Chem.* **58** 2213–2217.

Cofer, W.R. III and Edahl, R.A. Jr. (1986) A new technique for collection, concentration and determination of gaseous tropospheric formaldehyde. *Atmos. Environ.* **20**(5) 979–984.

Coutant, R.W. (1991) Theoretical evaluation of stability of volatile organic chemicals and polar volatile organic chemicals in canisters. Final report on EPA contract No. 68-DO-0007 to Battelle Columbus. Available from the Atmospheric Research and Exposure Assessment Laboratory, Research Triangle Park, NC 27711, USA.

Coutant, R.W., Callahan, P.J., Kuhlman, M.R. and Lewis, R.G. (1989) Design and performance of a high-volume compound annular denuder. *Atmos. Environ.* **23**(10) 2205–2211.

Coutant, R.W., Lewis, R.G. and Mulik, J.D. (1985) Passive sampling devices with reversible adsorption. *Anal. Chem.* **57** 221–223.

Dasgupta, P.K., Dong, S., Hwang, H., Yang, H.-C. and Genfa, Z. (1988) Continuous liquid-phase fluorometry coupled to a diffusion scrubber for the real-time determination of atmospheric formaldehyde, hydrogen peroxide and sulfur dioxide. *Atmos. Environ.* **22** 949–964.

Durham, J.L., Wilson, W.E. and Bailey, E.B. (1978) Application of an SO_2-denuder for continuous measurement of sulfur in submicrometric aerosols. *Atmos. Environ.* **12**(4) 883–886.

Edner, *et al.* (1992) Remote sensing of pollutant fluxes using differential absorption LIDAR. In *Proceedings of the International Speciality Conference on Optical Remote Sensing and Applications to Environmental and Industrial Safety Problems*, Houston, TX, 5–8 April, 1992.

Evans, G.F. (1991) Final report on the operations and findings of the Toxic Air Monitoring System (TAMS). Draft Report, Atmospheric Research and Exposure Assessment Laboratory (AREAL), US Environmental Protection Agency, Research Triangle Park, NC 27711, May 1991.

Gormley, P.G. and Kennedy, M. (1949) Diffusion from a stream flowing through a cyclindrical tube. *Proc. R. Irish Acad. Sect. A.* **52** 163–169.

Grant, W.B., Kagann, R.H. and McClenny, W.A. (1992) Optical remote measurement of toxic gases. *J. Air Waste Manage. Assoc.* **42**(1) 18–30.

Grix, R. (1991) A time-of-flight mass spectrometer for high-speed chromatography. In *Proceedings of the 39th ASMS Conference on Mass Spectrometry and Allied Topics*, Nashville, TN, May 19–24, 1991.

Grix, R., Kutscher, R., Li, G., Gruner, U. and Wollnik, H. (1988) A time-of-flight mass analyzer with high resolving power. *Rapid Commun., Mass Spectrom.* **2**(5) 83–86.

Grix, R., Gruner, U., Li, G., Stroh, H. and Wollnik, H. (1989) An electron impact storage ion source for time-of-flight mass spectrometers. *Int. H. Mass Spectrom. Ion Processes.* **93** 323–330.·

Hanst, P.L. and Hanst, S.T. (1992) Infrared measurement of environmental gases. In *Proceedings of the International Speciality Conference on Optical Remote Sensing and Applications to Environmental and Industrial Safety Problems*, Houston, TX, 5–8 April, 1992.

Hazlett, J.M. II, and Bailey, R.E. (1991) A 24-hour air toxics monitoring station—Tracing air pollutants—700 times a month. Hewlett-Packard, *PEAK*, Number 3/1991.

Hsu, J.P., Miller, G. and Moran, V. III (1991) Analytical method for determination of trace organics in gas samples collected by canister. *J. Chrom. Sci.* **29** 83–88.

Hudson, J. *et al.* (1992) An overview assessment of the intercomparability and performance of multiple open-path FTIR systems as applied to the measurement of air toxics. *Proceedings of the International Speciality Conference on Optical Remote Sensing and Applications to Environmental and Industrial Safety Problems*, Houston, TX, April 5–8, 1992.

Hunt, R.N. (1992) Continuous monitoring of atmospheric pollutants by remote sensing FTIR. In *Proceedings of the International Speciality Conference on Optical Remote Sensing and Applications to Environmental and Industrial Safety Problems*, Houston, TX, 5–8 April, 1992.

International Symposium on Environmental Sensing, 22–26 June 1992, Congress Hall Alexanderplatz, Berlin, Germany.

Jarvis, H.L., Wohltjen, J. and Lint, J.R. (1991) Surface acoustic wave (SAW) personal monitor for toxic gases. In *Proceedings of the Second International Symposium on Field Screening Methods for Hazardous Wastes and Toxic Chemicals*, Las Vegas, NV, February 12–14, 1991, pp. 73–82.

Ke, H., Levine, S.P. and Berkley, R. (1992) Analysis of complex mixtures of vapors in ambient air by fast-gas chromatography. *J. Air Waste Manage. Assoc.* **42** 1446–1452.

Kelly, T.J. and Barnes, R.H. (1991) Development of real-time monitors for gaseous formaldehyde. *USEPA Report EPA/600/S3-90/088*, January 1991. Available from W. McLenny, USEPA, Research Triangle Park, NC 27711, USA.

Kelly, T.J., Kenny, D.V., Spicer, C.W. and Sverdrup, G.M. (1989) Continuous analysis of human breath using atmospheric pressure ionization MS/MS with a novel inlet design. In *Proceedings of the 1989 EPA/A&WMA International Symposium on Measurement of Toxic and Related Air Pollutants*, Raleigh, NC, May 1989, pp. 478–483.

Koutrakis, P., Wolfson, J.M., Bunyaviroch, A., Froehlich, S.E., Hirano, K. and Mulik, J.D. (1993) Measurements of ozone using a nitrite coated filter. *Anal. Chem.* **65**(3) 209–214.

Lane, D.A., Johnson, N.D., Barton, S.C., Thomas, G.H.S., and Schroeder, W.H. (1988) Development and evaluation of a novel gas and particle sampler for semivolatile chlorinated organic compounds in ambient air. *Environ. Sci. Technol.* **22**(8) 941–947.

Lawson, D.R., Groblicki, P.J., Stedman, D.H., Bishop, G.A. and Guenther, P.L. (1990) Emissions from in-use motor vehicles in Los Angeles: A pilot study of remote sensing and the inspection and maintenance program. *J. Air Waste Manage. Assoc.* **40** 1096–1105.

Lewis, R.G., Mulik, J.D. and Coutant, R.W. (1985) Thermally desorbable passive sampling device for volatile organic chemicals in ambient air. *Anal. Chem.* **57** 214–219.

Li-Shi, Y. and Levine, S.P. (1989) Evaluation of the applicability of Fourier transform infrared (FTIR) spectroscopy for quantitation of the components of airborne solvent vapors in aïr. *Am. Ind. Hyg. Assoc. J.* **50**(7) 360–365.

Liu, Z. and Phillips, J.B. (1989) High speed gas chromatography using an on-column thermal desorption modulator. *J. Microcol. Sep.* **1–5** 250.

McAllister, R.A. *et al.* (1989) Non-Methane Organic Compound Monitoring Program, Final Report, Volume II: Urban Air Toxics Monitoring Program. *EPA-450/4-89-005*, Radian Corporation, Research Triangle Park, NC, USA, April 1989.

McLane, R.D. *et al.* (1991) High speed analysis of complex mixtures by gas chromatography/time-of-flight mass spectrometry. In *Proceedings of the 39th ASMS Conference on Mass Spectrometry and Allied Topics*, Nashville, TN, May 19–24, 1991.

McClenny, W.A., Pleil, J.D., Holdren, M.W. and Smith, R.N. (1984) Automated cryogenic preconcentration and gas chromatographic determination of volatile organic compounds. *Anal. Chem.* **56** 2947–2951.

McClenny, W.A., Oliver, K.D. and Pleil, J.D. (1989) A field strategy for sorting volatile organics into source-related groups. *Environ. Sci. Technol.* **23**(11) 1373–1379.

McClenny, W.A. *et al.* (1991) Canister-based method for monitoring toxic VOCs in ambient air. *J. Air Waste Manage. Assoc.* **41**(10) 1308–1318.

McClenny, W.A. *et al.* (1992) Superfund innovative technology evaluation, the Delaware SITE study, 1989. *EPA Report EPA/600/S3-91/071*, January 1992.

McLuckey, S.A., Glish, G.L., Asnao, K.G. and Grant, B.C. (1988) Atmospheric sampling glow discharge ionization source for the determination of trace organic compounds in ambient air. *Anal. Chem.* **60**(20) 2220–2227.

Mouradian, R.F. (1989) *Fast Gas Chromatography for Industrial Hygiene Analysis and Monitoring.* Ph.D. Thesis, University of Michigan.

Mulik, J.D. *et al.* (1991) Using passive sampling devices to measure selected air volatiles for assessing ecological change. *Proceedings of the 1991 EPA/A&WMA International Symposium on Measurement of Toxic and Related Air Pollutants*, Durham, NC, May 1991.

Nielsen, N.B., Uthe, E.E., Livingston, J.M. and Scribner, E.J. (1991) Compact airborne LIDAR mapping of lower atmospheric ozone distributions. Presented at the International Conference on Lasers '91, San Diego, California, 8–13 December 1991.

OAQPS (1993) Revision of the ambient air quality surveillance regulations (40 CFR Part 58) of the United States Combined Federal Register, January 1993. Available from W. Hunt, Office of Air Quality Planning and Standards, US Environmental Protection Agency, Research Triangle Park NC 27711, USA.

Ogle, L.D. *et al.* (1992) Moisture management techniques applicable to whole air samples analyzed by Method TO-14. *Proceedings of the International Symposium on Measurement of Toxic and Related Air Pollutants*, May 3–8, 1992, Durham, NC.

Overton, E.B., Grande, L.H., Sherman, R.W., Collard, E.S. and Steele, C.F. (1989) Rapid and reliable analysis of volatile organic compounds with a field deployable gas chromatograph. In *Proceedings of the 1989 EPA/A&WMA International Symposium on the Measurement of Toxic and Related Air Pollutants*, pp. 13–18. Raleigh, NC, May 1989.

Penrose, W.R., Stetter, J.R., Findlay, M.W., Buttner, W.J. and Cao, Z. (1991) Arrays of sensors and microsensors for field screening of unknown chemical wastes. In *Proceedings of the Second International Symposium on Field Screening Methods for Hazardous Wastes and Toxic Chemicals*, Las Vegas, NV, February 12–14, 1991, pp. 85–89.

Plane, J.M.C. and Nien, Chia-Fu (1992) Differential optical absorption spectrometer for measuring atmospheric trace gases. *Rev. Sci. Instrum.* **63**(3) 1867–1876.

Platt, Y. and Perner, D. (1980) Direct measurements of atmospheric CH_2O, HNO_2, O_3, NO_2, and SO_2 by differential optical absorption in the near UV. *J. Geophys. Res.* **85** 7453–7458.

Pleil, J.D. (1991) Demonstration of a valveless injection system for whole air analysis of polar VOCs. *Proceedings of the 1991 EPA/A&WMA International Symposium on Measurement of Toxic and Related Air Pollutants*, Durham, NC, May 1991.

Pleil, J.D., McClenny, W.A. and Oliver, K.D. (1988) Wind direction dependent whole-air sampling for ambient VOCs. Paper 88.150.6, *Air Pollution Control Association 81st Annual*

Meeting and Exhibition, June 19–24, 1988, Dallas, Texas.

Possanzini, M., Febo, A. and Liberti, A. (1983) New design of a high-performance denuder for the sampling of atmospheric pollutants. *Atmos. Environ.* **17**(12) 2605–2610.

Puig, L. (1990) *High Speed Vacuum Outlet Capillary Gas Chromatography with Selective Detection.* Ph.D. Thesis, University of Michigan.

Purdue, L. (1991) The Atlanta ozone precursor study. Paper No. 91-69.8, Air and Waste Management Association, 84th Annual Meeting, Vancouver, BC, June 16–21, 1991.

Purdue, L.J., Dayton, D.P., Rice, J. and Bursey, J. (1991) Technical assistance document for sampling and analysis of ozone precursors, October 1991. Available from Purdue, L.J. Atmospheric Research and Exposure Assessment Laboratory, US Environmental Protection Agency, Research Triangle Park, NC 27711, USA.

Russwurm, G.M. (1992) Quality assurance and the effects of spectral shifts and interfering species in FTIR analysis. In *Proceedings of the International Speciality Conference on Optical Remote Sensing and Applications to Environmental and Industrial Safety Problems*, Houston, TX, 5–8 April, 1992.

Russwurm, G.M., Kagaan, R.H., Simpson, O.A., McClenny, W.A. and Herget, W.F. (1991) Long-path FTIR measurements of volatile organic compounds in an industrial setting. *J. Air Waste Manage. Assoc.* **41**(8) 1062–1066.

Smith, B. (1992) Differential absorption LIDAR for monitoring stratospheric ozone. In *Proceedings of the International Speciality Conference on Optical Remote Sensing and Applications to Environmental and Industrial Safety Problems*, Houston, TX, 5–8 April, 1992.

Smith, D.L. (1991) Closed Cycle Cooler for VOC Preconcentration. Unpublished project report on USEPA Contract 68-DO-0007 to Battelle, September 1991. Available from Branch Chief, AMRB, MD 44, US Environmental Protection Agency, Research Triangle Park, NC 27711, USA.

Spartz, M.L. *et al.* (1989) *Evaluation of a mobile FT-IR system for rapid VOC determination—Part 1: Preliminary qualitative and quantitative calibration results.* American Environmental Laboratory, November 1989, pp. 18–30.

Spellicy, R.L. (1991) Spectroscopic remote sensing—addressing requirements of the Clean Air Act. *Spectroscopy* **6**(9) 24–34.

Stevens, R.K. *et al.* (1990) Evaluation of a differential optical absorption spectrometer as an air quality monitor. *Proceedings of the 1990 EPA/A&WMA International Symposium on Measurement of Toxic and Related Air Pollutants*, Raleigh, NC, April 30, 1990.

Superfund Admendments and Reauthorization Act (SARA) of 1986, Emergency Planning and Community Right to Know (Title III). Section 313, US Congress.

Tanner, R.L., Ferreri, G.Y. and Kelly, T.J. (1986) Sampling and determination of gas-phase hydrogen peroxide following removal of ozone by gas-phase reaction with nitric oxide. *Anal. Chem.* **58** 1857.

Van Hese, V.G. and Grant, D. (1988) *Automatic Multidimensional Capillary GC/On-Line Pre-Separation and Sample Pretreatment.* American Laboratory, December 1988.

Vecera, Z. and Dasgupta, P.K. (1991) Measurement of ambient nitrous acid and a reliable calibration source for gaseous nitrous acid. *Environ. Sci. Technol.* **25**(2) 255–260.

Vecera, Z. and Janak, J. (1987) Continuous aerodispersive enrichment unit for trace determination of pollutants in air. *Anal. Chem.* **59**(11) 1494–1498.

Vossler, T.L., Stevens, R.J., Paur, R.J., Baumgardner, R.E. and Bell, J.P. (1988) Evaluation of improved inlets and annular denuder systems to measure inorganic air pollutants. *Atmos. Environ.* **22**(8) 1729–1736.

Walters, R.B. (1983) Automated determination of formaldehyde in air without the use of tetrachloromercurate(II). *Am Ind. Hyg. Assoc. J.* **44**(9) 659–661.

Winberry, W.T. Jr., Murphy, N.T. and Riggan, R.M. (1988) Method TO-14 in *Compendium of Methods for the Determination of Toxic Organic Compounds in Ambient Air*, US Environmental Protection Agency, Research Triangle Park, NC 27111, EPA-600/4-9-017, June 1988.

Wise, M.B., Hurst, G.B., Thompson, C.V., Buchanan, M.V. and Guerin, M.R. (1991) Screening volatile organics by direct sampling ion trap and glow discharge mass spectrometry. In *Proceedings of the Second International Symposium on Field Screening Methods for Hazardous Wastes and Toxic Chemicals*, Las Vegas, NV, February 12–14, 1991, pp. 273–288.

Yefchak, G.E. (1991) Using time-of-flight mass spectral data to deconvolute unresolved chromatographic peaks. In *Proceedings of the 39th ASMS Conference on Mass Spectrometry and Allied Topics*, Nashville, TN, May 19–24, 1991.

7 VOCs and occupational health

G. de MIK

7.1 Introduction

VOCs, including aliphatic and aromatic hydrocarbons, chlorinated hydrocarbons, alcohols, ethers, esters and ketones, have a wide range of applications in industry. Their applications are so widespread that they are found in nearly every workplace, although the extent may be different. VOCs are used as fuels, solvents for paints, varnishes, lacquers, inks, glues and pesticides, cleaning and drying of metals and instruments, intermediates for the synthesis of other chemicals, etc.

Due to their volatile character the main route of exposure in industry is via the respiratory route, although in some cases absorption via the skin cannot be excluded. In many cases exposure at the workplace is not to a single compound but to a mixture of compounds, since either the compound itself is a mixture (for instance white spirit), or several compounds are used at the same moment at the same workplace (for example in the painting industry). The main health effects of interest are on the nervous system, mucous membranes and skin.

7.2 Sources

7.2.1 Importance

Nearly everyone is occupationally exposed to VOCs. The degree of exposure, however, may differ to a large extent because in some industries VOCs are the main products, while in other industries they are only of minor importance.

The widespread use of VOCs is a consequence of their properties, especially their ability to solve a tremendous number of water-insoluble organic and inorganic compounds. Millions of tons are produced worldwide.

7.2.2 Applications

As already mentioned, VOCs have a wide range of applications. It is not possible here to give an overview of all the compounds and all their applications. Instead, a summary of the application of some compounds will be given.

7.2.2.1 Acetone. Acetone is used as a solvent in paints, lacquers and varnishes, and as a solvent for cements in the leather and rubber industries. It is also used for cleaning and drying purposes (ACGIH, 1986).

7.2.2.2 Benzene. The main uses of benzene are as a constituent of gasoline (up to 5% v/v) and as a raw material in the chemical industry for the production of ethylbenzene, styrene, cumene, cyclohexane, nitrobenzenes, alkylbenzenes, maleic anhydride and chlorobenzenes (CONCAWE, 1987; Wallace, 1989).

7.2.2.3 Butanols. 1-Butanol is used as a solvent for paints, lacquers, coatings, natural and synthetic resins, gums, vegetable oils, dyes, alkaloids and camphor. It is a starting material in the synthesis of butyl acetate, dibutyl phthalate and dibutyl sebacate. It is used as an extractant in the manufacture of antibiotics, hormones and vitamins, and provides an excellent diluent for formulating brake fluids.

2-Butanol is used as a solvent and in hydraulic brake fluids, industrial cleaning products, polishes, paint removers and penetrating oils. It is used in the preparation of ore-flotation agents, fruit essences, perfumes and dyestuffs. Its largest use is in its conversion to methyl ethyl ketone.

t-Butanol is used for the removal of water from products, in the manufacture of perfumes, flotation agents, flavours, cellulose esters, plastics and lacquers, and in paint removers and extraction of drugs. It is used as a solvent and as a denaturant for other solvents (ethanol). It has been patented for use as a gasoline antiknock agent (DECOS, 1991).

7.2.2.4 Chloroform. In the past chloroform was used as an anaesthetic and as a solvent. Its main use now is as a raw material for the preparation of refrigerants and resins, and as an extractant solvent for pharmaceuticals (ACGIH, 1986).

7.2.2.5 Ethylbenzene. Ethylbenzene is used as a precursor for styrene and acetophenone. It is also used as a solvent in paints, glues and ink, and is a constituent of gasoline, naphtha and asphalt. Mixed xylenes are used as diluents in the paint industry, in agricultural sprays for insecticides and in gasoline blends (Sittig, 1985).

7.2.2.6 Gasoline. The production figures for automotive and aviation gasoline in the EC in 1985 were estimated at 99 569 000 and 232 000 tons, respectively (IARC, 1989).

7.2.2.7 Methyl chloride. The current principal use of methyl chloride is in the production of methyl silicone polymers and resins, and in the manufacture of tetramethyl lead antiknock compounds for gasoline. To a lesser extent it is used as a chemical intermediate and as a solvent. It is also used as a blowing agent of polystyrene foams, and has been used as an intermediate in the production of plastics, pharmaceuticals, herbicides, dyes, disinfectants, methyl ethers and dichloromethane. To a limited extent it has been used as a local anaesthetic (Ahlstrom and Steeler, 1979).

7.2.2.8 Methyl ethyl ketone. Methyl ethyl ketone is used in the manufacture of colourless synthetic resins, as a solvent and in the surface coating industry (ACGIH, 1986).

7.2.2.9 Pyridine. Pyridine is used as a solvent in, for example, pharmaceutical and polycarbonate resin industries. A main use is in the production of agricultural chemicals, such as the herbicides paraquat and diquat, and the insecticide chlorpyrofos. Other uses include the production of piperidine, the manufacture of pharmaceuticals, and the dyeing of textiles. Minor quantities are used for the denaturation of alcohol and antifreeze mixtures, and as a flavouring agent (Reed, 1990).

7.2.3 *Multiple exposure*

A complicating factor in studying occupational exposure is the multiplicity of compounds. In the painting industry different solvents are used at the same time, such as white spirit, toluene, xylenes, alcohols, ketones and aldehydes. Some of the solvents are themselves mixtures, for example, white spirit consists of hundreds of compounds. Since many solvents have general as well as specific toxic effects, it is difficult to describe dose–effect relationships when multiple exposure exists.

7.3 Exposure

7.3.1 *Differences from the general population*

To some extent the occupational exposure is different from the exposure of the general population:

- Some compounds are only used in industry so the general population will not be exposed to these compounds
- The level of exposure in industry will usually be much higher than in the general environment due to dilution in the atmosphere
- The duration of exposure in industry is mostly shorter than for the

general population. In most modern societies the exposure will be limited to eight hours a day and five days a week. However other time schedules are possible. The general population may be exposed during 24 hours a day. One should not forget that, outside the factory, the worker is part of the general population, so in most cases the occupational exposure is superimposed on the background exposure of the general population

- The exposure route may be different. Due to the volatility of VOCs the respiratory toute is dominant, although for the general population the oral route may also be important. At the workplace the oral route is negligible under normal hygienic conditions. As many compounds are used in the liquid stage, dermal absorption cannot be excluded. The relevance of this route is dependent on the absorption rate. Phenol, for example, is absorbed at a high rate so in this case skin exposure must be prevented.
- The age of the working population varies between 20 and 65 years. The general population includes all ages. The consequence of this is that sensitive people such as children and the elderly are not found at the workplace. One should realize however that children at the most sensitive period of their development can be exposed at the workplace during the pregnancy of the mother. After birth the child can be exposed via contaminated mother's milk.
- A common finding in epidemiologic studies is the so-called 'healthy worker effect', which means that the occupational population is a rather healthy subpopulation of the general population.

The most predominant difference between occupational exposure and exposure of the general population is the level of exposure. Acute toxicity, which is rather seldom in the general population (suicide excluded), may occur at the workplace. Peripheral neuropathy caused by exposure to *n*-hexane or methyl butyl ketone will only be seen in industry. The same holds for leukaemia caused by benzene, and haemangiosarcoma caused by vinylchloride. Toluene intoxications, however, are also described due to abuse of the solvent.

Health effects due to VOCs are described in chapter 2.

7.4 Sampling

Sampling methods are described in chapter 1. As outlined there activated charcoal is historically the method of choice but other methods are also introduced, for example, passive diffusion badges. The sampling strategy is dependent both on the aim of the measurements, and on the properties of the particular substance. The ultimate goal is to prevent adverse health effects. The basic principle at the workplace is that exposure should be as low as possible. In cases where exposure has been shown to be far below the limit that causes adverse effects, the air can be routinely sampled as a control measure once a week or once a month.

Often a sampling stragegy is used to ensure that exposure is in compliance with air quality standards. In these cases the recommendations for particular substances can be followed. The ACGIH list of *threshold limit values* (TLVs), or preferably a hygienist, give indications if a continuous measurement—a time-weighted average over 15 min or over 8 h—is necessary. The decision will largely depend on the process characteristics involved and on the exposure pattern. The strategy also depends on the toxicity of the substance. For a compound with irritating effects and short peak exposure a sampling time of, for example, 15 min will be recommended, since the actual concentration, and not the average over the day, is relevant. For compounds with systemic effects and long half-lives in the body (e.g. perchloroethylene) the average over the day or even over a week is recommended.

In general for substances with a very long half-life in the body time-weighted averages over 8 h are measured. Even when peak exposure occurs, this will average out to a certain extent in the body.

In conclusion no general rules can be given; only a hygienist having observed the workplace can outline the adequate strategy.

7.5 Health-based recommended exposure limits

7.5.1 Introduction

In 1977 the International Labour Organization adopted the generic term *occupational exposure limit* (OEL) for the various operational chemical limits for workplace air, for instance threshold limit value (TLV), maximum allowable concentration (MAC) and maximale arbeitsplatzkonzentration (MAK). This term has also been adopted by the World Health Organization. The OEL is defined as "the maximum permissible concentration of a chemical compound as gas, vapour, particulate matter present in the air within a working area which, according to current knowledge, generally does not impair the health of the employee nor causes undue annoyance, under the conditions that exposure can be repeated and in long-term duration over a daily period of 8 hours constituting an average work week of 40 hours" (Zielhuis *et al.*, 1991).

The first OELs were published in Germany. The first ACGIH–TLV list was published in 1947. After that, in a large number of both Western and Eastern European countries, policies were developed for setting OELs (Dutch Expert Group in 1976, Swedish Expert Group in 1978 and Nordic Expert Group in 1977) (Zielhuis *et al.*, 1991). As a consequence, several criteria documents were prepared, often on the same chemical substances, to underpin the national OELs. In 1989 a workshop on harmonization of criteria documents used for the establishment of health-based occupational exposure limits was organized by the Dutch Directorate General of Labour and the Commission of the European Communities, in order to harmonize the OELs in the community and to start a procedure for setting EC–TLVs. The workshop decided that criteria documents should contain all available information relevant to identification of health risks for exposed workers, critically examined to assess their validity. The following items are to be explicitly discussed in the criteria documents:

- identity, physical and chemical properties, monitoring (environmental and biological);
- sources of exposure;
- environmental levels and human exposure;
- toxicokinetics;
- toxicodynamics;
- evaluation of human health risks: (a) guidelines and standards from national and international bodies; (b) previous evaluation by international bodies; (c) assessment of human health risks (dose–response relationships (NOAEL—see section 7.5.1.2); (d) groups at risk;
- recommendations for research;
- references.

The establishment of an occupational exposure limit is a two-step process, described by the WHO (1981) as follows: "The first step is the development of recommended health-based exposure limits derived from data on exposure–effect and exposure–response relationships. The second step is the translation of these health-based limits into operational limits (or standards) by the responsible authorities. In reaching a decision on these operational limits, policy makers may have to take a variety of factors into consideration, e.g. the views of governments, employers, employees and the social, cultural, economic and technological background."

For the derivation of a health-based occupational exposure limit the existing toxicological data must be taken into consideration.

7.5.1.1 Human data. These data can be obtained from observations on volunteers, industrial experience or epidemiological studies. They are preferred to experimental animal data when there is a clear dose–effect or dose–response relationship.

Data on volunteers are available only on a limit number of substances. Usually the exposures are of short duration, e.g. hours or one day, with low doses to prevent adverse health effects and normally under well-defined conditions. Mostly the volunteers are healthy young people with no pre-employment status. These data are of limited value.

Data from industrial experience, obtained from periodical examination of exposed workers, may give information for assessing the adequacy of exposure limits and detecting or excluding the occurrence of early effects in man.

Epidemiological studies are very useful for setting health-based limits if they are valid. The main limitation is that, as regards long-term effects, the relevant airborne concentrations in the past may not have been adequately measured and therefore have to be estimated.

7.5.1.2 Animal data. For many compounds no or insufficient data on humans are available. Therefore experimental animal data are used. Since there are well-known differences and variations in toxicokinetics and toxicodynamics between species, the results from experiments in animals must be thoroughly evaluated to derive a limit to protect man.

When establishing health-based recommendations, the aim is to choose a value which minimizes the likelihood of the occurrence of effects caused by exposure of humans to a substance. To determine the threshold in animals, parameters of harmful effect (such as growth inhibition, change in organ weight, lesions in organs, enzyme leakage, etc.) are investigated after medium- or long-term exposure to the substance. The mean values of these parameters in the group of treated animals are compared with those of the untreated control group. The highest exposure (dose) at which there is no statistically significant difference for the relevant parameter of the effect regarded as critical is recorded as the *no-observed-adverse-effect-level (NOAEL)*. To extrapolate this NOAEL to man a factor is introduced to allow for the difference in body size between laboratory animals and human beings. A safety factor is then introduced, which should compensate for the influence of species-specific differences in biologic availability and sensitivity between laboratory animals and man.

The way in which animal data can be extrapolated, and extrapolation and safety-factors can be applied is described by Zielhuis and van der Kreek (1979), and Arlman-Hoeke and van Genderen (1988).

7.5.2 Threshold limit values

In 1946, the ACGIH published its first annual list of recommended *maximum allowable concentrations (MACs)* for 144 substances. Since then the list has been updated nearly every year. Many countries have adopted the TLV list or have incorporated it in their own national list, for instance The Netherlands in 1977. In recent years, however, the TLV list has been discussed, and not everyone accepts the values as safe (Ziem and Castleman, 1989).

TLVs refer to airborne concentrations of substances and represent conditions under which it is believed that nearly all workers may be repeatedly exposed day after day without the adverse health effects stated by ACGIH in the introduction of the TLV list. For many compounds, however, no acceptable data exist to derive human dose–effect or dose–response curves. In these cases the limits are based on experimental animal data, using extrapolating and safety factors to derive a limit value. However, one can never exclude the fact that workers are more sensitive then rats or mice. For this reason the ACGIH warns that exceptions are possible:

> Because of wide variation in individual susceptibility, however, a small percentage of workers may experience discomfort from some substances at concentrations at or below the threshold limit; a smaller percentage may be affected more seriously by aggravation of a pre-existing condition or by development of an occupational illness.

Individuals may also be hypersusceptible or otherwise unusually responsive to some industrial chemicals because of genetic factors, age, personal habits (smoking, alcohol or other drugs), medication or previous exposures. Such workers may not be adequately protected from adverse health effects from certain chemicals at concentrations at or below the threshold limits.

TLVs for gases and vapours are usually expressed in terms of parts per million (ppm) of substance in air by volume. At normal atmospheric conditions the number of ppm can be converted to mg/m^3 by multiplying with the molecular weight of the substance, and dividing by 24.45.

Three categories of TLV can be distinguished:

(i) Threshold limit value–time-weighted average (TLV–TWA), the time-weighted average concentration for a normal 8 h working day and a 40 h working week.

(ii) Threshold limit value–short-term exposure limit (TLV–STEL), the concentration to which workers can be exposed for a short period of time, provided that the daily TLV–TWA is not exceeded. TLV–STELs are used where acute toxic effects occur after high short-term exposures. Mostly this is the case with substances with potential irritating properties.

(iii) Threshold limit value–ceiling (TLV–C), the concentration that should not be exceeded during any part of the working exposure. For many substances an instantaneous monitoring is not possible. In these cases the substances are sampled over a 15 min period.

In most countries the carcinogenic substances are treated in a different way. In the TLV list the carcinogens are categorized as:

A1—confirmed human carcinogens. Substances, or substances associated with industrial processes, recognized to have carcinognic potential.

A2—suspected human carcinogens. Chemical substances, or substances associated with industrial processes, which are suspected of inducing cancer, based on either limited epidemiological evidence or demonstration of carcinogenesis in one or more animal species by appropriate methods.

Examples of category A1 are vinylchloride and benzene. Examples of category A2 are acrylonitrile, 1,3-butadiene, carbon tetrachloride, chloroform, chloromethyl methylether, ethylene dibromide, ethylene oxide, formaldehyde, methylene chloride and xylidine.

7.5.3 *Other occupational exposure limits*

As mentioned in section 7.5.1, different countries have developed their own OELs, which in some ways are different from the TLVs. In Western European countries such as Germany, The Netherlands, Sweden and Norway the OELs are established by means of the two-step procedure outlined in section 7.5.1. The first step is to derive a health-based recommended OEL. In the second step the national authorities derive a standard taking into account economic and social considerations. The health-based recommended value is documented in a criteria document and sometimes published in international journals (see Bos *et al.* (1991) for a critical review of the toxicity of methyl *n*-butyl ketone). The main differences compared with the TLV list are due to the carcinogenic compounds. In The Netherlands no OEL is given for genotoxic carcinogens. Instead, the risk of developing cancer due to lifetime exposure to a substance is calculated. In Germany a figure is given for carcinogenic substances, indicating which level is technical reasonable (DFG, 1992).

7.6 Biological exposure limits

7.6.1 *Introduction*

The ultimate goal of occupational health practice with respect to the use of chemicals at the workplace is the prevention of adverse health effects. To reach this goal occupational exposure limits are developed. If the concentration of an airborne chemical is in compliance with the exposure limit there will be no adverse health effects to the exposed worker. The general assumption behind this statement is that the amount of chemical monitored in the air is the same amount as absorbed by the worker. Generally this is not the case due to: (i) workplace-related factors such as differences in intensity of the physical workload, fluctuations in the exposure depending on the location, differences in temperature and humidity, multiple exposure; (ii) person-related factors such as body weight, diet, biotransformation capacity, age, sex, pregnancy, medication, and disease state; and (iii) individual life-style factors such as after-work activities, personal hygiene, eating habits, smoking, alcohol and drug intake.

The advantage of biological monitoring over exposure monitoring is that biological monitoring includes all routes of exposure. However, one should be aware that no distinction can be made between uptake at the workplace, and uptake via water and food or exposure to chemicals from hobbies. For the ultimate effect, however, the source of exposure is irrelevant. An extensive overview of biological monitoring is given by Lauwerys (1983).

The ACGIH is one of the few bodies publishing BM values. The TLV list of 1990–1991 includes the compounds shown in Table 7.1, together with their determinants. In the ACGIH-TLV list remarks are also given for the use of biological exposure indices. One should not use these indices without consulting the specific documentation and without sufficient expertise in this field. An example is the measurement of *m*-methylhippuric acid in urine, which is useful for the biological monitoring of workers exposed to *m*-xylene. If a worker is exposed to xylene and has taken aspirin then the methylhippuric acid concentration will be artifactually reduced (Campbell *et al.*, 1988).

Table 7.1 Compounds for which biological exposure indices have been established.

Airborne chemical	Determinants
Benzene	Total phenol in urine Benzene in exhaled air
Carbon disulphide	TTCA in urine
N,N-Dimethylformamide	N-methylformamide in urine
Ethylbenzene	Mandelic acid in urine
n-Hexane	2,5-Hexanedion in urine *n*-Hexane in end-exhaled air
Methyl chloroform	Methyl chloroform in end-exhaled air Trichloroacetic acid in urine Total trichloroethanol in urine Total trichloroethanol in blood
Methyl ethyl ketone	Methyl ethyl ketone in urine
Perchloroethylene	Perchloroethylene in end-exhaled air Perchloroethylene in blood Perchloroethylene in urine
Phenol	Total phenol in urine
Styrene	Mandelic acid in urine Styrene in mixed-exhaled air Phenylglyoxylic acid in urine Styrene in blood
Toluene	Hippuric acid in urine Toluene in venous blood Toluene in end-exhaled air
Trichloroethylene	Trichloroacetic acid in urine Trichloroacetic acid and trichloroethanol in urine Free trichloroethanol in blood Trichloroethylene in end-exhaled air
Xylenes	Methylhippuric acids in urine

7.7 Combined exposure to chemicals

7.7.1 *Introduction*

In any situation, humans are exposed to a variety of chemicals present in the environment, in food and in drinking water. In addition, as already mentioned, most workplace situations involve exposure not to a single, but to several chemicals. There may be prior, coincidental or successive exposure to a multiplicity of chemicals. The nature and conditions of exposure may lead to a significant alteration of the toxicity of the single chemical. Evaluation of health hazards from multiple chemical exposure is much more complex then evaluation of those from a single exposure. Many interactions are possible before and after absorption. In addition to chemicals, physical (heat, radiation, noise) and biological factors (increased risk of infectious disease) may be involved.

7.7.2 *Interactions in the exposure phase*

The working atmosphere may contain gases, vapours, fibres and liquid or solid aerosols. The route of entry is mostly by inhalation or by skin contact. For skin contact, liquids present the greatest hazard. The composition of workplace air may be different from the liquids from which it is produced, so there may be a difference between the hazard after absorption by skin contact and by inhalation. Special attention must be paid when heating or combustion is involved in a process, since thermal degradation products of unknown toxicity may be released in the atmosphere through photolysis, photosynthesis, pyrolysis, hydrolysis or adsorption to dust particles.

7.7.3 *Toxicokinetic interactions*

The rate of absorption through biological membranes is dependent both on the physicochemical properties of the chemical, and on the integrity of the membrane. Contact of the skin with organic solvents may lead to removal of the surface film, thus facilitating the entrance of other chemicals. Increased temperature may cause hyperaemia, with consequent increase in the rate of transport of the penetrated chemical into the bloodstream. The absorption of pesticides may be enhanced by special formulations.

Interference with drug distribution cannot be excluded, for instance by competition for the same binding site in plasma proteins. Special attention must be payed to the interaction of work-related chemicals with drugs which have a high protein binding capacity, since in a normal working population as many as 25% of the workers may use drugs. Many drugs are bound to plasma albumin, and by competing for the same binding site, the drug may displace a chemical, increasing the free plasma concentration. An important feature is that drugs are present in the body in physiologically active concentrations.

Many chemicals undergo biotransformation by a variety of oxidative, reductive and conjugating mechanisms such as hydroxylation, dealkylation, deacetylation and conjugation (see chapter 2). As the duration of toxicity of a chemical agent depends on the body's effectiveness in transforming and eliminating the chemical, any interaction will lead to an increase or decrease of toxicity.

Apart from biotransformation, elimination is effected mainly by excretion via urine, bile or exhaled air. Alterations in the pH of urine affects excretion.

7.7.4 *Toxicodynamic interactions*

Toxicodynamic interactions are based on the fact that interference between two or more chemicals occurs at the target or receptor level. An example of this type of interaction is the potentiation of the anxiolytic action of diazepam in the rat by toluene.

7.7.5 *Consequences of interactions*

The consequence of interactions is that the response of exposure to two or more chemicals given at the same time or after each other cannot be explained by the action of the single chemical. The overall resulting response may be:

1. Independent—each chemical produces a different effect owing to a different mode of action
2. Additive—the magnitude of the combined effect is equal to the sum of the effects produced by each chemical separately
3. Potentiating—the magnitude of the combined effect is more than the sum of the effects produced by each chemical separately
4. Antagonistic—the magnitude of the combined effect is less than the sum of the effects produced by each chemical (de Mik *et al.*, 1988).

When it is known that components of a mixture have similar effects (irritation, neurotoxic effects) the additive rule is used (ACGIH, 1990). When no data on interactions exist it is assumed that the compounds have independent effects.

References

Ahlstrom, R.C. and Steeler, J.M. (1979) Methyl chloride. In Grayson, M. and Eckroth, D. (eds) *Kirk-Othmèr Encyclopedia of Chemical Technology*, 3rd edn., Vol. 5., pp. 677–685. John Wiley and Sons, New York.

ACGIH (1986) *Documentation of the Threshold Limit Values and Biological Exposure Indices.* American Conference of Governmental Industrial Hygienists Inc. 5th edn., Cincinnati, Ohio, USA.

ACGIH (1990) *Threshold Limit Values for Chemical Substances and Physical Agents and Biological Exposure Indices.* American Conference of Governmental Industrial Hygienists Inc. Cincinnati, Ohio.

Arlman-Hoeke, M. and van Genderen, H. (1988) Workshop on new approaches in extrapolation procedures and standard setting for non-carcinogenic substances in human exposure. *Reg. Toxicol. Pharmacol.* **8** 381–437.

Bos, M.J. Mik, G. de and Bragt, P.C. (1991) Critical review of the toxicity of methyl *n*-butyl ketone: Risk from occupational exposure. *Am. J. Ind. Med.* **20** 175–194.

Campbell, L., Wilson, H.K., Samuel, A.M. and Gompertz, D. (1988) Interactions of *m*-xylene and aspirin metabolism in man. *Br. J. Ind. Med.* **45** 132–137.

CONCAWE (1987) A survey of exposures to gasoline vapour. *Report No. 4/87*, The Hague.

Dutch Expert Committee on Occupational Standards (DECOS) (1991) *Health Based Recommended Occupational Exposure Limit for 1-Butanol, 2-Butanol, t-Butanol.* Directorate-General of Labour, The Hague.

International Agency for Research on Cancer (IARC) (1989) Gasoline. In *IARC Monographs on the Evaluation of Carcinogenic Risks to Humans. Occupational Exposures in Petroleum Refining; Crude Oil and Major Petroleum Fuels*, Vol. 45, pp. 159–201. Lyon.

Lauwerys, R.R. (1983) *Industrial Chemical Exposure; Guidelines for Biological Monitoring*: Biomedical Publications, Davis.

DFG (1992) *MAK- und BAT-Werte-Liste 1991.* Deutsche Forschungsgemeinschaft. Senatskommission zur Prüfung Gesundheitsschädlicher Arbeitsstoffen. Mittcirlung 28. VCH, Weinheim.

Mik, G. de, Henderson, P.Th. and Bragt, P.C. (1988) In Notten, W.R.F., Herber, R.F.M., Hunter, W.J., Monster, A.C. and Zielhuis, R.L. (eds) *Health Surveillance of Individual Workers Exposed to Chemical Agents.* pp. 54–62. Springer Verlag, Berlin.

Reed, R.L. (1990) Pyridine, In: Buhler, D.R. and Reed, D.J. (Eds) *Ethel Browning's Toxicity and Metabolism of Industrial Solvents*, 2nd edn. Vol. 2, pp. 259–267. Elsevier Science Publishers B.V., Amsterdam.

Sittig, M. (1985) Handbook of toxic and hazardous chemicals, 2nd edn., pp. 412–414. Noyes Publications, New Jersey, USA.

Wallace, L.A. (1989) Major sources of benzene exposure. *Environ. Health Perspect.* **82** 165–169.

World Health Organization (1981) Recommended health based limits in occupational exposure to selected organic solvents. *Technical Report Series No. 664*, Geneva.

Zielhuis, R.L. and van der Kreek, F.W. (1979) Use of a safety factor in setting health based permissible levels for occuptional exposure. *Int. Arch. Occup. Environ. Health* **42** 191–201.

Zielhuis, R.L., Noordam, P.C., Maas, C.L., Kolk, J.J. and Illing, H.P.A. (1991) Harmonization of criteria documents for standard setting in occupational health: a report of a workshop. *Regul. Toxicol. Pharmacol.* **13** 241–262.

Zeim, G.E. and Castleman, B.I. (1989) Threshold limit values: Historical perspectives and current practice. *J. Occ. Med.* **31** 910–918.

Index

DATE DUE

NOV. 0 6 1996

GAYLORD

PRINTED IN U.S.A.